Lecture Notes in Energy

Volume 104

Lecture Notes in Energy (LNE) is a series that reports on new developments in the study of energy: from science and engineering to the analysis of energy policy. The series' scope includes but is not limited to, renewable and green energy, nuclear, fossil fuels and carbon capture, energy systems, energy storage and harvesting, batteries and fuel cells, power systems, energy efficiency, energy in buildings, energy policy, as well as energy-related topics in economics, management and transportation. Books published in LNE are original and timely and bridge between advanced textbooks and the forefront of research. Readers of LNE include postgraduate students and non-specialist researchers wishing to gain an accessible introduction to a field of research as well as professionals and researchers with a need for an up-to-date reference book on a well-defined topic. The series publishes single- and multi-authored volumes as well as advanced textbooks.

Indexed in Scopus and EI Compendex The Springer Energy board welcomes your book proposal. Please get in touch with the series via Anthony Doyle, Executive Editor, Springer (anthony.doyle@springer.com).

Muhammad Hussain · Tieling Zhang

Big Data Analytics in Energy Pipeline Integrity Management

Springer

Muhammad Hussain
School of Mechanical, Materials,
Mechatronic and Biomedical Engineering
Faculty of Engineering and Information
Sciences
University of Wollongong
Wollongong, NSW, Australia

Tieling Zhang
School of Mechanical, Materials,
Mechatronic and Biomedical Engineering
Faculty of Engineering and Information
Sciences
University of Wollongong
Wollongong, NSW, Australia

ISSN 2195-1284 ISSN 2195-1292 (electronic)
Lecture Notes in Energy
ISBN 978-981-96-8018-4 ISBN 978-981-96-8019-1 (eBook)
https://doi.org/10.1007/978-981-96-8019-1

This Springer imprint is published by the registered company Springer Nature Singapore Pte Ltd.
The registered company address is: 152 Beach Road, #21-01/04 Gateway East, Singapore 189721, Singapore

Executive Summary

The energy sector is the backbone of modern infrastructure, and pipelines play a critical role in ensuring the safe and efficient transportation of oil, gas, and other energy resources. However, maintaining pipeline integrity is a significant challenge due to aging infrastructure, environmental risks, safety concerns, and increasing regulatory requirements. The integration of Big Data Analytics (BDA) into Pipeline Integrity Management Systems (PIMS) presents a transformative approach to improving monitoring, predictive maintenance, and risk management in the energy pipeline sector.

This book provides a comprehensive exploration of how Big Data Analytics is revolutionizing pipeline integrity management by enabling real-time monitoring, predictive modeling, anomaly detection, and risk assessment. It examines the traditional approaches to Pipeline Integrity Management (PIM), their limitations, and how data-driven technologies such as artificial intelligence (AI), Internet of Things (IoT), cloud computing, and blockchain are addressing these challenges.

Key Themes and Contributions

1. **The Importance of Pipeline Integrity Management (PIM)**
   - Pipelines are vital for global energy security, and failures can lead to catastrophic economic, environmental, and human safety consequences.
   - Traditional Pipeline Integrity Management Systems (PIMS) have relied on manual inspections, hydrostatic testing, and in-line inspection (ILI) tools, which are often costly and reactive rather than proactive.
2. **The Role of Big Data Analytics in Pipeline Integrity**
   - BDA enables real-time data collection, integration, and advanced analytics for decision-making.

- Technologies such as predictive analytics, machine learning (ML), and AI-driven automation enhance early detection of pipeline threats.
- Real-time monitoring using IoT sensors, satellite surveillance, drones, and acoustic monitoring improves data collection accuracy.

3. **Modern Data Collection and Processing in PIMS**

   - Traditional methods of data collection are being replaced by automated systems that reduce human error and increase efficiency.
   - IoT-based data collection, AI-driven automated inspections, and cloud-based data storage enhance operational safety and predictive maintenance.
   - Data preprocessing techniques, including data cleaning, normalization, and integration, are critical for ensuring data quality and consistency.

4. **Machine Learning for Predictive Maintenance and Anomaly Detection**

   - Machine learning models help predict pipeline degradation, corrosion growth, and failure risks.
   - Techniques such as random forest regression, K-nearest neighbors (KNNs), and deep learning models are employed to analyze vast datasets from ILI tools, pressure sensors, and environmental monitoring systems.
   - AI-powered automated anomaly detection significantly reduces manual labor and enhances pipeline safety.

5. **Risk Assessment and Decision Support Systems**

   - Big Data Analytics enables risk-based decision-making by integrating historical data, sensor inputs, and real-time field conditions.
   - Techniques such as Monte Carlo simulation, Markov process models, and dynamic linear models (DLMs) improve the accuracy of risk assessments and failure probability predictions.

6. **Emerging Technologies in PIMS**

   - Blockchain ensures data security, enhances transparency, and streamlines compliance reporting.
   - Drones and robotics provide non-intrusive inspection methods for remote and hazardous areas.
   - Edge computing and 5G connectivity improve real-time data transmission and decision-making.
   - Quantum computing offers potential breakthroughs in complex risk assessments and predictive modeling.

7. **Data Visualization and Reporting**

   - Effective data visualization enhances anomaly detection, real-time monitoring, and strategic decision-making.
   - Tools such as Geographic Information Systems (GIS), risk heatmaps, and AI-powered dashboards improve the interpretability of pipeline integrity data.

8. **Challenges in Technology Adoption**

   - Despite the advantages, challenges such as data silos, integration complexity, cybersecurity risks, and regulatory hurdles slow down technology adoption in pipeline integrity management.
   - The book discusses strategies to overcome these barriers, ensuring scalable, and industry-compliant Big Data solutions.

Impact on the Energy Industry

This book provides energy companies, policymakers, engineers, and researchers with the knowledge and frameworks needed to leverage Big Data Analytics for improved pipeline integrity management. By adopting data-driven decision-making, predictive analytics, and emerging technologies, the energy sector can reduce operational risks, enhance sustainability, and achieve cost efficiencies.

With a detailed methodology, real-world case studies, and forward-looking perspectives, this book serves as a definitive resource for those seeking to understand and implement Big Data Analytics in the energy pipeline industry.

Muhammad Hussain
Tieling Zhang

Contents

List of Figures

List of Tables

Chapter 1
Introduction

1.1 Purpose Statement

The energy pipeline industry is the backbone of contemporary economies, guaranteeing the uninterrupted flow of natural gas, oil, and other essential resources over enormous geographic distances. The energy business faces tremendous hurdles in maintaining pipeline integrity as the world's infrastructure ages and energy demand rises. Energy pipeline failures can result in disastrous environmental events, monetary losses, and worries for public safety. To mitigate such hazards, a sophisticated approach to pipeline integrity management (PIM) is necessary.

Conventional techniques are used for monitoring and repairing energy pipelines. The manual methods of managing and overseeing energy pipelines that were previously in use are unable to handle the sophisticated pipeline systems of today. Pipeline integrity management can be revolutionized thanks to much richer datasets throughout the pipeline system and modern technology. In the energy pipeline industry, integrating Big Data Analytics (BDA) provides a potent tool for increased decision-making, real-time monitoring, and predictive maintenance.

The potential of integrating Big Data Analytics into Pipeline Integrity Management Systems (PIMS) needs to be studied. The goal is to provide a framework for improving PIMS using BDA, which will increase the safety, effectiveness, and economy of pipeline operations. Businesses may optimize maintenance schedules, maintain regulatory compliance, and discover potential issues early by utilizing the massive amounts of data generated by pipeline monitoring systems.

This book's goal is to give readers a thorough understanding of how Big Data Analytics is used in PIMS. It seeks to close the knowledge gap between established integrity management techniques and the cutting-edge digital innovations that have the potential to completely change the sector. This book will also explore the constraints of implementing BDA in the oil and gas sector, notably in terms of data quality, infrastructural restrictions, and worker preparation.

M. Hussain and T. Zhang, *Big Data Analytics in Energy Pipeline Integrity Management*, Lecture Notes in Energy 104, https://doi.org/10.1007/978-981-96-8019-1_1

1.2 Key Concepts

Oil and gas are the primary resources that supply over 50% of the world's energy today (Overholt 2016). Pipelines are among the most critical conduits for energy transportation. Among many pipeline systems, gas pipelines are the most critical and essential. Figure 1.1 illustrates the significance of pipelines across several sectors (Gracon 2023).

Natural gas transmission and gathering pipelines total more than 528,000 km (328,000 km), crude oil transmission and gathering pipelines total 119,000 km (74,000 mi), and hazardous liquid transmission pipelines total 132,000 km (82,000 mi). In the late 1800s, Australia's inaugural pipeline was established to provide water to the Coolgardie gold fields. Australia possesses around 42,000 km of natural gas transmission pipelines that transport high-pressure gas from producing sites to the peripheries of both big and local settlements (APGA 2021). In 1998, a total of $63.1 billion was invested by all natural gas pipeline companies, of which $13.6 billion was generated in income.

1.2.1 Energy Pipeline Systems

Energy pipelines are critical infrastructure that transmit oil, natural gas, and refined products from production sites to refineries, distribution centers, and end customers. These pipelines traverse thousands of kilometers across diverse topographies and climatic conditions. They comprise several components, including distribution pipelines that provide refined products to end consumers, transmission pipelines that convey oil and gas over extensive distances, and gathering pipelines that transport crude oil and gas from production wells to processing facilities.

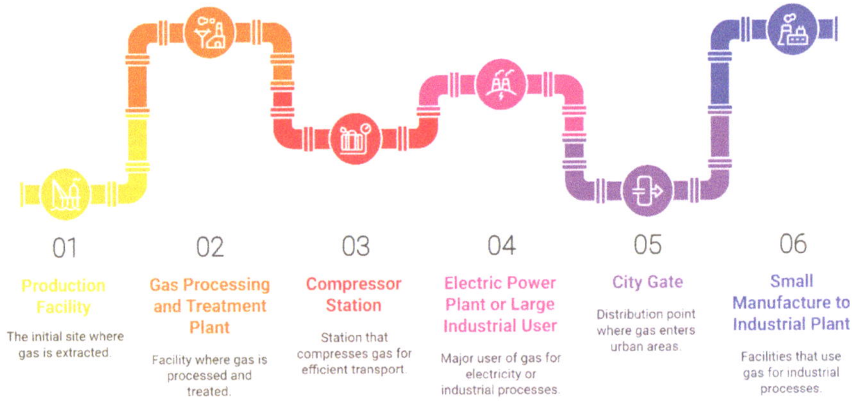

Fig. 1.1 The role of pipelines in different industries

Pipeline systems encompass complexities beyond their mere physical layout. Compressor stations, valves, metering points, and leak detection systems are integral components of pipelines, and all are crucial for the efficient and safe transportation of energy resources. To ensure pipeline integrity, these components generate substantial volumes of data that require monitoring and analysis. A multitude of factors, including internal and external corrosion, mechanical damage, material defects, and environmental conditions, can result in pipeline failures. Pipeline Integrity Management Systems are essential for the comprehensive monitoring, maintenance, and repair of pipelines required for hazard management.

1.2.2 Importance of Energy Pipelines in the Global Infrastructure

One of the most effective ways to move big amounts of gas, oil, and other hydrocarbons over long distances is via pipelines. Pipelines are a more affordable and environmentally friendly choice for transportation than roads or trains since they run constantly and require little human involvement. Huge pipeline networks spanning tens of thousands of kilometers cross borders, seas, and different geological terrains in nations like the US, Canada, and Russia.

Pipelines are the backbone of the world's energy infrastructure. Refineries can send refined oil via pipelines to distribution facilities, which subsequently provide the final consumer with gasoline, diesel, and other goods. Pipelines link extraction sites to power plants, factories, and even residential buildings in natural gas markets so that energy can be generated for heating and lighting.

1.2.3 Economic Significance of Pipelines

Energy pipelines are an economic asset worth billions of dollars since they enable the daily exchange of millions of barrels of oil and trillions of cubic feet of natural gas. These pipelines guarantee the continuous flow of energy resources and uphold the equilibrium between supply and demand in the world's energy markets. Disruptions to pipeline operations, including those caused by leaks or accidents, can lead to price increases, energy shortages, and financial losses for the operators.

1.2.4 Environmental and Safety Considerations

Pipelines are an effective mode of transportation; however, there are serious environmental hazards associated with them. A rupture or leak in a pipeline can result in disastrous releases of hazardous materials, such as natural gas or oil, contaminating water supplies, endangering wildlife, and endangering public health. Pipeline failures can create long-term ecological destruction, expensive cleanups, and detrimental effects on the company's reputation in addition to other environmental damage (Lu et al. 2023). Ensuring the structural and operational integrity of pipelines is therefore not only a matter of economic interest but also a key environmental and safety concern.

1.2.5 Energy Pipeline and Connected System

The foundation of the world's energy supply chain is made up of energy pipelines, which carry enormous volumes of refined goods, natural gas, crude oil, and even renewable energy sources over great distances. These pipelines, which can be thousands of kilometers long, are among the most important pieces of infrastructure in the energy sector. Both operators and regulators place a high premium on maintaining the integrity of pipelines due to the intricacy and significance of these systems. Energy pipeline maintenance errors can have serious negative effects on the environment, result in loss of human lives, incur large financial penalties, and damage the corporate reputations. A system of interconnected components, known as an energy pipeline, moves goods from production sites to distribution centers, refineries, or end consumers. These components are presented in Table 1.1.

Table 1.1 Pipeline system components

Component	Description
Transmission pipelines	Tasked with transporting crude oil, natural gas, and refined products over extended distances, often spanning states or countries. These pipelines are generally large in diameter (up to 48 inches) and function under high pressure
Gathering pipelines	Smaller pipelines that collect oil or gas from multiple wells and deliver it to processing facilities or larger transmission pipelines. Typically located in production areas like oil fields or offshore platforms
Distribution pipelines	Carry refined oil and gas products to end-users. These smaller pipelines serve as the final stage in the energy transportation network, delivering natural gas to homes or petroleum to gas stations
Compressor and pump stations	Facilities located along pipelines to maintain pressure are necessary for efficient product movement. Compressor stations are used for natural gas, and pump stations for liquids like oil
Valves and metering stations	Valves control the flow of product through the pipeline, allowing for emergency shut offs or maintenance. Metering stations provide data on the product's amount and pressure for integrity management

1.2.6 *Pipeline Threats and Problems*

Every year, billions of dollars are invested to ensure that gas and petroleum liquids are provided to consumers in a secure, reliable, and effective way. There are various potential challenges to the pipeline's integrity, as shown in Fig. 1.2. Recognizing these threats and setting up a program to evaluate their importance are two of the first hurdles that must be overcome to implement effective corrective measures as part of an all-encompassing integrity management program.

ASME B31.8S (Zheng et al. 2012; ASME 2001) lists nine failure type categories, three time-related fault types, and 21 risks that could jeopardize the integrity of energy pipelines. Internal corrosion, exterior corrosion, and stress corrosion cracking are the three main dangers to pipelines (NACE 2011). These nine classifications of failure types are:

- Manufacturing defect
- Materials and construction defects
- Stress corrosion cracking
- Internal corrosion
- External corrosion
- Equipment and related defects
- Third-party perpetrated damage
- Incorrect operations, maintenance, and procedures
- Weather-related, earth-related, and other outside unforeseen forces.

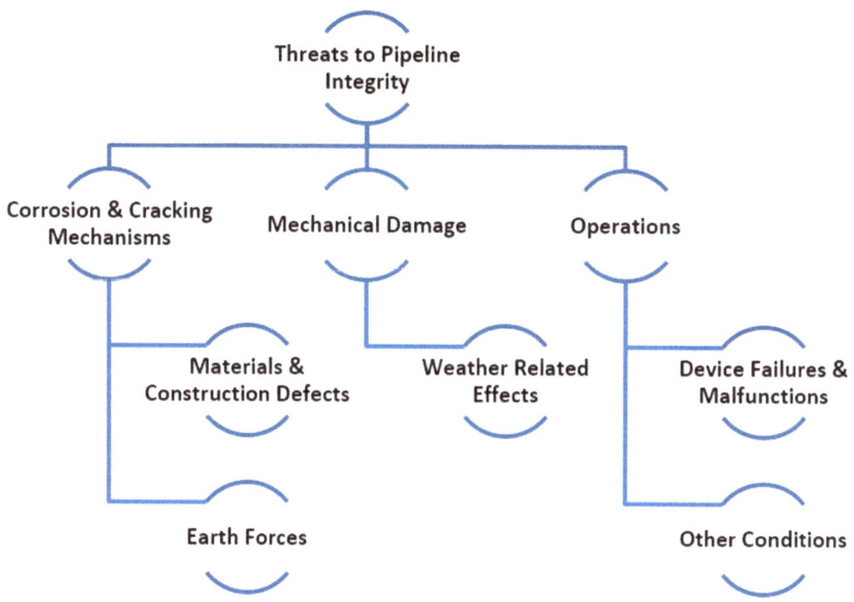

Fig. 1.2 Potential threats to pipeline integrity

Energy pipelines are vulnerable to a variety of challenges that could jeopardize their integrity and have disastrous effects. These risks can be divided into different categories, including material flaws, mechanical harm, corrosion, and environmental influences. Each one of these threats poses different challenges, therefore, it is critical to be able to monitor and reduce these risks by using data-driven decision-making approaches. Some of the damage mechanisms associated with pipelines are given in Table 1.2.

1.3 Defining Pipeline Integrity Management

To guarantee the safe, dependable, and effective functioning of pipelines, pipeline integrity management is an integrated method that incorporates risk management, engineering, and operational procedures (El-Akruti et al. 2016). It covers a pipeline's whole lifecycle, from planning and construction to operation, maintenance, and eventual decommissioning (Singh 2017) (see Fig. 1.3). In essence, pipeline integrity management is about ensuring that a pipeline is running as it was intended, without causing undue risk to humans, the environment, or the energy supply chain.

1.3.1 The Integrity Management Process

A continual process of inspection, monitoring, data analysis, and remedial action is involved in pipeline integrity management (Hussain et al. 2021). Regulatory frameworks, such as those instituted by the European Pipeline Research Group (EPRG) in Europe or the Pipeline and Hazardous Materials Safety Administration (PHMSA) in the United States, often govern this procedure. Operators must monitor potential threats, carry out regular integrity assessments, and put mitigation plans in place. The integrity management process is built around four core activities:

1. **Inspection**: Pipelines should undergo routine and periodic interior and external inspections to look for possible damage or deterioration. Non-destructive testing (NDT) methods including magnetic flux leakage (MFL), ultrasonic testing (UT), and smart pigs—high-tech instruments that move inside pipelines to find anomalies—may be used during these inspections.
2. **Data Collection and Analysis**: To evaluate the health of a pipeline, data from sensors, inspection instruments, and field activities must be gathered. In order to detect patterns, abnormalities, or possible threats, enormous volumes of data must be handled and examined in real-time, which is where Big Data Analytics come into play.
3. **Risk Assessment**: Evaluating the hazards associated with different pipeline sections according to elements such as material, age, operating pressure, and

Table 1.2 Pipeline damage mechanisms

Damage mechanism	Explanation/Details	Types
Corrosion	Corrosion is an important factor of pipeline failures. It occurs when metal reacts with its environment, causing the material to deteriorate over time. Corrosion in pipelines can take place both internally and externally	**Internal Corrosion**: This form of corrosion occurs inside the pipeline, typically due to the composition of the product being transported (Muthukumar 2014). For example, sour gas (natural gas containing hydrogen sulfide) can be highly corrosive, causing internal damage to the pipeline walls. Similarly, the presence of water and other contaminants in crude oil can accelerate internal corrosion **External Corrosion**: External corrosion occurs when the outside of the pipeline is subjected to moisture, oxygen, and various environmental elements. Soil composition, pH levels, and moisture content all influence the rate of external corrosion. Pipelines buried in certain types of soil, particularly those with high chloride content, are more prone to corrosion (Fessler 2008). External corrosion can also result from damage to the protective coatings that are applied to pipelines during installation
Mechanical damage	Mechanical damage to pipelines can result from a variety of sources, including construction activities, third-party interference, or natural disasters. Third-party damage, often referred to as "dig-ins," occurs when construction crews accidentally strike a buried pipeline during excavation work (Groysman 2014). Such incidents can cause immediate pipeline ruptures or create weaknesses that lead to failures over time. Significant damage can also result from natural disasters like earthquakes, floods, and landslides to pipelines, particularly in regions with seismic activity or unstable terrain	

(continued)

Table 1.2 (continued)

Damage mechanism	Explanation/Details	Types
Material defects and pipeline aging	Material defects in pipelines can occur during manufacturing, installation, or operation. These defects may include issues such as weld imperfections, cracks, or inclusions that compromise the strength and durability of the pipeline As pipelines age, they become more susceptible to failures due to material fatigue and the cumulative effects of corrosion and other threats. Aging pipelines often require more frequent inspections, repairs, and maintenance to ensure their continued safe operation. Big Data Analytics plays a significant role in tracking the condition of aging pipelines and predicting when and where failures are most likely to occur (Hussain et al. 2023)	
Environmental factors	Pipelines are subjected to several environmental conditions that can affect their reliability. Severe weather phenomena, including subzero temperatures, hurricanes, and substantial precipitation, might compromise pipelines or induce alterations in the adjacent soil. In frigid climates, pipelines may undergo frost heave, a phenomenon where the ground freezes and expands, applying pressure on subterranean pipelines. Additionally, soil movement, erosion, and seismic activity can all affect the stability of pipelines. Monitoring environmental conditions in real-time and incorporating this data into pipeline integrity management systems is essential for preventing failures caused by these external factors (Wasim and Djukic 2022; Niaz and Hussain 2021)	

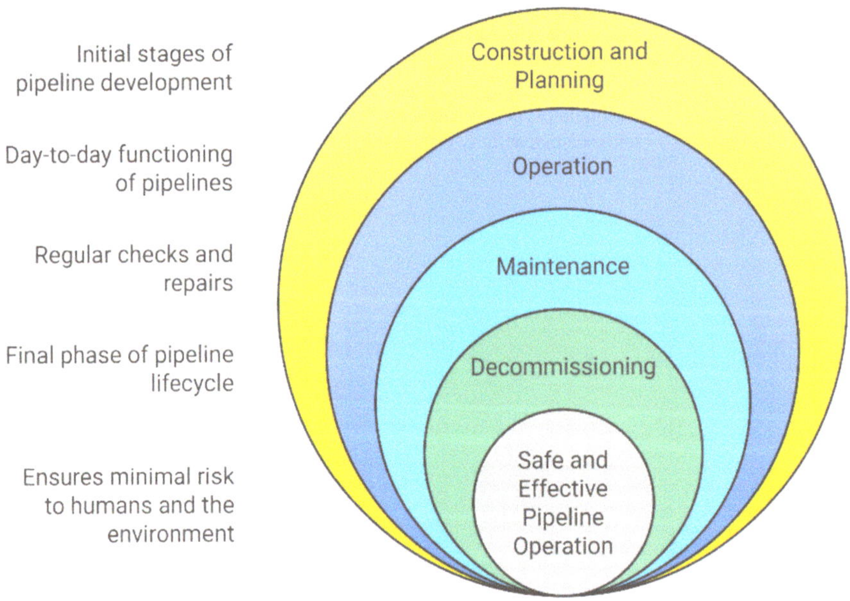

Fig. 1.3 Pipeline integrity management

environmental circumstances. Risk analyses aid in setting priorities for upkeep and repairs, guaranteeing effective resource allocation.

4. **Maintenance and Mitigation**: To guarantee long-term operational safety, proactive steps such as replacing infrastructure, applying protective coatings, and repairing damaged areas should be taken. Preventive maintenance is also a part of this, like cathodic protection to stop corrosion.

1.3.2 Lifecycle Approach to Integrity Management

The fact that pipeline integrity management must encompass the pipeline's whole lifecycle is a crucial component (Palkar 2015). Appropriate materials must be selected during the design and construction phases to ensure durability under anticipated working conditions, such as pressure, temperature extremes, and environmental influences. Continuous monitoring is required to identify any indications of deterioration or external damage during the operational period. Pipelines must be safely retired during decommissioning to guarantee that they do not present any risks to the environment or public safety in the future.

1.4 Traditional PIMS Approaches and Their Limitations

In the past, pipeline operators managed pipeline integrity using manual inspection procedures, recurring data collection, and reactive maintenance. Although these conventional methods have had some success, they have several serious drawbacks, as shown in Table 1.3.

Table 1.3 Traditional approaches and their limitations

Issue	Description
Reactive maintenance	Traditional Pipeline Integrity Management Systems (PIMS) often follow a reactive maintenance model, addressing repairs only after failure or degradation has been detected. This reactive approach can result in costly repairs, unplanned downtime, and, in extreme cases, catastrophic failures
Manual data analysis	The reliance on manual data collection and analysis hampers the ability to detect trends and patterns in pipeline data. Human errors and delays in data processing can lead to missed opportunities for early intervention
Data fragmentation	Data is usually gathered from multiple sources (e.g., ILI tools, SCADA systems, third-party contractors), but these datasets are often fragmented and siloed. This separation makes it challenging to integrate them into a comprehensive risk assessment framework
Infrequent inspections	Inspections that occur periodically (e.g., every few years) may fail to capture issues developing between inspection intervals, such as corrosion, cracks, or mechanical damage, which may go unnoticed and worsen over time

As the energy pipeline sector grows and becomes more complex, these limitations highlight the requirements for more innovative, data-driven approaches to maintaining pipeline integrity management. This is where Big Data Analytics becomes important today.

1.5 Big Data Analytics

Big Data refers to enormous and complicated datasets that traditional data management tools are unable to process (Hussain and Zhang 2023). In the context of pipeline integrity management, Big Data is generated from a wide range of sources, as shown in Table 1.4.

Big Data Analytics involves the processing and analysis of vast information with advanced algorithms and machine learning models. Big Data Analytics can improve the safety and efficiency of pipeline operations by aiding operators in making informed decisions through the identification of concealed patterns and correlations (Jan et al. 2019).

1.5.1 Data-Generating Technologies in Energy Pipelines

Numerous sensors and data collection technologies are installed in energy pipelines, generating a continuous stream of data. This information is essential for preserving

Table 1.4 Data sources from pipeline

Data source	Description
In-Line Inspection (ILI) tools	Devices that travel through the pipeline, collecting data on wall thickness, corrosion, cracks, and other structural defects
Supervisory Control and Data Acquisition (SCADA) systems	Real-time data from sensors along the pipeline, monitoring critical parameters like pressure, flow rate, and temperature
Remote sensing technologies	Technologies such as satellite imagery and drones that provide external data on environmental conditions surrounding the pipeline
Historical maintenance records	Records of past inspections, repairs, and maintenance activities, useful for identifying patterns and trends in pipeline integrity
External data	Factors like weather patterns, seismic activity, and market demand, which can influence pipeline operations and need to be integrated into assessments

pipeline integrity, guaranteeing secure operations, and averting malfunctions. Key technologies include those shown below:

- **Supervisory Control and Data Acquisition (SCADA) Systems**: These technologies are used to track and manage pipeline operations in real time. They gather information on pressure, temperature, flow rate, and valve positions from sensors positioned throughout the pipeline. Control centers receive this data so that their operators may keep an eye on the pipeline's condition and identify any irregularities.
- **In-Line Inspection (ILI) Tools**: ILI instruments, also known as "smart pigs," are devices that move through pipelines to gather data about their status. These instruments use a variety of sensors (such as ultrasonic sensors and magnetic flux leakage) to find anomalies including corrosion, wall thinning, and cracks. The data provided by ILI instruments is useful for identifying probable failure sites and planning maintenance tasks.
- **Remote Sensing Technologies**: Apart from the sensors positioned all along the pipeline, remote sensing technologies like satellites, drones, and aerial photography can offer external data on the surrounding environmental conditions. These devices are particularly valuable for monitoring pipelines in remote or difficult-to-access sites.

1.5.2 Role of Big Data in Pipeline Integrity Management

The incorporation of Big Data Analytics into pipeline integrity management is changing the way energy companies oversee and predict pressures, excavations and moderate defects in pipeline systems. Big Data are the large and unstructured datasets that are generated from sensors, inspection systems, monitoring devices, and operational databases which offer current and past state of pipeline health and performance indicators. Such a shift from a reactive to a proactive approach has the ability to shave millions off the repair bills and bring down the downtime factor significantly.

1.5.3 Real-Time Monitoring and Anomaly Detection

A key advantage of Big Data Analytics is its capability to offer real-time pipeline monitoring. IoT-enabled sensors can continuously stream data to centralized systems, which can then apply machine learning algorithms to detect anomalies (Erhan et al. 2021). These anomalies could be a slight change in pressure, a rise in temperature, or an increase in vibrations—any of which could signal a problem. Detecting these anomalies early allows operators to investigate and resolve potential issues before they become major failures (Jaramillo-Alcazar et al. 2023).

1.5.4 *Risk Assessment and Decision Support*

Big Data Analytics can enhance risk assessment and decision-making support in Project Information Management Systems. The integration of data from various sources, including ILI tools, SCADA systems, and other remote sensors, enables operators to develop advanced risk models that can incorporate multiple variables. These models can be utilized to determine the frequency and position of the subsequent pass, ensuring that resources are allocated where they are most required (Singh 2017). Furthermore, decision support systems based on Big Data Analytics assist operators in making informed decisions on pipeline operations.

1.5.5 *Environmental Impact Monitoring*

Big Data Analytics can be used to monitor environmental conditions in real-time, such as changes in soil moisture and seismic activity. Pipelines are frequently located in environmentally sensitive areas, where failures can result in severe ecological damage, or weather patterns (Jaber et al. 2022). When compiled in the pipeline integrity management systems, operators can evaluate environment effects on pipeline operations and prevent the resulting impacts.

1.5.6 *Predictive Maintenance Through Machine Learning*

Predictive maintenance is a game-changer for pipeline integrity management (Gama et al. 2024). By utilizing past data and machine learning models, operators can detect the location and time of a failure. Predictive analytics can recommend the best time to replace or maintain pipeline sections by looking for patterns in the data, such as trends in exterior damage, material fatigue, or corrosion rates. Because of this, maintenance may be planned for off-peak hours, lessening the impact on operations and preventing expensive unplanned downtime.

1.6 Emerging Trends in Pipeline Integrity Management

Utilizing cutting-edge technologies to improve monitoring, analysis, and repair capabilities is where pipeline integrity management is headed.

1.6.1 Artificial Intelligence and Automation

Artificial intelligence (AI) is being utilized to pipeline integrity management vastly. Massive volumes of data can be processed by AI algorithms, which can also learn from prior mistakes and make recommendations or judgments with little to no human input (Sattari et al. 2022). AI systems, for instance, are capable of monitoring pipelines on their own, anticipating possible faults, and even suggesting mitigation techniques. Drones and robots with AI capabilities are also being utilized for remote inspections, which eliminates the need for human personnel to enter dangerous areas.

1.6.2 Drones and Robotics

Aerial pipeline inspections are being carried out by drones fitted with high-definition cameras, infrared sensors, and even LIDAR systems, especially in hazardous or difficult-to-reach locations (Asadzadeh et al. 2022). These drones can capture images in the visible spectrum, as well as thermal ones, the material from which the latter can be further analyzed using specialized software and identify corrosion, leaks, or physical damage. Likewise, it is eased to transport robotic systems known as 'crawler pigs,' which move through pipelines to assess internal features such as corrosion, deformations, or hindrances.

1.6.3 Blockchain for Data Security

The increase in data in the energy industry means that data security and integrity are highly crucial as the industry evolves. Blockchain helps to maintain and share integrity and safety related records through a decentralized and immutable record keeping method (Ajao et al. 2019). This ensures that records are tamper proof, that adds another layer of reliability and ensures documentation that is credible in integrity management processes.

1.7 Challenges in Technology Adoption

The oil and gas industry has been modest in gravitating toward Big Data solutions, principally among the pipeline franchises, even though it is an energetic purveyor of superior technologies in exploration and production, see Table 1.5 for more details.

Table 1.5 Challenges in technology adoption

Challenge	Description
Legacy systems	Many pipeline operators depend on outdated legacy systems that are not designed to manage the vast volumes of data generated by modern sensors and monitoring technologies. Integrating these older systems with new Big Data platforms is often complex and expensive
Lack of technical expertise	Implementing Big Data Analytics demands specific skills in data science, machine learning, and software engineering. However, the oil and gas industry has traditionally focused on expertise in engineering and geology, leading to a shortage of personnel capable of effectively managing and analyzing Big Data
Data silos	A key obstacle in adopting Big Data Analytics is the fragmentation of data across various departments and systems within an organization. Pipeline operators collect data from multiple sources, like in-line inspection tools and SCADA systems, but this data is often isolated in silos, complicating integration and analysis
Cost and infrastructure	Implementing Big Data solutions necessitates substantial investment in IT infrastructure, including data storage, processing capabilities, and analytics software. Smaller pipeline operators, in particular, may struggle with the financial burden, especially given the volatility of oil prices

1.8 Expected Contributions to the Industry

The followings are expected to be some of the most important findings of this book to the oil and gas industry mainly focusing on pipeline integrity and analysis. The key contributions include:

- **Advancement of Predictive Maintenance Techniques**: By identifying and validating the pipeline maintenance risks prediction methods; the present study will aid pipeline operators in shifting from reactive to predictive maintenance approaches.
- **Improved Data Integration and Analysis**: The integration of data proposed in the framework to operate the systems will provide the needed remedies to sweeping tendencies of data fragmentation and enable real time analysis of large volumes of data by operators.
- **Enhanced Decision-Making Processes**: The creation of decision support systems will thereby help the pipeline operators get better and more efficient advice on pipeline operation.
- **Compliance with Regulatory Standards**: Through understanding of regulatory needs and safety issues in the LTOs, this study will help operators comply with the regulatory requirements while creating more benefits for them from the Big Data Analytics.

References

Ajao LA et al (2019) Crypto hash algorithm-based blockchain technology for managing decentralized ledger database in oil and gas industry. J 2(3):300–325

Asadzadeh S, de Oliveira WJ, de Souza Filho CR (2022) UAV-based remote sensing for the petroleum industry and environmental monitoring: state-of-the-art and perspectives. J Petrol Sci Eng 208:109633

ASME (2001) ASME Standard B31. 8S-2001 Managing system integrity of gas pipelines. American Society of Mechanical Engineers

Australian Pipelines and Gas Association (APGA) (2021). Australia's gas transmission pipeline system. https://apga.org.au/pipeline-facts-and-figures/. Accessed 3 July 2025

El-Akruti K, Zhang T, Dwight R (2016) Maintaining pipeline integrity through holistic asset management. Eur J Industr Eng 10(5):618–638

Erhan L et al (2021) Smart anomaly detection in sensor systems: a multi-perspective review. Inf Fusion 67:64–79

Fessler RR (2008) Pipeline corrosion. Report, US Department of Transportation Pipeline and Hazardous Materials Safety Administration, Baker, Evanston, IL

Gama J et al (2024) From fault detection to anomaly explanation: a case study on predictive maintenance. J Web Semantics 81:100821

Gracon (2023) Types of pipelines [cited 3 June 2023]. Available from: https://graconllc.com/types-of-pipelines/

Groysman A (2014) Corrosion in systems for storage and transportation of petroleum products and biofuels: identification, monitoring and solutions. Springer Science & Business Media

Hussain M, Zhang T (2023) Potential of Big Data analytics for energy pipeline integrity management. Corros Manage 2023(Jan/Feb):32

Hussain M et al (2021) Application of big data analytics to energy pipeline corrosion management. Corros Manage 2021:28–29

Hussain M, Zhang T, Seema M (2023) Adoption of big data analytics for energy pipeline condition assessment. Int J Pressure Vessels Piping:105061

Jaber MM et al (2022) Predicting climate factors based on big data analytics based agricultural disaster management. Phys Chem Earth, Parts a/b/c 128:103243

Jan B et al (2019) Deep learning in big data analytics: a comparative study. Comput Electr Eng 75:275–287

Jaramillo-Alcazar A, Govea J, Villegas-Ch W (2023) Anomaly detection in a smart industrial machinery plant using IoT and machine learning. Sensors 23(19):8286

Lu H, Xi D, Qin G (2023) Environmental risk of oil pipeline accidents. Sci Total Environ 874:162386

Muthukumar N (2014) Petroleum products transporting pipeline corrosion—a review. In: The role of colloidal systems in environmental protection, pp 527–571

NACE (2011) Pipeline corrosion integrity management. AMPP (NACE)

Niaz U, Hussain M (2021) The threat to pipeline integrity from soil corrosion. Corros Manage 2021(Nov/Dec):25

Overholt M (2016) The importance of oil and gas in today's economy. Tiger General

Palkar AM (2015) Sustain cycle pipeline integrity management system. Proceedings of the Thirty-Ninth Annual Convention & Exhibition, Indonesian Petroleum Association

Sattari F et al (2022) A theoretical framework for data-driven artificial intelligence decision making for enhancing the asset integrity management system in the oil & gas sector. J Loss Prev Process Ind 74:104648

Singh R (2017) Pipeline integrity: management and risk evaluation. Gulf Professional Publishing

Wasim M, Djukic MB (2022) External corrosion of oil and gas pipelines: a review of failure mechanisms and predictive preventions. J Nat Gas Sci Eng 100:104467

Zheng H et al (2012) Research on oil and gas station integrity management system. In: Proceedings of the 9th International Pipeline Conference, vol 1, IPC2012-90261, pp 387–393. September 24–28, 2012. Calgary, Alberta, Canada

Chapter 2
Fundamentals of Big Data Analytics in the Energy Sector

2.1 Introduction

Advancements in data collection, processing, and analytical technologies are propelling the digital transformation of the energy sector. The efficacy, dependability, and security of energy systems can be enhanced by the application of Big Data Analytics. Big Data offers unprecedented opportunities for the oil and gas industry, especially in pipeline operations, to improve decision-making, optimize processes, and mitigate risks.

2.2 Defining Big Data in the Energy Sector

The amalgamation of advanced technologies such as cloud computing, artificial intelligence (AI), the Internet of Things (IoT), and Big Data has emerged as a disruptive influence on the energy sector. The link facilitates the collection, storage, and analysis of substantial data volumes generated by various sources, including smart meters, sensors, and grid management systems. Leveraging this data is essential for improving operational efficiency, optimizing energy usage, and facilitating the shift to renewable energy sources (Yan et al. 2022; Rusitschka and Curry 2016).

Big Data in the energy sector is characterized by its substantial volume, rapid velocity, diverse variety, and high veracity, which collectively enhance predictive analytics and improve decision-making. The application of machine learning techniques enhances performance and reliability through real-time monitoring and evaluation of power systems (Hossain et al. 2019). Big Data Analytics is crucial for addressing the issues associated with the integration of renewable energy, as variable generation sources replace traditional generation resources (de Mergelina 2020). This transition necessitates sophisticated data-driven methodologies to accurately forecast energy production and demand trends.

M. Hussain and T. Zhang, *Big Data Analytics in Energy Pipeline Integrity Management*, Lecture Notes in Energy 104, https://doi.org/10.1007/978-981-96-8019-1_2

The energy sector is experiencing a paradigm shift towards sustainability, where Big Data and environmental, social, and governance metrics are gradually identified for its synergistic potential in driving sustainable investments. Case studies illustrate how organizations leverage Big Data to improve precision and liability in energy management, thereby adopting a culture of sustainability (Olanrewaju et al. 2024). The challenges associated with this transition include regulatory considerations, the need for skilled personnel, and the development of vigorous data governance frameworks to ensure data privacy and security (Almeida 2018).

Besides operational enhancements, Big Data also facilitates strategic initiatives in the oil and gas sector. The implementation of smart grid technologies enhances connectivity and communication within the energy infrastructure, allowing for the integration of renewable energy sources and enhancing demand response capabilities (Hossain et al. 2019; Alexakis et al. 2022). The increasing dependence on geospatial Big Data Analytics enhances the modeling and evaluation of long-term energy transitions, therefore guiding policy decisions and investment strategies.

2.2.1 Big Data's Role in Energy Operations

Pipeline owners can derive significant insights from the extensive data they collect through Big Data Analytics. The integration of Big Data Analytics into energy systems facilitates predictive maintenance, enhances decision-making, and optimizes resource allocation. This is particularly apparent in predictive maintenance, where pipeline firms may minimize downtime and extend the lifespan of essential aging assets by employing advanced analytics to foresee equipment issues (Judijanto et al. 2024). Reliable energy systems necessitate the capability to evaluate the substantial data generated by smart grids and other digital infrastructures (Liao et al. 2023). Figure 2.1 demonstrates that Big Data provides operators with several advantages in pipeline integrity control.

Conventional techniques frequently fail to derive meaningful insights from the diverse and unstructured data common in energy operations. Guo et al. (2018) emphasize that the utilization of Big Data technologies and machine learning algorithms

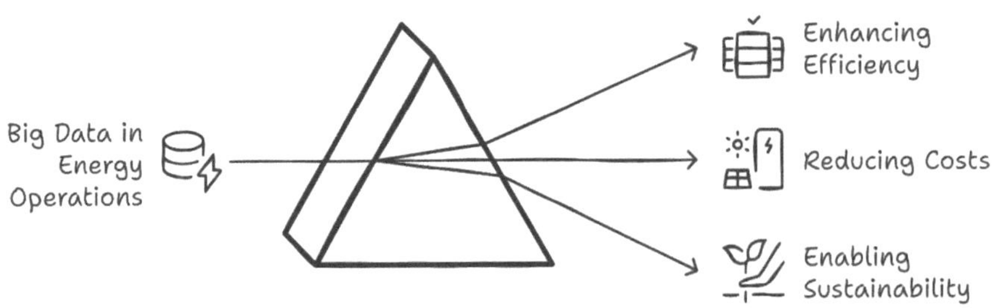

Fig. 2.1 Unveiling Big Data's impact on energy

is crucial for the real-time monitoring and assessment of power system conditions. Angadi et al. (2020) underscore the necessity of sophisticated analytics, highlighting the significance of utilizing extensive data analysis to improve system stability and operational efficiency in the electricity industry.

Big Data's role transcends maintenance and operational efficiency; it also includes sustainability initiatives in the energy sector. For example, Big Data Analytics can enhance grid visualization and administration, aiding the sustainability objectives of smart grids (Ponnusamy et al. 2021). The amalgamation of Big Data and artificial intelligence (AI) technologies has demonstrated encouraging outcomes in enhancing energy consumption efficiency and augmenting the overall efficacy of energy systems. The capacity of Big Data to facilitate green development projects is widely acknowledged, as it allows for more rational resource and energy utilization (Wei and Zhang 2023).

Big Data Analytics not only improve operations but also strengthen security protocols in the energy industry. Given the possible dangers of data breaches or tampering, energy data protection is crucial. According to research, protecting national interests and preserving the integrity of energy operations require an emphasis on secure data management procedures. One strategy being investigated to improve the security of energy data centers is the merging of blockchain technology with Big Data.

2.3 Data Collection in the Energy Sector

The first step in leveraging Big Data is data collection. In energy pipeline integrity management, data is gathered from various sources to provide a comprehensive picture of the pipeline's health, environmental conditions, and operational performance.

2.3.1 Traditional Data Collection Methods

In the past, pipeline operators evaluated the state of pipes using planned maintenance and recurring manual inspections. Despite their value, the frequency, precision, and breadth of these examinations were constrained. Manual data collection techniques, including hydrostatic testing or field inspections, took a lot of time and frequently only gave a momentary picture of the pipeline's state.

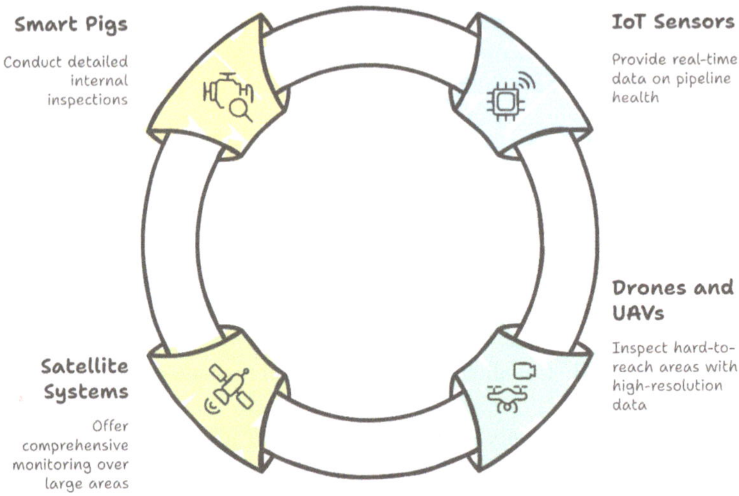

Fig. 2.2 Modern data collection tools

2.3.2 Modern Data Collection Tools

The advent of modern technologies has revolutionized how data is collected from pipelines. Today, advanced tools as shown in Fig. 2.2 being utilized to monitor pipeline condition.

2.3.3 Data Transmission and Storage

The enormous volumes of data being gathered in real time make storage and transmission crucial issues. Constant connectivity can be difficult in remote areas where several pipelines operate. Low-power wide-area networks, 5G networks, and satellite uplinks are examples of advanced communication technologies that are being used more and more to guarantee dependable data transmission.

Once transmitted, the data is often stored in cloud platforms that offer the flexibility and scalability needed to handle Big Data. Cloud storage solutions allow operators to store and access large datasets, while cloud-based analytics tools provide the processing power required for complex analyses (Pothukuchi 2022).

2.4 Types of Data in Energy Pipeline Integrity Management

A clear understanding of the data types available for analysis is critical to efficient pipeline integrity management, refers to Fig. 2.3.

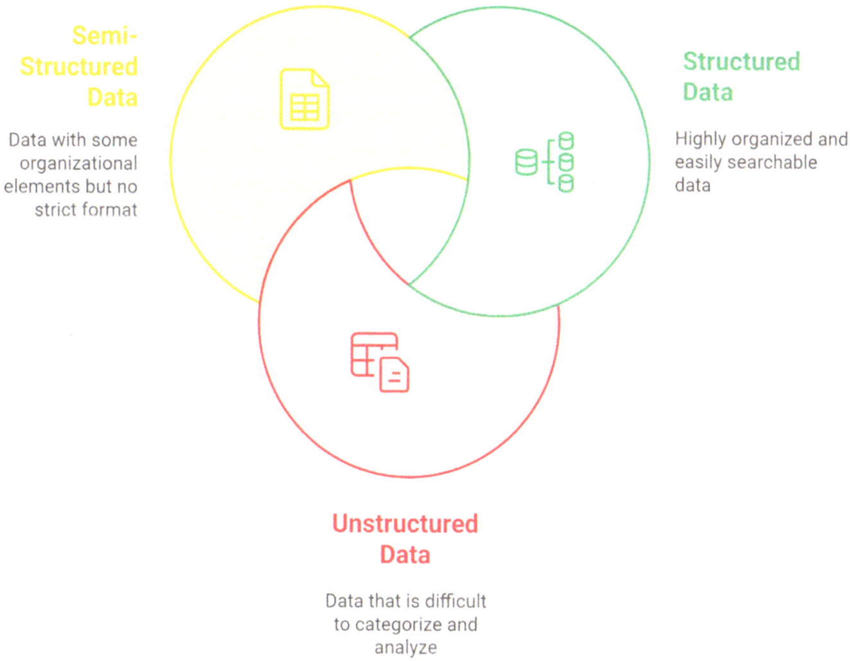

Fig. 2.3 Types of data

Research has shown that full-scale tests, such as fatigue and burst tests on girth welded pipeline sections, provide critical insights into the material properties and behaviors under operational conditions (Dakhel et al. 2023). These experimental findings are essential for developing integrity management systems and conducting residual lifetime calculations, thereby informing guidelines for safe operations (Dakhel et al. 2023). Furthermore, material databases derived from such experiments are pivotal for engineering critical assessments and integrity management plans (Lukács et al. 2010).

Another essential element of pipeline integrity management is data monitoring, especially from corrosion assessments. AI models can use the extensive database generated by corrosion monitoring to prioritize maintenance tasks and identify high-risk pipes. In addition to improving safety, this proactive strategy maximizes resource allocation and lessens the environmental effects of pipeline failures (Lukács et al. 2010). Moreover, the integration of advanced monitoring technologies such as Distributed Acoustic Sensing enables the real-time detection of leak-induced vibrations, which is crucial for timely action. The utilization of machine learning algorithms, capable of analyzing vast datasets to predict potential failures and informing maintenance strategies, significantly enhances the effectiveness of these monitoring systems (Awuku et al. 2023).

Predictive analytics, supported by data from in-line inspections and historical performance, is increasingly being utilized to enhance pipeline integrity management. Operators can make more informed decisions about maintenance and upgrades and have a better understanding of the risks involved in pipeline operations by using models that analyze inspection data. It has been suggested to use Bayesian survival analysis models to forecast girth weld failures, highlighting the significance of data-driven methods in risk assessment (Feng et al. 2019). The accuracy of risk assessments and the integrity management framework as a whole can be greatly increased by including other data types, such as operating circumstances and environmental elements, into predictive models (Zhang et al. 2022).

2.5 The Role of Big Data in Decision-Making

The oil and gas sector has historically generated substantial data quantities, but the advent of Big Data technologies has enabled companies to optimize the utilization of this data. While the oil and gas sector has utilized data since its establishment, Sumbal et al. (2017) indicate that technical improvements facilitating the incessant production of substantial data volumes from diverse sensors and instruments deployed in the field have contributed to the emergence of the phrase "Big Data". A thorough approach to data management and analytics is essential for this transformation to generate value and facilitate informed decision-making.

The benefit of data analytics in the pipeline sector is particularly critical, as it allows for the optimization of operations, risk management, and predictive maintenance. For instance, Tan et al. emphasize that successful project management in the industry can be significantly enhanced through data analytics, which provides fresh insights that lead to improved business value and profitability (Tan et al. 2022). The ability to anticipate failures and address potential issues proactively is vital in pipeline operations, where even minor disruptions can lead to substantial financial losses and environmental hazards. Companies monitor pipeline integrity, predict maintenance requirements, and optimize resource allocation by incorporating sophisticated analytics into decision-making processes. This improves overall operational efficiency, as shown in Fig. 2.4.

By going beyond simple digitization to include the development and tracking of data-driven insights, the idea of "Oil and Gas 4.0" emphasizes the significance of intelligence in decision-making (Majstorović 2022). The need for oil and gas companies to implement integrated platforms that enable the smooth exchange of information across several operational domains is highlighted by this paradigm shift. To fully utilize the promise of Big Data in the pipeline business, the difficulties involved in putting such systems into place, such as integrating various technologies and data sources, must be resolved.

It is difficult to overrate how Big Data can improve environmental sustainability and safety. Ogbu et al. (2024) point out that data science is essential to enhancing oil recovery techniques while reducing negative environmental effects. Businesses can

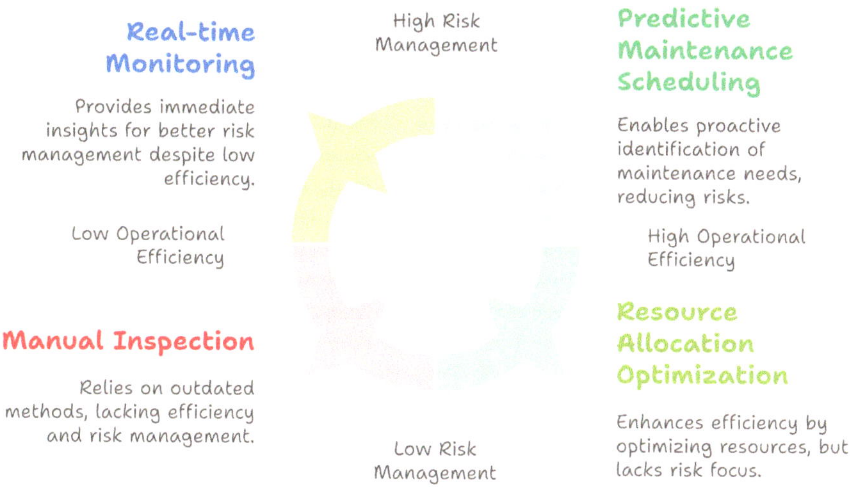

Fig. 2.4 Enhancing pipeline operations with Big Data Analytics

find patterns and trends that guide safer and more sustainable practices by examining the enormous data sets produced during exploration and production. This is especially important when it comes to pipeline operations, as the possibility of spills and leaks presents serious environmental problems. By using predictive analytics, pipeline conditions may be monitored in real time, allowing operators to effectively reduce risks and react quickly to possible problems.

Nguyen et al. (2020) conduct a comprehensive analysis that underscores the significance of Big Data Analytics in the digitization efforts of the oil and gas sector. Analyzing extensive data enhances operational efficiency and fortifies decision-making capabilities. Analyzing seismic data is particularly crucial, since advanced analytics can provide deeper insights into subsurface conditions, aiding in decisions about pipeline route and drilling. The integration of machine learning algorithms into these processes enhances predictive capabilities, facilitating more accurate forecasting of operational and production challenges.

For the oil and gas industry to create value, the interaction between knowledge management and Big Data is also essential. Sumbal et al. (2017) emphasize that fostering innovation and gaining a competitive edge requires the efficient use of data for knowledge discovery. Businesses can improve their capacity to react to operational difficulties and shifts in the market by cultivating a culture of data-driven decision-making. This is especially crucial in the pipeline sector, where it is critical to be able to adjust to changing environmental regulations and standards. Big Data Analytics is crucial for financial decision-making in the oil and gas industry, in addition to operational effectiveness and safety. Businesses may make better financial decisions by using the insights gained from data analytics to guide investment plans, risk assessments, and market forecasts. IoT and Big Data integration can unlock the potential of digital oil fields by supplying real-time data that improves decision-making capabilities, as mentioned by Albar et al. (2019). In addition to enhancing operational effectiveness, this all-encompassing approach to data management supports the organization's overall financial stability.

The challenges associated with Big Data implementation in the oil and gas industry must also be acknowledged. Alıguliyev and İmamverdiyev (2017) highlight that while many companies are beginning to explore Big Data technologies, the application of these tools is still in its infancy. The complexity of data management, coupled with the need for specialized skills and technologies, presents significant hurdles for organizations seeking to fully leverage Big Data. However, the potential benefits far outweigh these challenges, as evidenced by the increasing prioritization of Big Data initiatives within the industry.

2.6 The Value of Real-Time Data Processing

The integration of real-time data processing technologies is instigating a transformative shift in the oil and gas sector. This advancement is crucial for enhancing environmental sustainability, safety, and operational efficiency; it is not merely a trend. The advent of advanced technologies such as Big Data Analytics, IoT, and AI has enabled businesses to leverage real-time data to improve operational efficiency and decision-making. To enhance resource recovery and reduce environmental impacts, these technologies offer real-time monitoring, predictive maintenance, and optimization of production processes (Arinze and Jacks 2024; Arinze et al. 2024).

The constant monitoring of vital factors like temperature, pressure, and flow rates is made possible by real-time data processing. According to Yue et al. (2018), this skill is especially useful in workflows related to production optimization and reservoir surveillance. Operators can improve production efficiency by making well-informed decisions based on the transformation of wellhead data into actionable information. Real-time monitoring of these parameters lowers the chance of operational failures considerably, improving overall safety and dependability in oil and gas operations.

The benefits of AI are amplified when integrated into real-time data processing. AI-driven predictive maintenance solutions enable operators to anticipate equipment failures, hence reducing maintenance costs and minimizing downtime. Artificial intelligence (AI) systems can identify trends and anomalies indicative of potential issues by analyzing historical performance data with real-time inputs. This proactive technique enhances asset integrity management while promoting sustainable operations through the reduction of waste and resource consumption. By allocating resources to areas of greatest need, artificial intelligence (AI) enhances operational efficiency, increasing production while reducing costs.

It is impossible to overestimate the importance of IoT in the oil and gas industry. IoT devices make it easier to gather enormous volumes of data from a variety of sources, such as sensors placed across production facilities and drilling sites. Real-time processing of this data can yield insights into environmental conditions and operational performance (Hossein Motlagh et al. 2020). Improved safety measures are made possible by the deployment of IoT technologies, which enable operators to continuously monitor environmental and equipment conditions. This capacity is essential for averting situations that can result in safety risks or environmental catastrophes (Hossein Motlagh et al. 2020; Alakbarov and Hashimov 2018). Additionally, a strong framework for prompt and well-informed decision-making is produced by combining IoT with AI and Big Data Analytics.

The integration of real-time data processing technology in the oil and gas industry is best illustrated by the Digital Oil Field (DOF) concept. DOF signifies a change from conventional operational models to a more data-driven and integrated strategy. It includes the application of cutting-edge software and data analysis methods to improve operational effectiveness and profitability (Qin and Han 2019). Operators may monitor important performance indicators and make well-informed decisions in real-time thanks to the DOF framework's facilitation of data visualization on dashboards.

Addressing the intricacies and difficulties that the sector faces today requires this shift to a proactive operational strategy. Furthermore, real-time data processing in the oil and gas industry has important environmental ramifications. Businesses can monitor pollutants and streamline operations to lessen their environmental impact thanks to advanced technologies. Operators may put plans into place that not only increase productivity but also support sustainability objectives by utilizing real-time data. The potential for technology innovation to lead to positive environmental outcomes is demonstrated by the incorporation of AI and data analytics into emission reduction initiatives (Arinze and Jacks 2024). Facing escalating criticism regarding its environmental impact, the dual emphasis on environmental stewardship and operational efficiency is becoming paramount. Real-time data processing enhances organizational resilience in the oil and gas sector, alongside operational and environmental benefits. Navigating the industry's intricacies requires teamwork and well-informed decision-making, which are fostered by the ability to share real-time data amongst several stakeholders. Businesses may adapt to difficulties and uncertainties more successfully when they have integrated operations, which are defined by data exchange and cross-disciplinary collaboration (Bento et al. 2021). In a sector that frequently experiences unstable market conditions and regulatory changes, this resilience is essential.

It is important to consider the difficulties involved in putting real-time data processing systems into practice. Adoption is significantly hampered by problems such as data availability, system integration, and the requirement for qualified staff to analyze and act upon data insights. To guarantee that staff members are prepared to use new technologies efficiently, the quick speed of technical improvement demands ongoing investment in training and development. To fully utilize real-time data processing in the oil and gas industry, these obstacles must be overcome. Figure 2.5 illustrates a few advantages of real-time data processing.

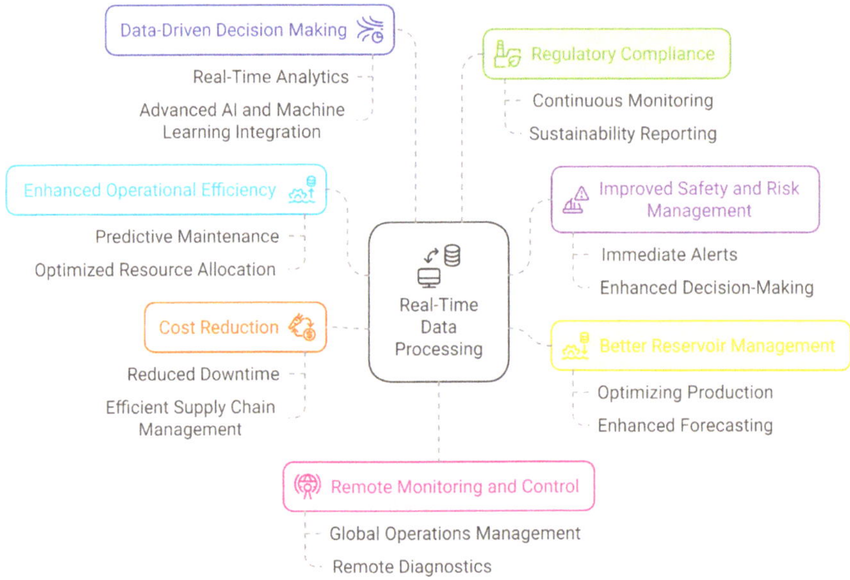

Fig. 2.5 Benefits of real-time data processing in oil and gas

References

Alakbarov RG, Hashimov MA (2018) Application and security issues of Internet of Things in oil-gas industry. Int J Educ Manage Eng 8(6):24

Albar A et al (2019) Combining the power of IoT and big data to unleash the potential of digital oil field. In: International petroleum technology conference. IPTC

Alexakis K et al (2022) Intelligent querying for implementing building aggregation pipelines. In: 2022 13th International conference on information, intelligence, systems & applications (IISA). IEEE

Aliguliyev RM, Imamverdiyev YN (2017) Conceptual big data architecture for the oil and gas industry. Probl Inf Technol:3–13

Almeida F (2018) Big data: concept, potentialities and vulnerabilities. Emer Sci J 2(1):1–10

Angadi RV, Venkataramu P, Daram SB (2020) Role of big data analytics in power system application. In: E3S web of conferences. EDP Sciences

Arinze C, Jacks B (2024) A comprehensive review on AIDRIVEN optimization techniques enhancing sustainability in oil and gas production processes. Eng Sci Technol J 5(3):962–973

Arinze CA et al (2024) Evaluating the integration of advanced IT solutions for emission reduction in the oil and gas sector. Eng Sci Technol J 5(3):639–652

Awuku B, Huang Y, Yodo N (2023) Predicting natural gas pipeline failures caused by natural forces: an artificial intelligence classification approach. Appl Sci 13(7):4322

Bento F, Garotti L, Mercado MP (2021) Organizational resilience in the oil and gas industry: a scoping review. Saf Sci 133:105036

Dakhel A et al (2023) Fatigue and burst tests of full-scale girth welded pipeline sections for safe operations. Weld World 67(5):1193–1208

de Mergelina P (2020) Effects of renewable energies and Big Data on the biggest Spanish energy power companies business models. Am J Econ Bus Admin 12(1):42–48

Feng Q, Sha S, Dai L (2019) Bayesian survival analysis model for girth weld failure prediction. Appl Sci 9(6):1150

Guo Y et al (2018) Complex power system status monitoring and evaluation using big data platform and machine learning algorithms: a review and a case study. Complexity 2018(1):8496187

Hossain E et al (2019) Application of big data and machine learning in smart grid, and associated security concerns: a review. IEEE Access 7:13960–13988

Hossein Motlagh N et al (2020) Internet of Things (IoT) and the energy sector. Energies 13(2):494

Judijanto L, Uhai S, Suri I (2024) The influence of business analytics and Big Data on predictive maintenance and asset management. Eastasouth J Inf Syst Comp Sci 1(03):123–135

Liao H, Michalenko E, Vegunta SC (2023) Review of big data analytics for smart electrical energy systems. Energies 16(8):3581

Lukács J, Nagy G, Török I (2010) The role of the external and internal reinforcing on the structural integrity of industrial and transporting steel pipelines. In: Materials science forum. Trans Tech Publ

Majstorović VD (2022) Application of Industry 4.0 model in oil and gas companies. J Eng Manage Competitiveness (JEMC) 12(1):77–84

Nguyen T, Gosine RG, Warrian P (2020) A systematic review of big data analytics for oil and gas industry 4.0. IEEE Access 8:61183–61201

Ogbu AD, Ozowe W, Ikevuje AH (2024) Data science's pivotal role in enhancing oil recovery methods while minimizing environmental footprints: an insightful review. Computer Science & IT Research Journal 5(7):1621–1633

Olanrewaju OIK, Ekechukwu DE, Simpa P (2024) Driving energy transition through financial innovation: the critical role of Big Data and ESG metrics. Comp Sci IT Res J 5(6):1434–1452

Ponnusamy VK et al (2021) A comprehensive review on sustainable aspects of big data analytics for the smart grid. Sustainability 13(23):13322

Pothukuchi AS (2022) A critical analysis of the challenges and opportunities to optimize storage costs for Big Data in the cloud. Asian J Multidiscip Res Rev 3(1):132–144

Qin H, Han Z (2019) Crude-oil scheduling network in smart field under cyber-physical system. IEEE Access 7:91703–91719

Rusitschka S, Curry E (2016) Big Data in the energy and transport sectors. In: New horizons for a data-driven economy: a roadmap for usage and exploitation of Big Data in Europe, pp 225–244

Sumbal MS, Tsui E, See-to EW (2017) Interrelationship between big data and knowledge management: an exploratory study in the oil and gas sector. J Knowl Manag 21(1):180–196

Tan V, Aziz KA, Razavi SH (2022) Data analytics for effective project management in the oil and gas industry. In: Proceedings of the international conference on technology and innovation management (ICTIM 2022)

Wei J, Zhang X (2023) The role of big data in promoting green development: based on the quasi-natural experiment of the big data experimental zone. Int J Environ Res Public Health 20(5):4097

Yan Q et al (2022) Exploring the promotion method and pathway of energy big data service business. IOP Conf Ser: Earth Environ Sci

Yue Y, Minh V, Xiao C (2018) Reservoir surveillance and production optimization workflow from real-time data advantage at a joint venture gas project. In: Proceedings of the international field exploration and development conference 2017. Springer

Zhang M et al (2022) A data-driven based method for pipeline additional stress prediction subject to landslide geohazards. Sustainability 14(19):11999

Chapter 3
Data Collection Methods in Pipeline Integrity Management

3.1 Introduction

Throughout the pipeline's lifecycle, monitoring, inspection, maintenance, and repair are all part of the holistic strategy that is known as pipeline integrity management. Data collection techniques are essential to this process because they give the information needed to evaluate pipeline status, anticipate possible breakdowns, and carry out efficient maintenance plans. The pipeline integrity management process is depicted in Fig. 3.1.

The methodical gathering and examination of pipeline condition data is the basis of an efficient PIM. Understanding the integrity of pipeline systems requires a variety of data sources, such as historical performance data, inspection results, and environmental conditions. In order to build integrity management systems and provide reliable integrity evaluations and residual lifespan predictions for pipelines, for example, Dakhel et al. (2023) emphasize the significance of experimental data. The development of policies and procedures that improve pipeline systems' operating effectiveness and safety is made possible by this data-driven methodology.

For pipeline management to be efficient in urban settings, geographical data integration is crucial. In order to increase the accuracy and completeness of data, Shen et al. (2024) discuss about a spatial data fusion model that integrates the physical, temporal, and semantic features of subterranean pipes. Rapid scenario modeling and pipeline system maintenance depend on effective data sharing and improved information content, both of which are made possible by this approach. In highly populated locations where pipelines may be exposed to a variety of environmental influences, the capacity to integrate multi-feature information is especially advantageous.

One of the biggest risks to pipeline integrity is corrosion, which calls for constant maintenance and observation. In order to prevent failures and anticipate possible places of failure through data collection and analysis, Usuoyibo and Obogai (2020) highlight the critical need for a systematic approach to pipeline monitoring and inspection. The authors contend that to guarantee the durability and security of

M. Hussain and T. Zhang, *Big Data Analytics in Energy Pipeline Integrity Management*, Lecture Notes in Energy 104, https://doi.org/10.1007/978-981-96-8019-1_3

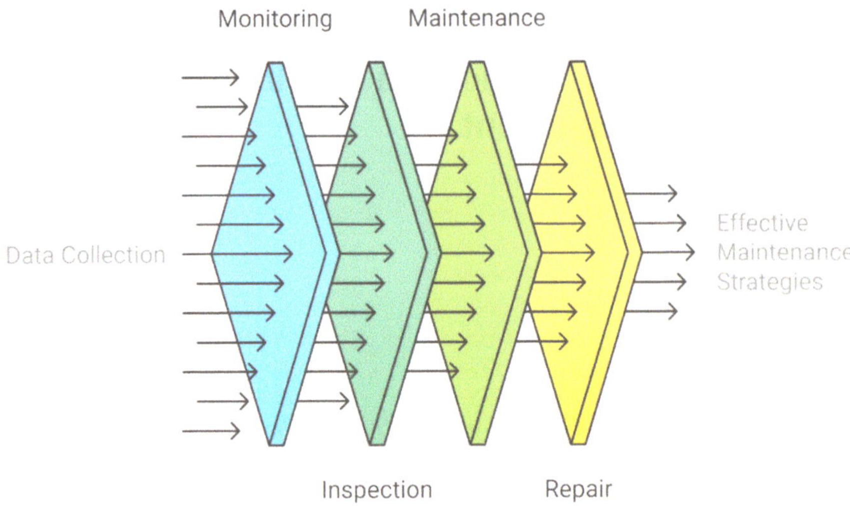

Fig. 3.1 Pipeline integrity management process

pipeline networks, a cradle-to-grave strategy to PIM is necessary. This method entails gathering information on operational characteristics, environmental factors, and corrosion rates to guide maintenance choices.

One of the best approaches for gathering information on pipeline integrity is in-line inspection (ILI). The significance of historical data in identifying high-risk pipelines is emphasized by Ma et al. (2021) as they outline different ILI techniques and data management tools. According to the authors, condition-based maintenance techniques can be applied considering the knowledge gathered from ILI data, enabling prompt actions that reduce the likelihood of pipeline failures. To increase the efficacy of ILI techniques, sophisticated sensors and data management systems must be integrated. Another crucial component of PIM that significantly depends on data collection is condition-based maintenance (CBM). The use of reinforcement learning in CBM for gas pipes that are susceptible to interior corrosion is covered by Mahmoodzadeh et al. (2020). Operators can make well-informed judgments about maintenance and repairs by consistently monitoring wall thickness and other important metrics. In addition to improving safety, this proactive strategy lessens the financial effect of pipeline failures.

Geometric data collected through ILI tools is essential for assessing pipeline defects such as dents. Lei et al. (2024) emphasize how important it is to model and analyze strain using geometric ILI data in order to assess pipeline integrity. The capacity to identify and evaluate dents enables prompt repairs, averting possible ruptures or leaks. The operational integrity of pipelines depends on this data-driven approach to defect management. PIM is increasingly using machine learning techniques to improve its predictive and data analysis capabilities. Comprehensive integrity management plans that take into account the unique hazards connected to long-distance pipelines are essential, according to Fan and Feng (2023). The authors emphasize the need of efficient data collection and analysis by pointing out that pipeline leaks can cause substantial economic losses as well as environmental harm.

3.2 Traditional Data Collection Methods

Historically, pipeline integrity was managed using manual and scheduled inspection techniques. While these methods provided operators with insights into pipeline health, they were limited by their infrequency, high cost, and difficulty in covering extensive pipeline networks.

3.2.1 Hydrostatic Testing

A pipeline is filled with water and pressurized to a pressure greater than its typical operating pressure in order to conduct hydrostatic testing. By using this test, operators can evaluate the pipeline's strength and integrity and look for any leaks or vulnerabilities. This approach is labor-intensive and usually carried very infrequently, therefore it cannot offer real-time data on pipeline condition, even though it is successful in detecting large-scale faults.

3.2.2 Manual Inspections

Field technicians regularly inspect above-ground pipelines for visible signs of damage, such as corrosion, dents, or leaks. These inspections are typically scheduled at regular intervals, and while they can detect obvious problems, they are subjective and dependent on the skill and attention of the inspector. Manual inspections are also time-consuming and may not be feasible for pipelines located in remote or hazardous areas.

3.2.3 In-Line Inspection (ILI) Tools

"Smart pigs" or in-line inspection (ILI) tools are among the most significant developments in pipeline inspection (Parlak and Yavasoglu 2023). Figure 3.2 illustrates some of the instruments and their uses. These robotic devices are placed inside the pipeline and move along its length, gathering information on wall thickness, corrosion, cracks, and other structural problems. Smart pigs use advanced sensors, such as magnetic flux leakage (MFL) or ultrasonic testing (UT), to detect anomalies within

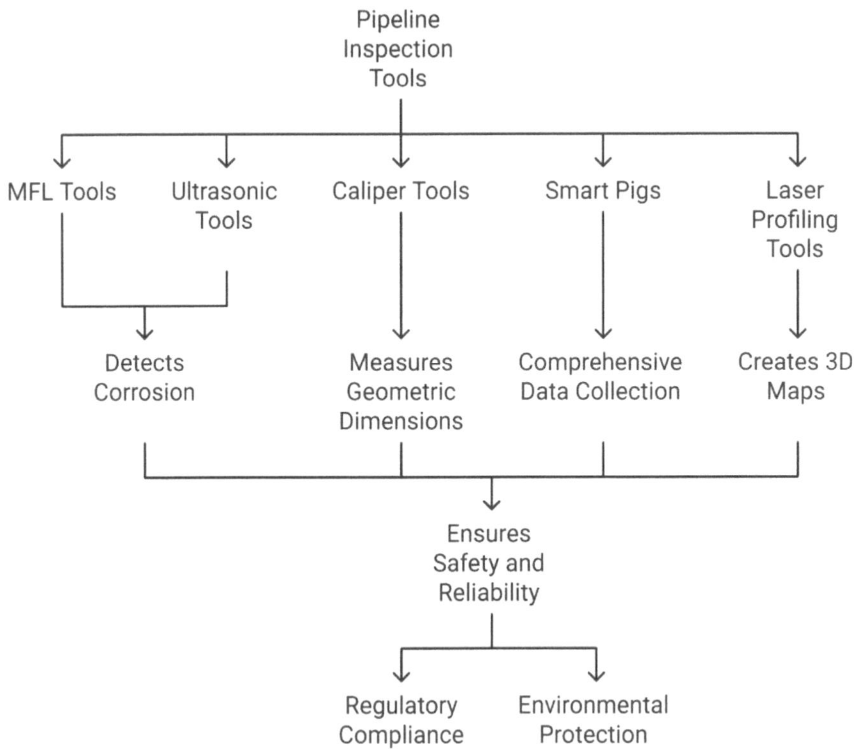

Fig. 3.2 Pipeline inspection tools and their importance

the pipeline (Xie and Tian 2018). However, smart pigging is generally conducted periodically and is limited in its ability to provide continuous real-time monitoring.

3.3 Modern Data Collection Techniques

Advancements in technology have revolutionized the way data is collected from pipelines. These innovations offer operators real-time, continuous data streams that enable predictive maintenance and proactive decision-making.

3.3.1 Internet of Things (IoT) Sensors

IoT sensors have transformed the way pipeline data is gathered, providing operators with a continuous flow of real-time information (Priyanka et al. 2021). These sensors are positioned thoughtfully throughout the pipeline to keep an eye on a variety of characteristics, such as:

- **Pressure**: Sudden pressure drops may indicate a leak or rupture, while gradual pressure changes could signal corrosion or pipeline degradation.
- **Temperature**: Variations in temperature, particularly in pipelines transporting gases or liquids, can indicate inefficiencies or potential damage.
- **Vibration**: Excessive vibration can be a sign of mechanical wear or external disturbances, such as nearby construction activities.
- **Flow Rate**: Changes in flow rate can reveal blockages, leaks, or issues with pipeline throughput.

IoT sensors transmit their data wirelessly to a centralized control center, where operators can analyze the information in real time. These sensors provide continuous monitoring, ensuring that potential issues are detected before they escalate into significant problems.

3.3.2 Drones and Aerial Surveillance

The use of drones, or unmanned aerial vehicles (UAVs), is becoming increasingly common in pipeline integrity management. Drones equipped with high-resolution cameras, thermal imaging, and other sensors can fly over long stretches of pipeline and capture detailed visual data, which is then analyzed for signs of damage or wear (Peddinti et al. 2023; Liang et al. 2023).

- **Thermal Imaging**: Thermal cameras can detect temperature anomalies that may indicate a leak or external damage to the pipeline.
- **High-Resolution Imagery**: Drones capture high-definition images of the pipeline, allowing operators to spot corrosion, dents, or other visible defects.
- **3D Mapping**: Drones can create 3D maps of the terrain around a pipeline, which helps in assessing potential geohazards, such as landslides or erosion.

In remote or challenging-to-reach locations, such hilly regions or offshore platforms, drones are particularly helpful for pipeline inspections. Additionally, especially in dangerous settings, drones offer a safer substitute for manual inspections.

3.3.3 Satellite Monitoring

Satellite imagery offers an unparalleled view of pipeline networks, covering large geographic areas that may be difficult or impossible to monitor on the ground (Roper and Dutta 2005). Satellites equipped with optical, radar, or infrared sensors can capture images of pipelines and the surrounding environment (Parker et al. 2021). Monitoring environmental changes, spotting encroachments, and estimating the likelihood of natural disasters like earthquakes or floods can all benefit from this data.

- **Radar Satellite Imaging**: Radar satellites can monitor ground displacement, identifying areas where subsidence or ground movement may compromise the structural integrity of a pipeline.
- **Optical Satellites**: Optical satellites provide high-resolution images of the Earth's surface, helping operators detect encroachments, construction activities, or unauthorized access near pipelines.

3.3.4 *Acoustic Monitoring*

Acoustic sensors are used to detect sounds generated by pipeline leaks or ruptures (Fan et al. 2022; Xu et al. 2013). These sensors capture sound waves produced by escaping gases or liquids and can pinpoint the location of a leak by analyzing the acoustic signal (Xu et al. 2013; Adnan et al. 2015). This technique is particularly effective for detecting small leaks that may not be identified by other methods.

3.4 Challenges in Data Collection

The oil and gas industry has numerous technical, operational, and legal obstacles when it comes to gathering data for pipeline integrity management. Because the commodities being transported are toxic, pipeline integrity is crucial, and improper management can result in catastrophic breakdowns. Since it influences risk evaluations, maintenance plans, and operational choices, data collecting is essential to guaranteeing the security and dependability of these systems (Dai et al. 2017; Wang et al. 2013).

The requirement for dependable, high-quality data is one of the main obstacles in data collection. Accurate historical data is essential for pipeline integrity management because it can be used to spot failure patterns and evaluate how well pipeline management systems are working (Dai et al. 2017; Wang et al. 2013). However, the variety of sources from which this data must be obtained—such as sensors, inspection reports, and operations logs—often makes data collection challenging. The formats, standards, and accuracy levels of these sources may vary, which could result in discrepancies that compromise the integrity evaluation procedure (Munappy et al. 2020).

Another level of difficulty is added by the intricacy of contemporary pipeline systems. The amount of data created grows exponentially as pipelines are increasingly integrated with sophisticated monitoring technologies, such as Supervisory Control and Data Acquisition (SCADA) systems. Strong data management techniques are therefore required to guarantee that the information gathered is timely, accurate, and pertinent for making decisions (Liu et al. 2022; Polyzotis et al. 2017).

Complex data pipelines that can manage the dynamic nature of pipeline operations are necessary for the integration of multiple data types, ranging from historical maintenance records to real-time operational data.

In addition to technical challenges, operational factors also play a significant role in data collection difficulties. For instance, the need for continuous monitoring and the implementation of condition-based maintenance strategies can strain existing data collection frameworks. Operators must balance the demands of real-time data acquisition with the need for comprehensive data analysis, which can lead to gaps in data collection if not managed effectively (Ma et al. 2021). The geographical dispersion of pipeline networks often complicates data collection efforts, as remote locations may lack reliable communication infrastructure, leading to delays in data transmission and potential data loss (Febaide and Uzedhe 2023).

Regulatory compliance is another critical aspect that influences data collection practices in pipeline integrity management. Operators are required to adhere to stringent regulations that mandate the collection and reporting of specific data types, including environmental impact assessments and safety audits. Navigating these regulatory requirements can be challenging, particularly when they change frequently or vary by jurisdiction. The need for compliance can also lead to an overemphasis on certain data types at the expense of others, potentially skewing the overall integrity assessment (Lukács et al. 2010).

There are advantages and disadvantages to integrating new technology for data collection in pipeline integrity monitoring, such as wireless data capture systems and sophisticated sensors. These technologies bring new challenges to data security and management, even while they can improve monitoring systems' accuracy and data collection capacities (Febaide and Uzedhe 2023; Haro-Olmo et al. 2021). To safeguard sensitive data, for instance, the deployment of wireless sensor networks may expose data to cybersecurity risks, requiring strong security measures. Traditional data management frameworks may not be able to handle the enormous volume of data produced by contemporary pipeline systems, which could result in processing errors and inefficiencies. Operators must deal with the difficulties of data integration and curation as they work to fully utilize Big Data Analytics, making sure that the information gathered is both actionable and useful (Lukács et al. 2010; Haro-Olmo et al. 2021).

Given these difficulties, pipeline operators must implement thorough data management plans that include every stage of the data lifecycle, from collection to analysis and reporting. This involves spending money on cutting-edge data processing tools that can improve the quality of data obtained and expedite data workflows (Polyzotis et al. 2017). Furthermore, companies can make sure that data gathering activities are in line with both regulatory requirements and more general operational goals by cultivating a culture of data-driven decision-making.

In the end, overcoming the difficulties in gathering data for pipeline integrity monitoring necessitates a multipronged strategy that blends innovative technology with efficient management techniques. Operators can improve pipeline systems' safety and dependability and reduce the risks of pipeline failures by utilizing advanced data analytics, streamlining data integration procedures, and making sure regulatory standards are followed.

3.4.1 Data Accuracy and Reliability

The accuracy of the data collected by sensors, drones, and other devices is critical to the success of a pipeline integrity management program. Inaccurate or unreliable data can lead to false positives or missed defects, undermining the effectiveness of predictive maintenance efforts (Achouch et al. 2022). Regular sensor calibration and data validation against other sources are necessary to guarantee data reliability. Safety and environmental protection depend on pipeline integrity and efficient monitoring. Risk assessment and decision-making procedures depend on precise data collection. However, several obstacles stand in the way of achieving high data reliability and precision.

The unpredictability and inconsistency of data gathered from different monitoring systems present a major obstacle. According to Mao et al. (2024), the reliability of risk assessments for subterranean oil and gas pipelines can be improved by efficiently managing irregular and changeable data. To glean valuable insights from the gathered data, sophisticated data processing and analysis techniques must be applied. Additionally, by enabling real-time monitoring and decision-making based on thorough data analytics, the integration of Industry 4.0 technology can promote enhanced data accuracy (Majstorović 2022). But there are drawbacks to the shift to such integrated platforms, such as the requirement for existing technologies to work together and the creation of new frameworks for data management. By automating the detection and removal of faulty data, machine learning techniques have become a viable way to improve data accuracy. For example, Doan and Vo showed how machine learning techniques like K-Means Clustering and Support Vector Regression can efficiently eliminate noisy data and outliers from production datasets, increasing the precision of production forecasts (Doan and Vo 2020). This automated method improves the dependability of the insights obtained from the data while also streamlining data processing.

In offshore environments, where conditions can be extreme, the reliability of equipment is further challenged by environmental factors. Rahimi et al. proposed a framework for reliability assessment that incorporates various types of data, including design and operational environment data, to evaluate equipment performance in challenging conditions (Rahimi et al. 2011). This comprehensive approach underscores the importance of accurate data collection in predicting equipment reliability and ensuring the integrity of offshore pipelines. The development of predictive models using advanced analytics is another avenue for enhancing data accuracy and reliability. Yusoff et al. (2025) highlight that trustworthy predictive models may greatly support decision-making in the oil and gas industry, especially when it comes to production optimization and maintenance. Operators can lower the risks involved in pipeline operations and increase the accuracy of their forecasts by utilizing Big

Data Analytics. Additionally, there are chances to improve data management procedures in the oil and gas industry through the combination of Big Data and cloud computing technology. Yang et al. (2019) talk about how these technologies might make it easier to manage and analyze vast amounts of data, which will ultimately increase operational effectiveness and the precision of decisions. However, in order to handle possible problems with data security and integrity, the shift to cloud-based systems needs to be handled cautiously.

3.4.2 Data Volume and Management

The substantial quantities of data generated during pipeline operations present significant challenges for data volume and management in pipeline integrity management within the oil and gas sector. To ensure safety, compliance, and operational efficacy, this data must be maintained proficiently. The vast volume of data produced by many sources, including sensors, monitoring systems, and operating records, constitutes a primary challenge. For example, generated water, a by-product of oil and gas extraction, creates significant volumes, frequently surpassing 70 billion barrels worldwide (Abdelhamid 2024). The substantial volume of wastewater, along with the data produced from monitoring its attributes, requires comprehensive data management systems to guarantee precise tracking and reporting (Dickhout et al. 2017). The amalgamation of data from several sources, including flowback and produced water, complicates management, as producers frequently report aggregated volumes without differentiating between the two fluid types (Kondash and Vengosh 2015).

The complexity of data types adds another layer of difficulty. Pipeline operators must manage diverse data sets, including pressure–volume-temperature (PVT) data, gas-to-oil ratios (GOR), and various operational parameters (Barry et al. 2018; Soliman et al. 2020). The integration of these data types into a cohesive management framework is essential for accurate monitoring and decision-making. For example, the accurate determination of parameters such as bubble point pressure and oil formation volume factor are critical for effective reservoir management and pipeline design (Soliman et al. 2020; Gholami 2016). The use of intelligent management systems that make use of multi-source data fusion is becoming more and more crucial to handle these issues. By combining and evaluating several data sources, these systems can improve pipeline management's effectiveness and security. Promising methods for handling massive amounts of data are also provided by machine learning techniques. Machine learning algorithms can detect patterns and abnormalities in sensor and equipment data, increasing monitoring system accuracy and enabling predictive maintenance (Jambol et al. 2024).

The implementation of Geographic Information Systems (GIS) can significantly enhance data management capabilities in pipeline integrity management. GIS can facilitate the visualization and analysis of spatial data related to pipeline construction, operation, and maintenance, thereby improving decision-making processes. The integration of GIS with real-time data acquisition systems can provide operators with comprehensive insights into pipeline integrity, enabling timely interventions to mitigate risks. In addition to technological advancements, regulatory compliance plays a crucial role in data management. Operators are required to adhere to stringent regulations regarding data reporting and monitoring, which can complicate data management efforts. The need for accurate and timely reporting of data related to environmental impacts, safety audits, and operational performance necessitates robust data management frameworks that can accommodate regulatory requirements (Dai et al. 2017).

3.4.3 Transmission and Connectivity

Transmission and connectivity in pipeline integrity management are critical components that ensure the safe and efficient transportation of oil and gas. The challenges associated with these aspects are multifaceted, encompassing technological, operational, and regulatory dimensions. Effective transmission and connectivity are essential for real-time monitoring, data collection, and communication between various components of the pipeline system, ultimately contributing to maintaining pipeline integrity.

The requirement for dependable communication systems that can function in a variety of challenging and frequently severe conditions is one of the transmission and connection problems. Usually, pipelines are found in isolated locations with inadequate traditional communication infrastructure. According to Ravishankar et al. (2022), to improve the resilience of pipeline systems, sophisticated technologies like drones and artificial intelligence must be adopted due to the limits of human observation-based inspection methods. By enabling real-time data transmission and monitoring, these technologies can raise pipeline operations' general level of safety and dependability.

A key component of efficient pipeline management is the integration of multiple data sources. The connectivity and data transfer between several monitoring systems can be greatly improved by using multi-source data fusion techniques. To increase the effectiveness and safety of gas pipeline management, transmission and connectivity practices are significantly shaped by the regulatory environment in addition to technological improvements. Strict rules pertaining to data reporting and monitoring must be followed by operators, which might make communication system management more difficult. Strong communication frameworks that can adapt to regulatory requirements are necessary for the correct and timely reporting of data pertaining to operational performance, safety audits, and environmental impacts. Ensuring that pipeline systems function in accordance with safety and environmental regulations requires this compliance.

Maintaining pipeline integrity requires the use of cutting-edge monitoring technologies, such as sensors and in-line inspection instruments. The significance of these technologies in identifying flaws and keeping an eye on pipeline status is covered by Ma et al. (2021). For analysis and decision-making, the data gathered from these monitoring devices need to be reliably sent to central management systems. Strong communication protocols and infrastructure that can handle the large amounts of data produced by these technologies are needed for this. The contribution of automation to improving connectivity and transmission is substantial and cannot be overlooked. Process parameter monitoring throughout the pipeline network can be made easier with the introduction of automation technologies such as programmable logic controllers (PLCs). Because automation enables real-time modifications and monitoring of operating parameters, Wang et al. (2017) emphasize the significance of automation in guaranteeing the safety and stability of pipeline transmission. When it comes to handling the complexity of long-distance pipelines that cross a variety of geographical terrains, automation is very important.

The reliability of communication systems may be impacted by the difficulties presented by environmental elements, such as severe weather and physical barriers. By offering alternate methods of data transmission in remote locations, the use of cutting-edge technologies like satellite communication and wireless sensor networks can assist in lessening these difficulties (Ho et al. 2020). By guaranteeing constant data transmission and monitoring, which is crucial for preserving pipeline integrity, these technologies improve the resilience of pipeline systems.

3.5 Role of Cloud Computing in Data Collection

In the field of data collection, cloud computing has become a game-changing technology, especially for sectors like oil and gas pipeline integrity management that need to manage enormous volumes of data. Cloud computing plays a variety of roles in supporting data collection, including storage, processing, and analysis, all of which are essential for sound decision-making and efficient operations.

The capacity of cloud computing to offer scalable storage solutions is one of its main benefits. According to Mohlamean and Ruxwana (2016), cloud computing enables businesses to store vast amounts of data without having to make substantial upfront expenditures in physical infrastructure. This is especially important in the oil and gas industry because pipeline monitoring systems can provide a significant amount of data. The ability to scale storage resources dynamically enables companies to accommodate fluctuating data needs without compromising performance or accessibility.

The processing power needed to analyze big datasets is improved by cloud computing. To make prompt decisions about pipeline integrity monitoring, real-time data processing is made possible by the integration of cloud-based analytics tools. According to Yeboah-Ofori et al. (2023), cloud computing makes it easier to process large data, which helps businesses extract useful insights from intricate datasets. This feature is especially helpful for predictive maintenance, as prompt sensor data analysis can assist in spotting possible problems before they become serious ones. Another important factor in data collection is cloud computing's collaborative nature. Cloud systems facilitate collaboration across pipeline management teams by granting numerous stakeholders' real-time access to and sharing of data. In the oil and gas sector, where several teams—from engineering to environmental compliance—must collaborate to guarantee pipeline integrity, this is especially crucial. Efficient data sharing between departments improves coordination and communication, which eventually results in better pipeline operations management.

Internet of Things (IoT) devices, which are being used more and more to monitor pipeline conditions, can be integrated with cloud computing. The cloud provides the necessary infrastructure required to manage the data produced by several sensors placed along pipelines. The integrity of the pipeline depends on the constant monitoring of important parameters like temperature, pressure, and flow rates. Because of the cloud's capacity to manage the volume of data generated by these devices, operators may manage pipeline systems proactively by having access to the most recent data.

In the context of cloud computing for data collecting, security and compliance are also important factors. Businesses need to make sure that the data they store on the cloud is safe and complies with all applicable laws. According to Alharbi (2017), worries about data security and privacy may prevent people from using cloud services. However, these worries are being lessened by developments in security protocols and encryption, which makes cloud computing a practical choice for managing sensitive data in sectors like gas and oil industry (Adee and Mouratidis 2022).

As regulatory frameworks develop, organizations must navigate these complexities to ensure compliance while leveraging cloud technologies for data collection. The cost-effectiveness of cloud computing is another compelling factor driving its adoption for data collection. Businesses can lower the capital costs of maintaining on-premises infrastructure by leveraging cloud services. Businesses can more effectively manage resources and invest in other crucial aspects of their business thanks to this financial flexibility. This cost effectiveness is further increased by the pay-as-you-go approach that many cloud providers offer, which allows businesses to only pay for the services they utilize.

3.5.1 Data Storage

Large data storage is made possible by cloud platforms, which eliminate the need for expensive on-premises infrastructure. For pipeline networks that produce terabytes of data every day, this is very crucial. Redundancy and backup features provided by cloud storage solutions guarantee that data is kept safe in the case of a system breakdown.

3.5.2 Data Analytics

Operators can evaluate and process their data in real time using cloud-based analytics tools. Cloud-based machine learning models and advanced algorithms can detect anomalies, predict errors, and propose solutions. Operators may proactively preserve the integrity of their pipelines and make data-informed decisions utilizing these analytical tools.

3.6 The Future of Data Collection in Pipeline Integrity Management

Given technology advancements, evolving laws, and the increasing need for efficiency and safety in the oil and gas sector, data collection in pipeline integrity management is poised for a significant transformation. Novel data collection techniques will be crucial for preserving the integrity and safety of pipeline networks as the industry contends with deteriorating infrastructure and the necessity for enhanced monitoring capabilities.

Combining cutting-edge sensor technology with the Internet of Things (IoT) is one of the most exciting advances in this area. Real-time pipeline condition monitoring is possible with IoT devices, enabling the ongoing gathering of information on temperature, pressure, flow rates, and other crucial factors. Zhang and Xie (2024) emphasizes that the use of IoT technology in pipeline systems can support clever safety management techniques, allowing operators to quickly identify leaks and other irregularities. This move to real-time data collection will improve the capacity to address possible problems before they become serious malfunctions. Furthermore, the way pipeline data is processed is about to undergo a revolution to the use of artificial intelligence (AI) and machine learning (ML) in data analysis. Large volumes of data gathered from multiple sources can be analyzed by AI algorithms, which can then find trends and more accurately forecast possible errors. Dakhel et al.'s work (Dakhel et al. 2023), for example, highlights the use of experimental data in integrity assessments and suggests that AI might improve the reliability of these assessments by more efficiently processing complicated datasets. The incorporation of AI technology into pipeline integrity management is expected to result in more proactive maintenance plans and enhanced safety results as these technologies develop further.

In the future, cloud computing will also play a bigger part in data collection. Cloud platforms give businesses the ability to easily handle and analyze massive amounts of data by offering scalable storage options and strong processing power. According to Haro-Olmo et al. (2021), attaining the required degree of usefulness in data collection requires the integration of IoT with Big Data in cloud environments. With the use of data from various sources, pipeline operators would be able to monitor pipeline integrity and make well-informed decisions based on thorough assessments. Alongside technical developments, regulatory frameworks will also change over time, impacting pipeline integrity management data gathering procedures. Operators will need to modify their data gathering techniques to adhere to new requirements as governments and regulatory agencies place an increasing emphasis on environmental preservation and safety. This could entail stricter reporting guidelines and the use of standardized data formats to make it easier for various entities to share and analyze data. Future data gathering efforts will be required to prioritize regulatory compliance, as highlighted by Paviglianiti and Shaw's (2018) study, which emphasizes the significance of consistent data review in maintaining safe pipelines.

The increasing focus on sustainability and environmental impact will drive the development of more sophisticated data collection methods. As the oil and gas industry faces pressure to reduce its carbon footprint, operators will need to implement technologies that not only monitor pipeline integrity but also assess environmental impacts. This may involve the integration of environmental sensors that track emissions and other ecological factors, providing a more holistic view of pipeline operations. The research by Huang et al. (2023) on using digital technologies for pipeline management underscores the potential for these advancements to support sustainable practices in the industry.

Improved stakeholder collaboration and data sharing will also define the future of data collection in pipeline integrity management. Data sharing across enterprises will become more and more crucial as the industry shifts more integrated methods to pipeline management. The creation of common databases and platforms that can compile information from various sources and enable more thorough analyses and insights will be made easier by this cooperative setting. According to Ma et al. (2024), integrating Geographic Information Systems (GIS) with cloud computing will improve teamwork by giving the gathered data a spatial context.

By bringing data processing closer to the location of data collection, edge computing will enhance cloud-based applications. By lowering latency and bandwidth needs, this method can facilitate quicker decision-making in urgent circumstances. According to Balouek-Thomert et al. (2019), the combination of cloud and edge computing will improve the responsiveness of pipeline integrity monitoring systems by fostering a more fluid ecosystem for data management. In remote locations with limited access to standard communication infrastructure, this move toward decentralized data processing will be especially advantageous.

3.6.1 AI-Driven Inspections

Artificial intelligence (AI) algorithms are being created to assess real-time data acquired by sensors, satellites, and drones. These algorithms can spot trends and abnormalities that can point to corrosion, cracks, or leaks in pipelines. By automating repetitive monitoring chores, AI-driven inspection systems would drastically minimize the need for manual inspections, freeing up operators to concentrate on high-risk regions.

3.6.2 Blockchain for Secure Data Management

Pipeline data can be securely stored and transmitted in a decentralized manner with blockchain technology. Operators may ensure the security and integrity of their data by utilizing blockchain technology, which protects against unauthorized access or modification. Blockchain technology might generate a transparent audit trail of maintenance and inspection activities, aiding operators in fulfilling regulatory requirements.

References

Abdelhamid C et al (2024) Wyoming's produced water: analysis and green hydrogen potential. IOP Conf Ser: Earth Environ Sci

Achouch M et al (2022) On predictive maintenance in industry 4.0: overview, models, and challenges. Appl Sci 12(16):8081

Adee R, Mouratidis H (2022) A dynamic four-step data security model for data in cloud computing based on cryptography and steganography. Sensors 22(3):1109

Adnan N et al (2015) Leak detection in gas pipeline by acoustic and signal processing-a review. IOP Conf Ser: Mater Sci Eng

Alharbi S (2017) An empirical investigation on the impact of trust mediated determinants and moderating factors on the adoption of cloud computing. Int J Inf Technol Comp Sci 9(11):12–22

Balouek-Thomert D et al (2019) Towards a computing continuum: enabling edge-to-cloud integration for data-driven workflows. Int J High Perform Computing Appl 33(6):1159–1174

Barry PH et al (2018) Noble gases in deepwater oils of the US Gulf of Mexico. Geochem Geophys Geosyst 19(11):4218–4235

Dai L et al (2017) Analysis and comparison of long-distance pipeline failures. J Petrol Eng 2017(1):3174636

Dakhel A et al (2023) Fatigue and burst tests of full-scale girth welded pipeline sections for safe operations. Weld World 67(5):1193–1208

de Haro-Olmo FJ et al (2021) Data curation in the Internet of Things: a decision model approach. Comput Math Methods 3(6):e1191

Dickhout JM et al (2017) Produced water treatment by membranes: a review from a colloidal perspective. J Colloid Interface Sci 487:523–534

Doan T, Van Vo M (2020) Using machine learning techniques for enhancing production forecast in north Malay Basin. In: International field exploration and development conference. Springer

Fan Y, Feng X-X (2023) Research on integrity management of long distance pipeline. In: E3S Web of conferences. EDP Sciences

Fan H, Tariq S, Zayed T (2022) Acoustic leak detection approaches for water pipelines. Autom Constr 138:104226

Febaide I, Uzedhe G (2023) Leveraging on low-cost devices for wireless data acquisition in remote pipeline networks. Nigerian J Technol Dev 20(3):154–162

Gholami A (2016) Oil formation volume factor determination through a fused intelligence. Acta Geophys 64:2510–2529

Ho M et al (2020) Inspection and monitoring systems subsea pipelines: a review paper. Struct Health Monit 19(2):606–645

Huang Y et al (2023) Using digital technologies to plan and manage the pipelines network in city. IET Smart Cities 5(2):95–110

Jambol D et al (2024) Leveraging machine learning to enhance instrumentation accuracy in oil and gas extraction. Comp Sci IT Res J 5(6):1335–1357

Kondash A, Vengosh A (2015) Water footprint of hydraulic fracturing. Environ Sci Technol Lett 2(10):276–280
Lei Z et al (2024) Modeling and strain analysis of pipeline dent based on geometric in-line-inspection data. In: Ninth international conference on energy materials and electrical engineering (ICEMEE 2023). SPIE
Liang H et al (2023) Towards UAVs in construction: advancements, challenges, and future directions for monitoring and inspection. Drones 7(3):202
Liu L et al (2022) Gas pipeline flow prediction model based on LSTM with grid search parameter optimization. Processes 11(1):63
Lukács J, Nagy G, Török I (2010) The role of the external and internal reinforcing on the structural integrity of industrial and transporting steel pipelines. In: Materials science forum. Trans Tech Publ
Ma Q et al (2021) Pipeline in-line inspection method, instrumentation and data management. Sensors 21(11):3862
Ma F et al (2024) An underground pipeline relocation decision support system based on BIM and GIS integration. J Phys: Conf Ser 2024
Mahmoodzadeh Z et al (2020) Condition-based maintenance with reinforcement learning for dry gas pipeline subject to internal corrosion. Sensors 20(19):5708
Majstorović VD (2022) Application of Industry 4.0 model in oil and gas companies. J Eng Manage Competitiveness (JEMC) 12(1):77–84
Mao B et al (2024) OPC UA application study in oil and gas pipeline network monitoring data forwarding. EAI Endorsed Trans Energy Web 11(1)
Mohlameane M, Ruxwana N (2016) The impact of existing South African policies and regulatory laws on cloud computing: a literature review networks. In: CS & IT conference proceedings
Munappy AR, Bosch J, Olsson HH (2020) Data pipeline management in practice: challenges and opportunities. In: Product-focused software process improvement: 21st international conference, PROFES 2020, Turin, Italy, November 25–27, 2020, Proceedings 21. Springer
Parker AL et al (2021) Applications of satellite radar imagery for hazard monitoring: insights from Australia. Remote Sens 13(8):1422
Parlak BO, Yavasoglu HA (2023) A comprehensive analysis of in-line inspection tools and technologies for steel oil and gas pipelines. Sustainability 15(3):2783
Paviglianiti J, Shaw S (2018) Analysis of the national energy board pipeline integrity performance measures data. In: Proceedings of the 12th International Pipeline Conference, vol 1, IPC2018-78413. September 24–28, 2018. Calgary, Alberta, Canada. American Society of Mechanical Engineers
Peddinti PR, Puppala H, Kim B (2023) Pavement monitoring using unmanned aerial vehicles: an overview. J Transp Eng, Part B: Pavements 149(3):03123002
Polyzotis N et al (2017) Data management challenges in production machine learning. In: Proceedings of the 2017 ACM international conference on management of data
Priyanka E, Thangavel S, Gao X-Z (2021) Review analysis on cloud computing based smart grid technology in the oil pipeline sensor network system. Petrol Res 6(1):77–90
Rahimi M, Rausand M, Wu S (2011) Reliability prediction of offshore oil and gas equipment for use in an arctic environment. In: 2011 International conference on quality, reliability, risk, maintenance, and safety engineering. IEEE
Ravishankar P et al (2022) Darts—drone and artificial intelligence reconsolidated technological solution for increasing the oil and gas pipeline resilience. Int J Disast Risk Sci 13(5):810–821
Roper WE, Dutta S (2005) Remote sensing and GIS applications for pipeline security assessment. In: 2005 ESRI user conference proceedings
Shen T et al (2024) Spatial data fusion model design and research for an underground pipeline in urban environment scene modeling. Appl Sci 14(15):6760
Soliman A, Attia A, Abdelwahab M (2020) Development of new models for predicting crude oil bubble point pressure, oil formation volume factor, and solution gas-oil ratio using genetic algorithm. J Petrol Min Eng 22(2):17–39

Usuoyibo A, Obogai L (2020) Corrosion analysis of API 5L Gr. three phase crude. NIPES-J Sci Technol Res 2(3)

Wang T et al (2013) Overview of oil and gas pipeline failure database. In: ICPTT 2013: Trenchless Technology, pp 1161–1167

Wang C-X et al (2017) The application of PLC control system in oil and gas pipeline transportation. In: Proceeding of the 2nd international conference on mechanical control and automation (ICMCA)

Xie M, Tian Z (2018) A review on pipeline integrity management utilizing in-line inspection data. Eng Fail Anal 92:222–239

Xu Q, Zhang L, Liang W (2013) Acoustic detection technology for gas pipeline leakage. Process Saf Environ Prot 91(4):253–261

Yang Z-F et al (2019) Cloud computing and big data for oil and gas industry application in China. J Comput 14(4):268–282

Yeboah-Ofori A, Darvishi I, Opeyemi AS (2023) Enhancement of Big Data security in cloud computing using RSA algorithm. In: 2023 10th International conference on future Internet of Things and Cloud (FICloud), 2023. IEEE

Yusoff M et al (2025) An in-depth review of predictive methods for oil and gas applications. 1(2):3

Zhang X, Xie X (2024) Model for the failure prediction mechanism of in-service pipelines based on IoT technology. Processes 12(8):1642

Chapter 4
Data Integration and Preprocessing Techniques

4.1 Introduction

Data integration and preprocessing techniques are crucial for the management and analysis of large data across various domains, including corporate analytics, social sciences, and healthcare. Utilizing these strategies enhances data quality, facilitates quick data analysis, and ensures proper formatting for machine learning (ML) and deep learning (DL) algorithms. The significance of preprocessing cannot be overstated, since it lays the groundwork for accurate data interpretation and decision-making processes.

Jukić et al. (2015) emphasize how Big Data can enhance already-existing analytical data, arguing that efficient integration of various data sources requires preparation. In the healthcare industry, where data from multiple modalities must be harmonized to produce insightful information, this integration is especially crucial.

The challenges associated with Big Data necessitate the development of specialized preprocessing methods. Traditional preprocessing techniques often fall short when applied to large datasets due to scalability issues. For example, Basgall et al. (2018) discuss the limitations of conventional methods in handling imbalanced classification tasks within Big Data contexts, indicating that new approaches are required to ensure effective data preprocessing. Elankavi et al. (2017) propose a hybrid approach to data mining that addresses the unique characteristics of Big Data, such as its volume, variety, and velocity. This hybrid approach underscores the need for innovative preprocessing techniques that can adapt to the complexities of Big Data environments.

Another essential component of preprocessing, especially when it comes to healthcare analytics, is data integration. According to Alcalde-Barros et al. (2019), incorporating streaming data for real-time analysis is crucial for prompt healthcare decision-making. Furthermore, Dhayne et al. (2019) stress that efficient healthcare analytics are made possible by good data integration, which improves the quality and utility of data. As Pedroni et al. (2019) noted in their work on EEG data, this integration

M. Hussain and T. Zhang, *Big Data Analytics in Energy Pipeline Integrity Management*, Lecture Notes in Energy 104, https://doi.org/10.1007/978-981-96-8019-1_4

procedure frequently uses automated pipelines that can expedite the preprocessing of big datasets.

The role of data quality in preprocessing cannot be overlooked. Lai and Leu (2017) propose a quality management procedure aimed at improving the efficiency and practicality of Big Data applications. They argue that ensuring data quality is paramount for achieving reliable outcomes in data-driven decision-making. This sentiment is echoed by García et al. (2016), who discuss various methods for preprocessing Big Data to enhance its quality and usability. The focus on data quality is particularly relevant in fields such as finance and healthcare, where inaccurate data can lead to significant consequences.

In the realm of machine learning, preprocessing techniques are critical for optimizing model performance. Zhu and Gao (2016) assert that effective data preprocessing can lead to improved classification performance and structural invariance within datasets. This is particularly important in scenarios where the data is noisy or contains outliers, as preprocessing can help mitigate these issues.

The evolution of cloud computing technologies has transformed the way data preprocessing is conducted. Dawelbeit and McCrindle (2014) describe a cloud-based elastic framework that facilitates efficient data preprocessing in Big Data environments. This framework allows for scalable and flexible data management solutions, which are essential for handling the vast amounts of data generated in today's digital landscape. The ability to preprocess data in a cloud environment enhances accessibility and collaboration among researchers and practitioners.

4.2 Challenges in Data Integration

Data integration in fact is essential to creating a thorough picture of pipeline health, especially when it comes to situations like resource allocation, infrastructure management, and environmental monitoring. A comprehensive understanding of pipeline systems is made possible by the integration of data from several sources, empowering stakeholders to make well-informed decisions on operational effectiveness, risk assessment, and maintenance. However, the process of data integration is fraught with challenges that can hinder the effectiveness of such initiatives, refer to Fig. 4.1 on some of the challenges associated with pipeline.

The dynamic nature of data sources further complicates the integration process on top of existing difficulties. It gets harder to maintain coherence and consistency among interconnected databases as data changes. For example, to accommodate changes in data gathering procedures and standards, the integration of data from several healthcare providers necessitates constant updates and modifications (McCarty et al. 2011). The creation of adaptive integration frameworks that can react to modifications in data sources and formats is required due to this dynamic feature of data integration.

The integration of data from disparate sources often involves addressing issues related to data semantics. Different data sources may use varying terminologies

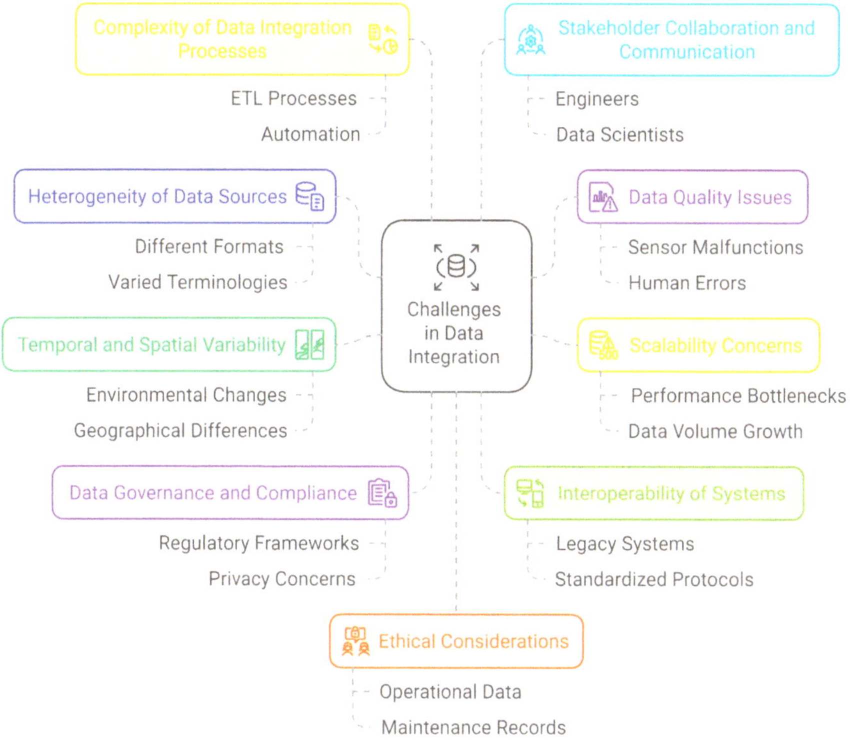

Fig. 4.1 Challenges in pipeline health data integration

and definitions, leading to potential misunderstandings and misinterpretations. For example, the integration of clinical data from multiple hospitals may require mapping different coding systems to a unified standard, as noted by Kreimeyer et al. (2022) in their discussion of pharmacovigilance data integration. Establishing common semantic frameworks is essential for ensuring that integrated data is meaningful and interpretable. The ethical ramifications of data usage and sharing are also called into question by the integration of Big Data. Data ownership, permission, and privacy issues become critical as businesses depend more and more on linked datasets for decision-making.

4.3 Data Preprocessing Techniques

In the realm of Big Data, characterized by vast quantities, diversity, and rapidity of information, data preparation is a crucial phase in the data analysis pipeline. The quality of data input into machine learning (ML) and deep learning (DL) algorithms significantly influences their effectiveness. This section examines several data preparation techniques essential for transforming raw data into a format suitable for analysis, highlighting their significance and application.

4.3.1 Data Cleaning

One of the most basic preprocessing methods is data cleaning, which looks for and fixes mistakes or inconsistencies in the dataset. This procedure entails dealing with missing values, eliminating duplication, and fixing errors. According to García et al. (2016), the enormous increase in data demands strong cleaning techniques to guarantee that the data is trustworthy and suitable for analysis. According to Luengo et al. (2012), who emphasize the significance of choosing suitable imputation approaches based on the classification methods being used, missing data can greatly distort findings. Various imputation methods have been developed to solve this issue.

The large amount of information in the Big Data context complicates the data cleansing process. Ding et al. (2022) underscore the importance of systematic data cleaning procedures in Internet of Things (IoT) scenarios, where time-series data often contains errors that must be rectified to enhance data quality. The incorporation of automated cleaning technology enhances precision and efficiency, facilitating real-time data processing and analysis.

4.3.2 Data Transformation

Data transformation refers to the process of converting data into a format suitable for analysis. This may include encoding categorical variables, standardization, and normalization. In machine learning, normalization is especially crucial since it guarantees that features make an equal contribution to the distance computations utilized by algorithms such as k-nearest neighbors (KNN). Furthermore, for algorithms that need numerical input, category variables must be encoded into numerical representations. To make this transition easier, methods like label encoding and one-hot encoding are frequently used.

4.3.3 Feature Selection and Extraction

Reducing the dataset's dimensionality through feature selection and extraction is essential for enhancing model performance and cutting down on computational expenses. While feature extraction generates new features based on preexisting ones, feature selection entails locating the most pertinent features within the dataset. For these goals, a variety of techniques are frequently employed, including Principal Component Analysis (PCA) and Recursive Feature Elimination (RFE). Specifically,

PCA works well for converting high-dimensional data into a lower-dimensional space while maintaining the greatest amount of variance. As Yang et al. (2018) highlight in their investigation of preprocessing strategies for Big Data, this strategy is essential for displaying complicated datasets and enhancing the interpretability of models.

4.3.4 Data Integration

Data integration involves consolidating information from multiple sources to create a unified dataset. This process is essential in Big Data environments, as data is often collected from diverse heterogeneous sources, including sensors, databases, and external APIs. Alcalde-Barros et al. (2019) assert that efficient data integration is crucial for achieving a comprehensive knowledge of data, particularly in streaming applications.

The integration process often requires resolving issues related to data format discrepancies, semantic differences, and data quality. Automated data integration tools can facilitate this process by providing standardized methods for merging datasets.

4.3.5 Data Reduction

The goal of data reduction strategies is to decrease the amount of data while preserving its accuracy and utility. This may entail techniques like dimensionality reduction, aggregation, and data sampling. To make it easier to examine and interpret big datasets without losing important information, sampling techniques can be used to create smaller, representative subsets. On the other hand, aggregation techniques summarize data to offer higher-level insights. Aggregating transaction data by time periods, for instance, can highlight trends and patterns in financial data analysis that would not be visible in raw data.

4.3.6 Data Encoding and Representation

When working with unstructured or semi-structured data, data encoding and representation are essential for getting the data ready for analysis. Tokenization, stemming, and lemmatization are examples of text preprocessing techniques that are crucial for converting textual data into an analysis-ready format. According to López-Miguel (2021), the quality of insights obtained from textual data can be greatly improved by applying natural language processing (NLP) approaches.

4.3.7 Handling Imbalanced Data

In machine learning, imbalanced datasets, where some classes are underrepresented, present serious difficulties. This issue can be addressed by techniques such as the Synthetic Minority Over-sampling Technique (SMOTE), which equilibrates the dataset by generating synthetic samples for the minority class. Basgall et al. (2018) discuss the importance of scalable oversampling methods for imbalanced classification tasks in Big Data contexts, emphasizing the need for effective preprocessing strategies to improve model performance.

4.3.8 Automation of Preprocessing

As the amount of data keeps expanding, it is becoming more and more crucial to automate data preparation methods. By streamlining the cleaning, converting, and merging of data, automated preprocessing can cut down on the time and effort needed for manual preparation. The creation of standardized preprocessing techniques for EEG data, as mentioned by Pedroni et al. (2019), is an example of how automation may improve data quality and analysis effectiveness.

4.3.9 Handling Time Series Data

Pipeline operations generate large volumes of time series data, that is collected at regular intervals over time. Preprocessing time series data requires special techniques to handle the sequential nature of the data, such as:

- **Smoothing**: Smoothing techniques, such as moving averages, are used to reduce the noise in time series data, making it easier to detect underlying trends (Verbesselt et al. 2010).
- **Lag Features**: Creating lag features, which represent previous time steps in the data, can help machine learning models capture temporal dependencies and make better predictions.
- **Resampling**: Resampling techniques are used to adjust the frequency of time series data. For example, minute-level data can be resampled to hourly or daily intervals, depending on the requirements of the analysis.

4.4 Data Quality and Its Importance

The integrity of data is crucial in the oil and gas sector, especially for pipeline management. Pipelines are essential infrastructure for the conveyance of oil and gas, and any malfunction or inefficiency in these systems can result in significant financial, environmental, and safety repercussions. This discussion highlights the significance of data quality concerning oil and gas pipelines, referencing multiple academic sources.

The integrity of pipeline operations is directly linked to the quality of data used for monitoring and maintenance purposes. Accurate and high-quality data enables real-time monitoring of pipeline conditions, which is essential for preventing leaks and failures. Chen et al. (2019) underscore the significance of risk assessments and data accuracy in predicting potential incidents and mitigating risks effectively in the oil and gas pipeline industry. Inaccurate or incomplete data can lead to erroneous assessments, which can compromise safety and increase the likelihood of incidents, potentially resulting in catastrophic consequences for both the environment and public safety. Predictive maintenance strategies that rely on data analytics are increasingly adopted in the oil and gas sector to enhance asset integrity management. Research by Arinze et al. (2024) discusses how integrating AI-driven analytics improves the ability of companies to perform predictive maintenance effectively. This integration is crucial for prolonging the lifespan and reliability of pipeline infrastructure, as poor data quality can severely hinder these analytics, leading to missed opportunities for timely maintenance interventions and resulting in higher operational costs and risk exposure.

The implications of data quality extend beyond operational efficiency; they also affect regulatory compliance and overall corporate sustainability. Regulatory bodies require accurate reporting and monitoring to ensure compliance with environmental and safety standards in oil and gas activities. Although specific studies may not directly address regulatory compliance, the need for high data quality is widely acknowledged as essential for maintaining trust with regulators and stakeholders. Non-compliance due to inadequate data can lead to significant penalties and damage the reputation of companies within the industry.

The dynamic nature of the oil and gas sector necessitates continuous updates and adjustments to data management practices. Pipeline integrity data must be regularly maintained and validated to reflect current conditions accurately. The importance of operational excellence in influencing health, safety, and environmental performance throughout the oil and gas sector is paramount and should guide data management practices to respond timely to changes in pipeline conditions. In the context of technological advancements, cloud computing and Big Data Analytics are becoming indispensable tools for improving data quality in the oil and gas industry. Yang et al. (2019) highlight that cloud computing can streamline data management processes, facilitating better aggregation and analysis of data from multiple sources. Improved data integration capabilities led by cloud technologies can enhance data quality by facilitating rapid data validation and cleansing processes, thereby supporting informed decision-making.

4.4.1 Data Accuracy

Data accuracy refers to how well the data reflects the true state of the pipeline (Stoianov et al. 2007). Ensuring accuracy requires regular calibration of sensors and validation of data against known benchmarks. Inaccurate data can lead to false conclusions, such as underestimating the severity of a potential defect.

4.4.2 Data Completeness

Incomplete data, such as missing sensor readings or unreported maintenance activities, can lead to gaps in the analysis (Teh et al. 2020). Data completeness is particularly important when conducting predictive maintenance or risk assessments, as missing data can skew the results (Burmeister et al. 2023).

4.4.3 Data Timeliness

Real-time data is essential for detecting anomalies and initiating corrective actions before a failure occurs (Reed et al. 2012). Delayed or outdated data can prevent operators from taking timely action, increasing the risk of failure.

4.5 The Role of Cloud Computing in Data Integration and Preprocessing

Particularly when it comes to Big Data, cloud computing has become a game-changing technology that is essential to data integration and preparation. Cloud computing helps businesses to effectively handle and process enormous volumes of data from many sources by offering scalable resources. The capacity of cloud computing to provide on-demand processing and storage is one of its main benefits. According to Shehab et al. (2022), Big Data's magnitude and dynamic nature pose serious problems for conventional processing and storage resources. By offering strong and scalable resources that can be used with different Big Data designs, cloud computing solves these issues. This flexibility allows organizations to adapt to changing data requirements and ensures that they can effectively manage large datasets without the need for extensive on-premises infrastructure.

A thorough understanding of the data landscape requires the integration of data from various sources, which cloud computing makes it possible. Cloud computing allows for unified management and scheduling of data processing operations by virtualizing scattered computing and storage resources. When data is gathered from several sources, including databases, IoT devices, and external APIs, this functionality is especially helpful. Organizations can improve the quality and usability of the combined datasets by using cloud-based integration solutions to expedite the data harmonization and merging process. In addition to integration, cloud computing significantly enhances data preprocessing capabilities. Cloud provides a range of tools and services that support various preprocessing tasks, including data cleaning, transformation, and feature selection. For instance, Wang et al. (2019) highlight the effectiveness of Hadoop, a cloud-based platform, in executing parallel algorithms for data mining, which can greatly reduce the execution time of preprocessing tasks. This efficiency is crucial for organizations dealing with large volumes of data, as it allows them to preprocess data in a timely manner, ensuring that it is ready for analysis. Cloud computing supports the implementation of advanced preprocessing techniques that may not be feasible with traditional computing resources. For example, the use of machine learning algorithms for feature selection and dimensionality reduction can be effectively executed in cloud environments. The cloud's computational power enables organizations to apply complex algorithms to large datasets, facilitating the extraction of meaningful insights and improving the overall quality of the data.

Another important factor in improving data preparation is cloud computing's scalability. Organizations need scalable solutions as data volumes continue to increase. Because cloud systems provide dynamic resource allocation, businesses can modify their processing capacity in response to changing demands. In large data environments, where data influx can be unpredictable, this scalability is especially crucial. By utilizing cloud resources, organizations can ensure that they have the necessary computational power to handle peak processing loads without incurring the costs associated with maintaining excess on-premises infrastructure.

When it comes to data integration and preprocessing, cloud computing improves accessibility and collaboration. Cloud-based solutions enable real time collaboration between data scientists, analysts, and stakeholders by enabling teams to access and share data from any location. For businesses that have many locations or need input from different departments, this accessibility is essential. According to Singh et al. (2021), cloud computing offers a platform for large data preprocessing, analysis, and visualization, facilitating more productive teamwork. This cooperative setting encourages creativity and speeds up the creation of data-driven solutions. Along with these advantages, cloud computing also solves security and regulatory issues related to preprocessing and data integration. Strong security features including encryption, access controls, and adherence to industry standards are provided by numerous cloud providers. For businesses that manage sensitive data, like financial institutions and healthcare providers, this is especially crucial. Organizations can improve their data security posture and guarantee regulatory compliance by utilizing cloud services.

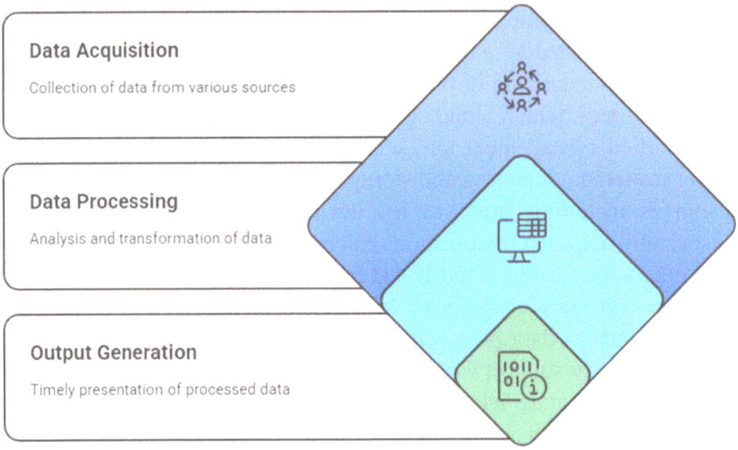

Fig. 4.2 Real-time processing framework

But it's important to understand that using cloud computing for preprocessing and data integration is not without its difficulties. Data privacy, security threats, and possible vendor lock-in are some of the things that organizations need to carefully think about. According to Gill et al. (2019), enterprises face new challenges when integrating cloud computing with other technologies like blockchain and the Internet of Things. Therefore, a strategic approach to cloud adoption is necessary to maximize the benefits while mitigating potential risks.

4.5.1 Scalability and Flexibility

Cloud platforms can scale up or down based on the volume of data being processed, making them ideal for handling the large datasets generated by modern pipeline monitoring systems. This flexibility allows operators to process real-time data from thousands of sensors without the need for expensive on-premises infrastructure.

4.5.2 Real-Time Processing

Many cloud platforms offer real-time data processing capabilities, allowing operators to preprocess and analyze data as it is collected. This is particularly important for pipelines that require continuous monitoring and rapid response to anomalies. Figure 4.2 shows real-time processing framework.

References

Alcalde-Barros A et al (2019) DPASF: a flink library for streaming data preprocessing. Big Data Analytics 4:1–17

Arinze CA et al (2024) Predictive maintenance in oil and gas facilities, leveraging ai for asset integrity management. Int J Front Eng Technol Res 6(1):16–26

Basgall MJ et al (2018) SMOTE-BD: an exact and scalable oversampling method for imbalanced classification in big data. In: VI Jornadas de Cloud Computing & Big Data (JCC&BD)(La Plata, 2018)

Burmeister N et al (2023) Exploration of production data for predictive maintenance of industrial equipment: a case study. IEEE Access 11:102025–102037

Chen X et al (2019) A methodology for overall consequence assessment in oil and gas pipeline industry. Process Saf Prog 38(3):e12050

Dawelbeit O, McCrindle R (2014) A novel cloud based elastic framework for big data preprocessing. In: 2014 6th Computer science and electronic engineering conference (CEEC). IEEE.

Dhayne H et al (2019) In search of big medical data integration solutions-a comprehensive survey. IEEE Access 7:91265–91290

Ding X et al (2022) IoT data cleaning techniques: a survey. Intell Converged Netw 3(4):325–339

Elankavi R, Kalaiprasath R, Udayakumar R (2017) Data mining with big data revolution hybrid. Int J Smart Sens Intell Syst 10(5):560

García S et al (2016) Big data preprocessing: methods and prospects. Big Data Analytics 1:1–22

Gill SS et al (2019) Transformative effects of IoT, Blockchain and Artificial Intelligence on cloud computing: evolution, vision, trends and open challenges. Internet of Things 8:100118

Jukić N et al (2015) Augmenting data warehouses with big data. Inf Syst Manag 32(3):200–209

Kreimeyer K, Spiker J, Botsis T (2022) Overcoming major barriers to build efficient decision support systems in pharmacovigilance. Advances in informatics, management and technology in healthcare. IOS Press, pp 398–401

Lai S-T, Leu F-Y (2017) Data preprocessing quality management procedure for improving big data applications efficiency and practicality. In: Advances on broad-band wireless computing, communication and applications: proceedings of the 11th international conference on broad-band wireless computing, communication and applications (BWCCA–2016) November 5–7, 2016, Korea. Springer

Lopez-Miguel ID (2021) Survey on preprocessing techniques for big data projects. Eng Proc 7(1):14

Luengo J, García S, Herrera F (2012) On the choice of the best imputation methods for missing values considering three groups of classification methods. Knowl Inf Syst 32:77–108

McCarty CA et al (2011) The eMERGE Network: a consortium of biorepositories linked to electronic medical records data for conducting genomic studies. BMC Med Genomics 4:1–11

Pedroni A, Bahreini A, Langer N (2019) Automagic: standardized preprocessing of big EEG data. Neuroimage 200:460–473

Reed M, Lee R, Coote B (2012) Development of pipeline integrity performance indicators for Canadian energy pipelines. In: International pipeline conference. 2012. American Society of Mechanical Engineers

Shehab N, Badawy M, Ali HA (2022) Toward feature selection in big data preprocessing based on hybrid cloud-based model. J Supercomput 78(3):3226–3265

Singh M et al (2021) Quantifying COVID-19 enforced global changes in atmospheric pollutants using cloud computing based remote sensing. Remote Sens Appl: Soc Environ 22:100489

Stoianov I et al (2007) Pipeneta wireless sensor network for pipeline monitoring. In: Proceedings of the 6th international conference on Information processing in sensor networks

Teh HY, Kempa-Liehr AW, Wang KI-K (2020) Sensor data quality: a systematic review. J Big Data 7(1):11

Verbesselt J et al (2010) Detecting trend and seasonal changes in satellite image time series. Remote Sens Environ 114(1):106–115

Wang Y, Liu Y, Jing W (2019) Hadoop-based parallel algorithm for data mining in remote sensing images. Int J Perform Eng 15(11):2860

Yang J, Wang H, Cao Y (2018) Tractable queries on big data via preprocessing with logarithmic-size output. Knowl Inf Syst 56(1):141–163

Yang Z-F et al (2019) Cloud computing and big data for oil and gas industry application in China. J Comput 14(4):268–282

Zhu C, Gao D (2016) Influence of data preprocessing. J Comput Sci Eng 10(2):51–57

Chapter 5
Literature Review

5.1 Overview of the Oil and Gas Sector

The oil and gas sector is making a substantial contribution to the global economy. This sector includes exploration, refining, and distribution, serving as a significant global energy source (Brewer 2023). In 2023, the global crude oil market had significant expansion, rising from $2747.78 billion in 2022 to $2904.09 billion, with a compound annual growth rate (CAGR) of 5.7%. The market is anticipated to increase further, attaining $3481.5 billion by 2027, with a projected CAGR of 4.6% (The Business Research Company 2023). In 2022, the global average daily oil output was 80.75 million barrels, including condensates (Bhutada 2023). Pipelines are among the most crucial energy transportation networks and constitute the safest method for moving oil and gas products (Liu et al. 2023; Shaik et al. 2022). They are anticipated to remain a dependable energy delivery source for several more centuries. In 1998, the aggregate investment for all natural gas pipeline companies amounted to $63.1 billion, generating total income of $13.6 billion. Liquid pipeline corporations received an investment of $30.2 billion, generating $6.9 billion in income (Khan et al. 2021a; Wikipedia 2015). The industry is often categorized into three primary sectors: upstream, midstream, and downstream, see Fig. 5.1 for further details. Each sector is integral to the exploration, production, transportation, and refinement of oil and natural gas.

5.2 The Importance of Pipeline Integrity in the Oil and Gas Sector

Pipeline integrity is a significant issue for firms engaged in the midstream and downstream segments of the oil and gas industry. Pipelines serve as the principal method for delivering oil and gas over extensive distances, and their secure functioning is

M. Hussain and T. Zhang, *Big Data Analytics in Energy Pipeline Integrity Management*, Lecture Notes in Energy 104, https://doi.org/10.1007/978-981-96-8019-1_5

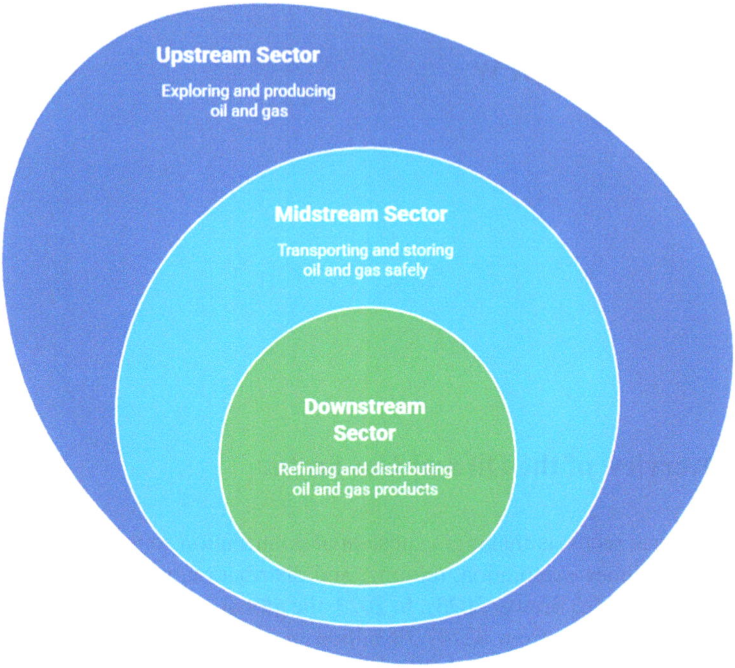

Fig. 5.1 The oil and gas sectors

essential for the industry's success. Pipeline failures, including leaks or ruptures, may lead to substantial economic losses, environmental pollution, regulatory sanctions, and harm to a company's reputation (Hussain et al. 2023).

As oil and gas pipelines deteriorate, they become progressively vulnerable to integrity concerns, such as corrosion, mechanical damage, and material fatigue. Consequently, sustaining pipeline integrity via efficient monitoring, inspection, and maintenance techniques is crucial for reducing failure risks and guaranteeing the long-term viability of pipeline operations.

The integration of Big Data Analytics (BDA) into pipeline integrity management signifies a fundamental transformation in the methodologies employed by pipeline operators for risk assessment, monitoring, and maintenance. Conventional techniques for pipeline inspection and maintenance are frequently labour-intensive, reactive, and constrained in their capacity to anticipate future problems. BDA equips pipeline operators with the necessary tools to adopt a proactive, predictive strategy for integrity management, therefore mitigating the risk of catastrophic failures and enhancing operational efficiency.

5.3 Pipeline Condition Assessment

The evaluation of pipeline condition is an essential element of Pipeline Integrity Management Systems (PIMS). The process entails assessing the physical state of pipelines to identify any indications of deterioration, corrosion, or mechanical impairment that may jeopardize their structural integrity. By 2040, worldwide demand for natural gas is projected to increase by 43% (Ahmed and Kabir 2021). These pipes are susceptible to degradation due to several factors, including aging, substandard construction, environmental alterations, and inadequate maintenance, along with external impacts such as climatic conditions, soil composition, seismic events, and flooding. Moreover, anthropogenic variables, including excavation methods and inadequate installation, exacerbate the susceptibility of pipelines to leakage (Yin et al. 2021). Although it may be infeasible to entirely eradicate pipe failures, the implementation of robust risk management and reliable maintenance procedures (Shahriar et al. 2012; Shipley et al. 2022; El-Shenawy and Shehadeh 2014) can substantially reduce the total risk of such failures to an acceptable threshold. In the past two decades, the increase in failure instances has intensified the emphasis on evaluating pipeline integrity and current conditions (Mora et al. 2016; Telford et al. 2011; Ahmad et al. 2022; Khan et al. 2021b). The integrity management program is a crucial instrument for assessing pipeline condition and estimating remaining lifespan. Consequently, utilizing modern analytics for pipeline condition evaluation would aid pipeline proprietors in preserving pipeline integrity and reliability (Spandonidis et al. 2022).

Pipeline condition assessment is a thorough review of the physical and operational state of pipelines utilized for the transfer of oil, gas, or other fluids. Figure 5.2 captures the fundamental principles of condition assessment, highlighting the significance of a comprehensive and proactive methodology for evaluating pipeline status.

There are 22 reasons regarded by the American Society of Mechanical Engineers (ASME) B31.8S as posing a "Threat" to pipeline integrity, and these threats have been divided into nine categories. ASME B31.8S, Canadian Standards Association (CSA) Z662, and the American Petroleum Institute (API) 1160 have integrity threat types that are listed in Table 5.2 (Mora et al. 2016). Conducting regular condition assessments of these pipelines is crucial to ensure its integrity, prevent leaks or failures, and maintain uninterrupted supply. The importance of pipeline condition is shown in Table 5.1.

The objective of these evaluations is to guarantee the integrity and safety of the pipeline and its associated systems, detect any faults or defects, and ascertain the necessity for maintenance or rehabilitation. The evaluation of defect risk encompasses several factors, including the current state of the pipeline, historical leakage occurrences, design attributes, operational and maintenance practices, and external influences such as population density, mechanical damage like dents, and environmental conditions.

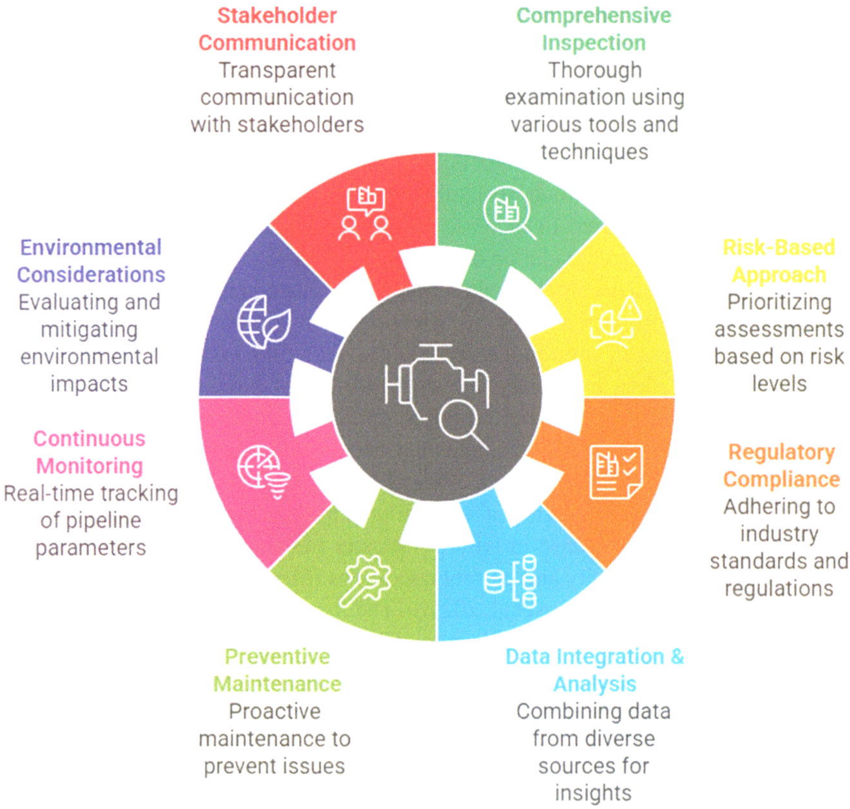

Fig. 5.2 Key principles of condition assessment

Table 5.1 Importance of pipeline condition assessment

Important aspect	Details
Safety	Assessing pipeline conditions helps identify potential integrity issues that could lead to accidents, spills, or explosions, thereby safeguarding human lives and the environment
Reliability	Regular assessments help identify areas of concern, allowing for timely repairs or replacements to minimize disruptions in the energy supply chain
Cost optimization	Effective condition assessment aids in prioritizing maintenance activities, reducing unnecessary expenses, and optimizing the lifespan of pipeline assets

5.3.1 Traditional Condition Assessment Methods

Historically, pipeline condition assessment has relied on a combination of periodic inspections, non-destructive testing (NDT) techniques, and visual inspections. While

Table 5.2 Correlation of pipeline integrity threat names from industry standards

ASME B31.8S Gas transmission pipelines	**CSA-Z662 (Canada)** Gas, liquid hydrocarbons, oilfield water and steam, liquid or dense phase carbon dioxide pipeline systems	**API 1160** Hazardous liquid pipelines	**UKOPA (United Kingdom)** Major accident hazard pipelines
External corrosion	Metal loss	External corrosion	External corrosion
Internal corrosion	Metal loss	Internal corrosion	Internal corrosion
Stress corrosion cracking—SCC (i.e., environmentally assisted cracking)	Cracking	Crack and crack-like	Other
Manufacturing-related (e.g., seam weld and pipe body)	Material or manufacturing	Design and material	Seam weld defect
Welding/fabrication related (e.g., girth weld, wrinkle bend, threads)	Construction	Construction	Girth weld and pipe defect, construction/material
Mechanical damage (e.g., immediate/delayed failure, vandalism/theft)	External interference	Third party damage	External interference
Incorrect operations (e.g., procedure)		Operation errors	Operator error
Weather-related and outside force (e.g., hydro-geotechnical, cold weather, lightning, heavy rains or floods, earth movements)	Weather	Ground movement	Ground movement
	Geotechnical failures		
Pipeline equipment (e.g., valves, seals)	Ancillary equipment	System operations	N/A

these methods are effective at identifying visible damage, they often fall short in detecting early-stage corrosion or cracks that could lead to future failures.

- **In-Line Inspection (ILI) Tools**: ILI tools, often referred to as "smart pigs," are widely used to assess the internal condition of pipelines. These tools travel through the pipeline and use sensors to detect anomalies such as corrosion, cracks, and dents. While ILI tools provide valuable data, they are typically used on a periodic basis, meaning that issues can develop between inspections without being detected.
- **Non-Destructive Testing (NDT)**: NDT techniques, such as ultrasonic testing and radiography, are used to assess pipeline condition without causing damage to the

pipeline itself. These techniques are commonly used in combination with visual inspections to detect defects on both the internal and external surfaces of pipelines.

- **Visual Inspections**: Visual inspections are often conducted manually by pipeline operators or inspectors. This method is useful for detecting obvious signs of damage, such as dents or coating disbondment. However, it is less effective at identifying subsurface defects or early-stage corrosion.

While traditional methods of condition assessment are valuable, they often lack the ability to provide continuous, real-time insights into the condition of pipelines. This limitation has led to the development of more advanced monitoring and assessment techniques that leverage the power of Big Data Analytics.

5.3.2 Big Data Analytics in Pipeline Condition Assessment

The utilization of Big Data Analytics in pipeline status assessment is revolutionizing how pipeline operators oversee and sustain their infrastructure. Through the analysis of extensive data from various sources, such as ILI tools, SCADA systems, and environmental sensors, BDA empowers pipeline operators to acquire profound insights into pipeline conditions and anticipate potential problems prior to their occurrence, refer to Fig. 5.3 for BDA benefits in pipeline integrity management.

5.4 Condition Monitoring Techniques for Pipeline Integrity

Condition monitoring techniques for pipeline integrity and reliability are vital elements in the efficient management of pipeline integrity programs (Telford et al. 2011). They act as the sensory system (Table 5.3 shows common sensors installed

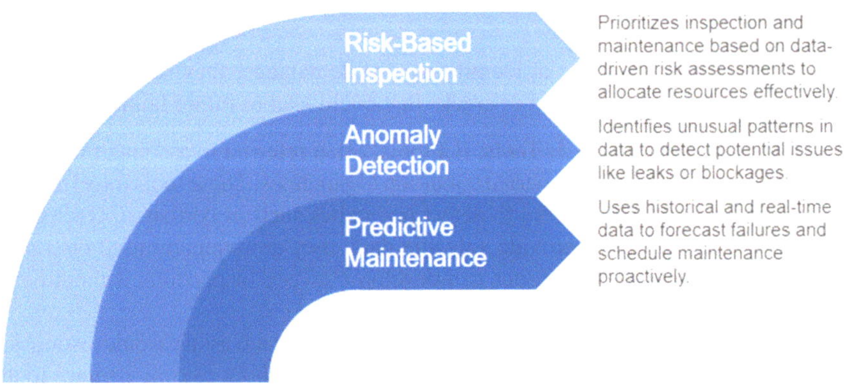

Fig. 5.3 Enhancing pipeline management with BDA

on oil and gas pipelines), offering immediate insights into the well-being, and functioning of complex pipeline systems. The significance of these techniques lies in their ability to detect anomalies, faults, or potential failures at an early stage, preventing catastrophic consequences and ensuring the uninterrupted flow of resources through the pipelines (Ahmad et al. 2022). In the era of pipeline integrity management, the implementation of a diverse array of monitoring techniques is essential to comprehensively address the complex challenges that pipelines face (Khan et al. 2021b).

One prominent condition monitoring technique is the use of advanced sensors strategically placed along the pipeline (Hancke et al. 2012). These sensors continuously collect data on parameters such as pressure, temperature, chemical composition, inhibitors, flow rates etc. (Isermann 2011). Analyzing this data allows operators to identify deviations from normal operating conditions, enabling proactive intervention before minor issues escalate into major problems (Hussain et al. 2021). Moreover, the integration of smart technologies, including the Internet of Things (IoT) devices, enhances the capabilities of these sensors, facilitating real-time data transmission and remote monitoring (Lampropoulos et al. 2019). This interconnected network of sensors forms a robust foundation for predictive maintenance strategies, optimizing the operational efficiency and longevity of the pipeline infrastructure.

In addition to sensor-based monitoring, another crucial technique involves the use of non-destructive testing (NDT) methods. Table 5.4 outlines common non-destructive testing (NDT) techniques used for oil and gas pipeline integrity management, along with their brief pros and cons. This table provides a brief overview of common NDT techniques for oil and gas pipeline integrity management, along with their respective advantages and disadvantages. The choice of the NDT method depends on factors such as the material, accessibility, and the specific types of defects or anomalies that need to be identified.

NDT techniques, such as ultrasonic testing, magnetic particle inspection, and radiographic testing, enable inspectors to assess the structural integrity of pipelines without causing any damage (Gupta et al. 2022). By employing these methods, operators can identify corrosion, weld defects, and material degradation, allowing for timely repairs or replacements (Soomro et al. 2022). The non-invasive nature of NDT ensures that the assessment process itself does not compromise the integrity of the pipeline, aligning with the overarching goal of maintaining the system's reliability (Gupta and Ahmad 2021).

Furthermore, the utilization of real-time monitoring and data analytics plays a pivotal role in enhancing the effectiveness of condition monitoring in pipeline management (Kumar et al. 2018; Davies 2012). Sophisticated algorithms can process vast amounts of data generated by sensors and NDT methods, identifying patterns and trends that may indicate potential issues (Baduge et al. 2022; Gupta et al. 2021). Machine learning algorithms, for instance, can learn from historical data and predict future failure probabilities, enabling a proactive approach to maintenance. This data-driven decision-making process empowers operators to make informed choices, optimizing resource allocation and minimizing downtime (Hussain et al. 2023).

Table 5.3 List of common sensors installed on oil and gas pipelines

Sensors type	Brief details/Applications
Pressure sensors	Monitoring pipeline pressure to ensure it remains within safe operating limits. Detects leaks, blockages, or anomalies
Temperature sensors	Measures the temperature of the pipeline to prevent overheating or freezing, ensuring the integrity of the pipeline
Flowmeters	Measures the flow rate of the transported fluid, aiding in volume calculation and ensuring optimal pipeline operation
Leak detection sensors	Identifies and locates leaks in the pipeline, preventing environmental damage and ensuring the safety of the infrastructure
Corrosion sensors	Monitors corrosive activity on the pipeline's surface, helping prevent corrosion-related failures and extend pipeline life
Acoustic sensors	Listening to abnormal sounds like vibrations or leaks, aiding in the early detection of potential issues
Hydrogen Sulfide (H_2S) sensors	Monitors the presence of H_2S, a corrosive and toxic gas, ensuring personnel safety and preventing pipeline damage
Ultrasonic thickness gauges	Measures pipeline wall thickness to detect thinning from corrosion or erosion, predicting, and preventing failures
Vibration sensors	Monitors vibrations in the pipeline, indicating issues such as mechanical failures or fluid turbulence
GPS (Global positioning system)	Tracks pipeline location, providing real-time information for monitoring and maintenance
Remote sensing technologies	Utilizes technologies like satellite imagery or drones to monitor pipeline sections for anomalies, encroachments, or threats
Oxygen sensors	Measures oxygen levels in the pipeline to prevent corrosion and maintain the integrity of the transported fluid
Cathodic protection monitors	Monitors the effectiveness of cathodic protection systems, preventing corrosion on buried pipelines
Density sensors	Measures fluid density, aiding in identifying changes in composition or contamination
Smoke and fire detectors	Detects smoke or fire along the pipeline, triggering immediate response actions to prevent accidents
Radiation sensors	Monitors radiation levels in and around the pipeline, crucial for pipelines transporting radioactive materials
Humidity sensors	Monitors humidity levels to prevent corrosion and assess environmental conditions affecting the pipeline

The integration of unmanned aerial vehicles (UAVs) or drones into condition monitoring practices has emerged as a cutting-edge technique in pipeline integrity management. Drones equipped with cameras and sensors can conduct aerial inspections of vast pipeline networks, reaching remote or challenging locations (Javaid et al. 2022). This aerial perspective provides a comprehensive view of the pipeline

Table 5.4 Common non-destructive testing (NDT)

NDT techniques	Application/Benefits	Pros	Cons
Ultrasonic Testing (UT)	Uses high-frequency sound waves to detect defects and measure thickness	High accuracy	Surface must be accessible for probe placement
		Can detect internal and external flaws	Requires skilled operators
		Suitable for various materials	
Radiographic Testing (RT)	Utilizes X-rays or Gamma rays to reveal internal structures and defects	Penetrates thick materials	Safety concerns due to radiation exposure
		Reveals internal and surface defects	Requires specialized equipment
		Provides permanent records	Limited to straight sections of pipelines
Magnetic Particle Testing (MPT)	Detects surface and near-surface defects using magnetic fields	Effective for ferrous materials	Limited to surface and near-surface defects
		Quick and relatively inexpensive	Requires magnetizable materials
		Can inspect complex shapes	
Liquid Penetrant Testing (LPT)	Identifies surface cracks and defects by applying a liquid penetrant	Detects fine surface cracks and defects	Limited to surface defects
		Inexpensive and simple to use	Not suitable for porous materials
		Works on various materials	
Eddy Current Testing (ECT)	Uses induced electrical currents to detect defects near the surface	Detects surface and near-surface defects	Limited penetration depth
		High sensitivity to cracks and corrosion	Requires electrically conductive materials
		Can be used for conductivity measurements	
Infrared Thermography (IRT)	Detect anomalies by measuring surface temperature variations	Rapid inspection of large areas	Limited to surface defects
		Non-contact and non-intrusive	Affected by environmental conditions
		Effective for detecting insulation issues	

infrastructure, aiding in the identification of physical damage, leaks, or encroachments. The use of drones not only enhances the speed and efficiency of inspections but also improves safety by minimizing the need for human inspectors to access potentially hazardous areas (Aiello et al. 2020).

The integrity of a pipeline can be assessed utilizing several instruments accessible to pipeline operators. The identification and elimination of all significant and near-critical SCC issues along the full pipeline length can solely be achieved through hydrotesting. Like excavations, visible exterior SCC characteristics can be identified and eliminated during excavations with 100% in-the-ditch NDE assessment. Regrettably, both alternatives possess considerable technological and financial constraints. Figure 5.4 shows a few listed ILI technologies that most of pipeline owners are using to detect SCC.

Table 5.5 summarizes (Canadian Energy Pipeline Association 2007) existing and developing crack tool technologies.

From sensor-based monitoring and non-destructive testing to real-time analytics and drone inspections, these techniques collectively contribute to a holistic approach in safeguarding pipeline systems (Rachman et al. 2021).

The proactive nature of condition monitoring not only prevents potential disasters but also enhances the overall reliability, safety, and efficiency of pipeline operations. As technology continues to advance, the synergy of these monitoring techniques will undoubtedly play a pivotal role in shaping the future of pipeline integrity management.

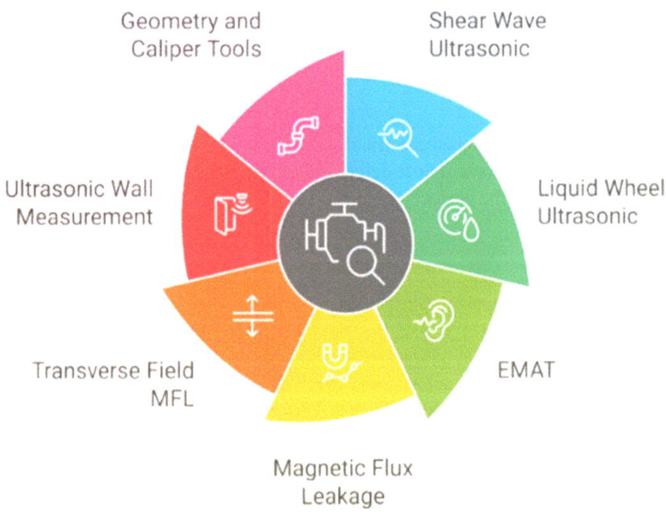

Fig. 5.4 ILI technologies to detect SCC

Table 5.5 Summary of in-line inspection technologies for crack detection

Technology	Pros	Cons	Run in operating	
			Oil	Gas
Shear wave (Liquid-coupled) ultrasonics	• Most proven detection • Description and sizing capabilitics	• Requires liquid slug to run in gas line	No	Yes
EMAT	• Run in either liquid or gas pipelines with higher resolution than LWUT • Reliability, detectability, and discrimination issues will improve as technology matures	• Mechanical and reliability issues • Discrimination and detectability performance not as good as traditional shear wave tools at the time of printing	Yes	Yes
FMFL	• Proven technology for metal loss and 'wide' opening axial defects • Range of 350 km+ • Secondary detection of corrosion and dents	• Only detects wide-opening axial defects—unreliable at detecting or sizing tight SCC cracks	Yes	Yes
LWUT	• Most operational kilometers in gas pipelines. Gas-bypass speed control is available. Tool reliability is good	• Lower resolution than traditional shear wave ILI. Signal/noise and discrimination issues	Yes	Yes
New technologies	• Evolving technologies and improvements to electronics, memory and interpretation algorithms will result in new tools and improvements to existing technologies	• Crack tool development is expensive and complex	Yes	Yes
SEEC (Self excited eddy current)	• Does not require a liquid medium • Can effectively manage higher pipeline product speeds	• New technology is currently being proven	Yes	Yes

5.5 Big Data and Pipeline Integrity Management

Key technologies and tools underpin the application of Big Data Analytics in pipeline management. Hadoop, distributed file systems, Apache Spark, and machine learning frameworks are instrumental in handling the massive datasets generated by pipeline sensor networks and IoT devices (Mantzoukas 2020; Buddhika et al. 2021). The O&G industry generates a significant quantity of structured and unstructured data. The algorithm (Raj et al. 2022) is utilized for analyzing the huge data in the O&G industry, resulting in the production of valuable information, see Fig. 5.5 (Nasurudeen et al. 2023).

The application of Big Data in pipeline management is multifaceted and brings about significant advantages (Valencia Parra 2019). One key aspect is predictive maintenance. By analyzing historical and real-time data from pipelines, predictive analytics can identify potential issues before they escalate into critical problems (Habeeb et al. 2019). This proactive approach allows for scheduled maintenance, minimizing downtime and reducing the likelihood of costly failures. Additionally, Big Data Analytics aids in optimizing pipeline operations by providing insights into efficiency and performance (Noshi et al. 2018). Through the analysis of operational data, organizations can fine-tune their processes, enhance resource allocation, and improve overall efficiency (Mohammadpoor and Torabi 2020).

To evaluate and forecast the causes, consequences, and likelihood of pipeline incidents, as well as to gauge the associated risks, contemporary data-driven methodologies for pipeline incident risk analysis heavily rely on statistical examination of historical incidents (Papadakis 1999; Lam and Zhou 2016; Bubbico 2018). Rachman et al. (2021) extensively reviewed the applications of machine learning in the realm of pipeline integrity management, underscoring the growing importance of machine learning techniques in enhancing the precision and efficiency of pipeline integrity management practices. Hussain et al. (2023) focused on the utilization of Big Data Analytics to empower energy companies to adopt a proactive strategy when it comes

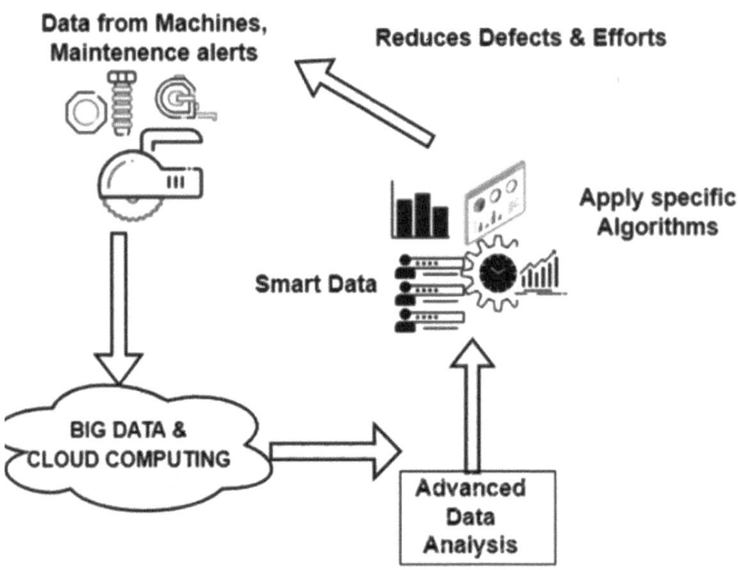

Fig. 5.5 Big Data in oil and gas sector (Nasurudeen et al. 2023)

to assessing the condition of their oil and gas pipelines and associated structure. Through the amalgamation of real-time data derived from sensors embedded within the pipelines, meteorological data, and maintenance records; it becomes feasible to identify potential concerns and forecast anomalies. Significant research efforts have been dedicated to the development of precise models for predicting corrosion growth rates. These are based on either theoretical generation mechanisms, observed physical appearance, such as the expansion of corrosion defects over time, or a combination of both. These degradation models have been analyzed by a combination of field data, which includes direct inspection and In-line Inspection (ILI), as well as data collected under controlled laboratory conditions. One practical approach is applying the maximum experienced measured corrosion growth rates to all defects. This approach is suggested in guidelines such as ASME B31.8S-2004, NACE RP0502, and GRI-00/0230. These recommended maximum corrosion growth rates range from 0.3 to 0.56 mm/year and mostly depend on the pipeline steel's Specified Minimum Yield Strength (SMYS). In order to reduce the uncertainty in the ASME B31G corrosion evaluation model, Leis et al. (2017) reported FEA numerical results. Heggab et al. (2023) conducted a numerical sensitivity investigation using Finite Element Analysis (FEA) to assess burst pressure in corroded pipelines.

Regarding corrosion prediction in oil and gas pipeline systems, the principal application scenarios, as illustrated in Fig. 5.6, encompass pipeline corrosion rate estimation, leakage detection, defect assessment, damage identification, fault diagnosis, pipeline erosion analysis, research on corrosion-resistant materials, corrosion image recognition, and the construction of corrosion maps.

Integration of deep machine learning into oil and gas pipeline systems will eventually enable the combination of intelligent sensing and physical state, data-driven and

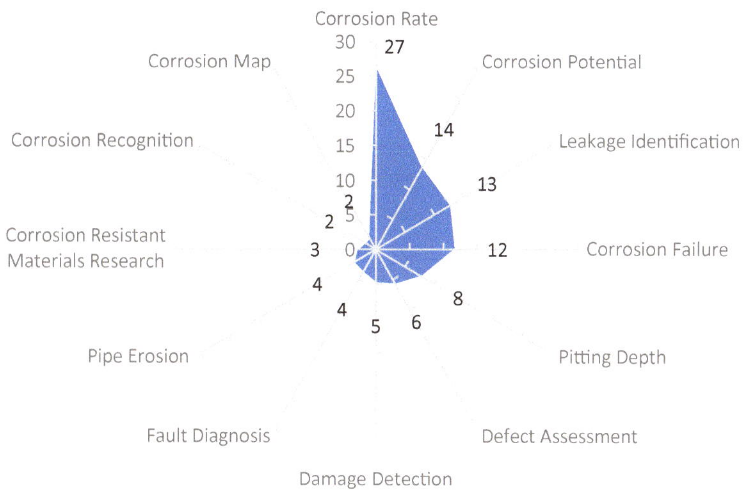

Fig. 5.6 Primary application scenarios of machine learning

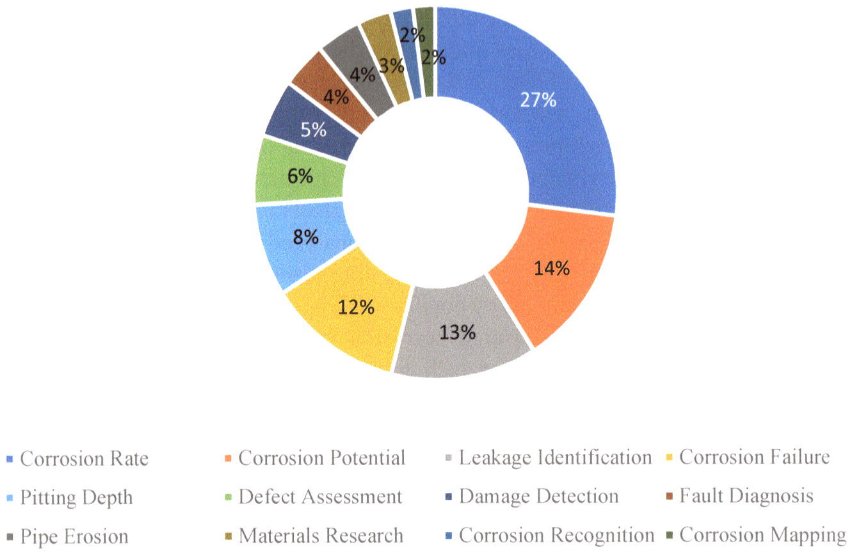

Fig. 5.7 Main application scenarios of machine learning in the field

simulation models, and auxiliary decision-making and operational control, effectively enhancing the ability to harness complex systems and enhancing the efficiency and safety of oil and gas pipeline system operations. The primary application scenarios for predicting corrosion in oil and gas pipeline systems are depicted in Fig. 5.7, and they primarily involve pipeline corrosion rate, leakage identification, defect assessment, damage detection, fault diagnosis, pipeline erosion, research into corrosion-resistant materials, corrosion image identification, and corrosion map construction.

The estimation of corrosion fault propagation requires further investigation to identify dependable solutions. Researchers still lack consensus on the patterns and expected timing of corrosion fault propagation. These concerns have prompted several scholars to utilize artificial intelligence as a viable method for assessing the corrosion behavior of pipes. The rate of metal corrosion significantly influences the actual service life of pipelines.

The literature encompasses various facets of Big Data applications in oil and gas pipelines, such as pipeline monitoring, predictive maintenance, safety improvement, operational efficiency, risk evaluation, data integration, maintenance expenses, and visualization. Currently, most of the oil and gas storage and transportation businesses employ a three-tiered strategy to oversee the security of their storage systems and pipelines. This method entails collecting temperature, pressure, and more data from long-distance pipelines and delivering it in real-time to a data monitoring center via the Internet, where it is subsequently analyzed and processed for potential early warning scenarios (Kong et al. 2020). In recent years, the rise in failure instances has heightened scrutiny on the integrity and condition assessment of gas pipelines (Hassani and Silva 2018; He et al. 2019).

Big Data encompasses a wide array of structured, semi-structured, and unstructured datasets, as implied by the term "Big Data." Big Data Analytics is the process of deriving significant conclusions, patterns, and insights from vast and complex databases. It involves utilizing advanced approaches and tools to analyze vast quantities of structured, semi-structured, and unstructured data to identify patterns, correlations, trends, and other critical insights that might facilitate decision-making, improve company operations, or drive innovation. Unstructured data refers to datasets lacking adequate categories or labels, in contrast to structured data, which possesses them. Many data in the oil and gas sector is unstructured, rendering analysis very tough. The technology for analyzing unstructured data is nascent, complicating the research and prediction of potential hazards. Despite 80% of data being unstructured, Gartner asserts that data generation would escalate by 800% in the next five years (Shah et al. 2022). Big Data Analytics projects can yield both intangible and concrete corporate value, referred to as benefits (Jensen et al. 2023). The oil and gas industry may save operational expenses and positively impact all stages of the oil and gas lifecycle by utilizing Big Data and analytics technologies (Holdaway 2014; Abdullahi et al. 2019).

Identifying abnormalities and precisely forecasting future behavior during operations will facilitate more effective decision-making, hence optimizing operational expenditures for risk mitigation. Numerous models have been created throughout the years to forecast pipeline failures and conditions. Nevertheless, most of these models employed corrosion characteristics as the sole criterion to assess the state of pipes. Regular inspections are essential to maintain the pipeline in a dependable and secure state. Over the past decade, many examination techniques, such as Ultrasound and Magnetic Flux Leakage (MFL), have been developed. These technologies furnish accurate and effective data for energy pipeline operators to detect pipeline faults that may result in catastrophic failure. In-line inspection (ILI) is an effective and proficient tool that identifies and/or forecasts pipeline irregularities.

Gu et al. (2019) introduced a pipeline lifetime state monitoring framework that offers lifetime analysis through machine learning, facilitating optimal decision-making to mitigate risks that could lead to severe consequences and jeopardize personal safety. Sarker et al. (2022) examined the opportunities and difficulties of Big Data across several businesses. They examine matters pertaining to data collection, storage, processing, and analysis, encompassing data quality, integration, privacy, and security challenges. They provide a comprehensive overview of the advantages of Big Data across several businesses, including energy pipelines; however, they did not address the utilization of Big Data inside the pipeline sector. Swetapadma and Yadav (2016) introduced a fault detection system utilizing decision trees. Layouni et al. (2017) developed a method for detecting metal loss by evaluating magnetic flux leakage data obtained from oil and gas pipeline scans, utilizing pattern-adapted wavelets and two machine learning algorithms: artificial neural networks and linear regression. They also deliberated on the quantitative and qualitative methods to ascertain the lengths, depths, and locations of metal loss in the oil and gas pipeline. Lv et al. (2017) examined data kinds, storage strategies, data security, and privacy, as well as the advantages of Big Data networks. Nguyen et al. (2020) examined the concept of Big Data within the oil and gas industry, including its opportunities and associated challenges. Nevertheless, they did not address the merits and drawbacks of each Big Data Analytics paradigm pertinent to Industry 4.0. Layouni et al. (2017) finds metal-loss issues from the MFL scans of oil and gas pipelines and assesses their dimensions utilizing machine-learning methodologies. The proposed approach, which attains elevated accuracy and computational efficiency, is founded on pattern-adapted wavelets and artificial neural networks.

Employing Big Data Analytics in the energy business diminishes costs and enhances implementations by introducing innovative solutions that optimize pipeline asset management, utilizing fewer resources while increasing delivery efficiency. As hardware-based tools necessitate increased engineering efforts, dedicating additional time to analysis is not a cost-effective strategy for numerous firms (Nguyen et al. 2020; Rawat 2014; Skrynkovskyy et al. 2018). The surveillance of energy pipelines has other applications beyond production assessment, such as security, preventive and predictive maintenance, leak detection, equipment regulation, and automation systems. Location identification and awareness of the pipeline trajectory are essential for pipeline monitoring. Employing geographic information systems and global positioning systems facilitates the identification of pipeline incidents and the tracking of reported events (Elijah et al. 2021; Eze et al. 2017). Corrosion, environmental factors, human error, installation and assembly, and manufacturing are among them. Currently, pipeline monitoring commonly uses wireless sensor networks (WSN) (Adegboye et al. 2019; Franconi et al. 2014; Wan et al. 2011). WSN, IoT, Big Data, and AI collaborate to enhance smart monitoring and facilitate remote access to data obtained from the pipeline (Wu et al. 2019; Yi et al. 2010; Barani and Lakshmi 2013; Seo et al. 2021; Seghier et al. 2021; Liu et al. 2019). To mitigate the risks of corrosion, erosion, and wear, the data obtained through IoT must be analyzed using an appropriate decision-making framework (Singh and Hetlevik 2017). Artificial intelligence has been employed to anticipate pipeline problems (Akram et al. 2014), model two-phase flow in pipes (Chaari et al. 2020), and project the deterioration rate of pipe fittings (Zahedi et al. 2018).

Numerous studies have demonstrated the benefits of Big Data utilization in pipeline asset management. A correlation was identified between the lifespan of ground pipes and extensive pipeline data using various model-fitting techniques (Fan et al. 2019). The methodology for evaluating pipeline performance with aggregated data statistics is widely recognized (Hussain et al. 2021; Ahmad et al. 2018). The incorporation of the safety factor parameter within established evaluation criteria is constrained due to the static nature of these parameters, whereas environmental factors, infrastructure elements, and material compositions differ from one pipeline to another (McMahon et al. 2020). Identifying abnormalities and precisely forecasting future behavior during operations will facilitate more effective decision-making, hence optimizing operational expenditures for risk mitigation. By employing Big Data Analytics, they can allocate less resources and enhance the decision-making process (Zhang et al. 2020).

Rouhanizadeh and Kermanshachi (2019) developed a prediction model employing multiple regression techniques to anticipate the performance of sewer pipes. The model utilizes historical data, encompassing previous failures and hydraulic information, as input. The model's results demonstrate a marked sensitivity to the age of the pipeline system and the type of pipes employed. The model facilitated pipeline owners in pinpointing pipe portions with a higher susceptibility to failure, hence allowing the formulation of a robust risk management system for sewer pipeline networks.

Cobanoglu et al. (2022) performed a thorough assessment of Big Data, offering an extensive analysis of the existing research landscape, evaluating diverse methodologies, and emphasizing future opportunities. Their discourse encompasses diverse methodologies and techniques employed for data collection, storage, processing, and analysis, including machine learning, data mining, natural language processing, and visualization across multiple sectors such as healthcare, banking, marketing, transportation, and social media. It examines the influence of Big Data Analytics on decision-making processes and underscores the potential advantages and hazards linked to its application. Consequently, leveraging Big Data facilitates the promotion of "digital energy" to improve operational and business performance.

5.5.1 Benefits of Big Data in Pipeline Integrity Management

Big Data Analytics offers several key benefits for pipeline integrity management, helping operators shift from reactive to proactive maintenance strategies:

- **Predictive Analytics**: One of the most significant benefits of Big Data Analytics is its ability to support predictive maintenance. By analyzing historical data on pipeline conditions, such as corrosion rates and material fatigue, machine learning models can predict when and where failures are likely to occur. This allows operators to schedule maintenance activities before issues arise, minimizing downtime and extending the lifespan of the pipeline.
- **Real-Time Monitoring**: Real-time monitoring systems, such as SCADA, provide continuous data on pipeline conditions, including pressure, flow rate, and temperature. Big Data Analytics can process this real-time data and detect anomalies, such as sudden pressure drops or temperature fluctuations, which may indicate a leak or rupture. By providing real-time alerts, BDA allows operators to take immediate action and prevent catastrophic failures.
- **Risk Assessment and Decision Support**: Big Data Analytics enhances risk assessment by providing operators with detailed information on the likelihood and consequences of pipeline failures. By integrating data from multiple sources, operators can develop more accurate risk models and prioritize maintenance activities based on the areas of greatest risk. Decision support systems powered by BDA provide operators with actionable insights that improve the overall efficiency and safety of pipeline operations.

- **Cost Savings**: The ability to predict failures and optimize maintenance schedules reduces the need for unplanned repairs and emergency shutdowns, leading to significant cost savings for pipeline operators. In addition, improved risk assessment and preventive maintenance can reduce the likelihood of environmental damage and regulatory penalties.

5.6 Defining Big Data in the Energy Sector

Big Data refers to datasets that are so large, complex, and fast-moving that traditional data processing methods are insufficient to handle them (Selmy et al. 2024). In the energy sector, Big Data typically encompasses a range of structured, semi-structured, and unstructured data collected from various sources, including sensors, industrial equipment, satellite systems, and external databases (Sarker et al. 2022; Ikegwu et al. 2022).

5.6.1 Big Data's Role in Energy Operations

Big Data Analytics enable energy companies to extract valuable insights from the massive volumes of data they collect. In the context of pipeline integrity management, Big Data helps operators in few aspects as shown in Fig. 5.8.

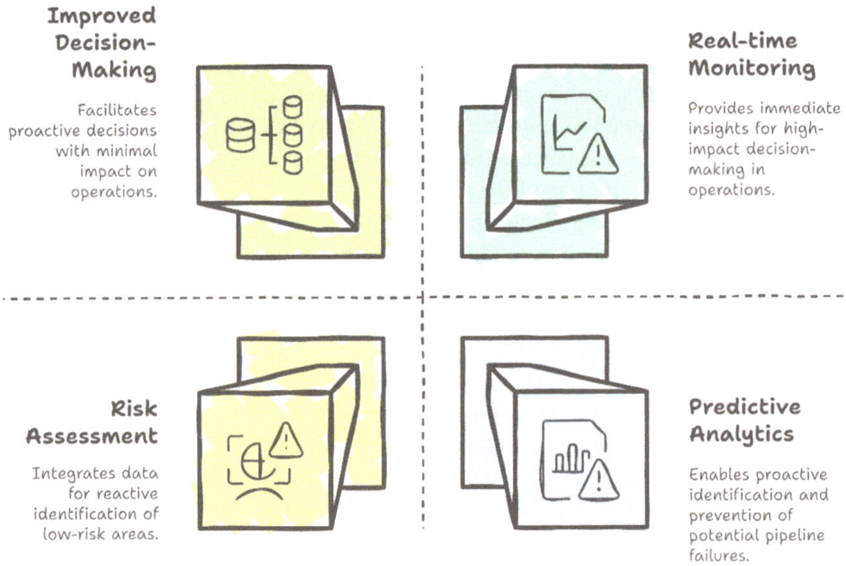

Fig. 5.8 Big Data in pipeline integrity management

5.7 Decision Making Models for Energy Pipeline Asset Management

To predict anomalies and understand existing conditions of the pipelines, multiple models have been developed over the years. Most of these models, however, used corrosion characteristics to determine the condition of pipelines (El-Abbasy et al. 2015). Frequent checks are mandatory to keep the pipeline in reliable and safe condition. In the last decade, numerous inspection techniques have been developed. These methods provide precise and efficient information for the energy pipeline operators to identify pipeline defects that could lead to a catastrophic failure. This section discusses a few models that are closely related to the prediction of pipeline degradation.

5.7.1 Neural Network (NN) Modeling

A set of training samples is used by NN models to "learn" patterns, which are then "applied" to new data to find related patterns. Figure 5.9 depicts a NN model's general structure. An input layer, an output layer, and one or more hidden layers are often present. As a unit, each input, output, or hidden node is indicated. The weights are represented by the connections between the nodes. Once the number of hidden layers is chosen, the main job of NN training is to establish these weights (Muñoz-Zavala et al. 2024). It should be emphasized that linkages flowing straight from inputs to outputs or feedback from outputs to inputs are both permitted.

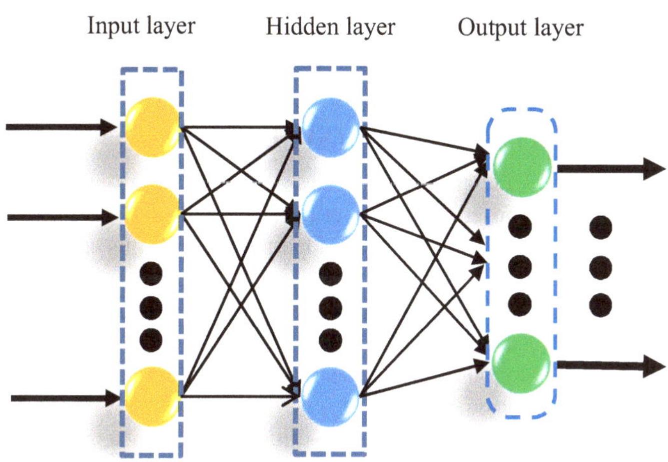

Fig. 5.9 Typical structure of a neural network (NN)

The neural network modeling approach was not applied in this study, as it is not suitable to predict future wall loss because the required data is not available from the current industrial practice. Using the neural network-based method requires plenty of data to train a model. Although in this study, the ILI datasets collected from two different pipelines and one presented in a research paper data, the amount of data at different positions is not enough to train a neural network for predicting an individual anomaly growth. It has been noticed that there is too little data along the timeline when referring to an individual anomaly, which is important for predicting future data points. Therefore, the neural network modeling approach is not an appropriate choice to predict corrosion in oil and gas pipelines using ILI data.

Zheng et al. (2011) conducted one attempt to calculate the corrosion growth-rate of a gas pipeline using a NN. Wang and He (2010) used the Adaptive Neuro-Fuzzy Interference System (ANFIS) to estimate the corrosion growth rate of a pipeline submerged in soil. Liao et al. (2011) used a backpropagation NN and a niche concept generic method to estimate a wet gas pipeline's internal corrosion growth rate and optimize the model. Their model utilized corrosion growth rate as both an input and an output, much like the models used by Wang and He (2010) and Zheng et al. (2011). In the above studies, laboratory tests were used to gather the data. Although the model can be built using the laboratory test data, it is not practically applicable.

It has been demonstrated that NN models can take measurement errors of ILI techniques into account. As an illustration, consider the work of Sinha and Pandey (2002), who employed probabilistic neural networks (NNs) to evaluate the reliability of oil and gas pipelines based on pipeline characteristics like yield strength and defect dimension with their associated uncertainties. The estimation of oil pipeline failure types, such as third-party damage, mechanical damage, natural hazard damage (Senouci et al. 2013); pipeline conditions (El-Abbasy et al. 2014); and potential control associated with cathodic protection, CP (AL-Shareefi 2009), are additional applications of NN models for trial testing.

5.7.2 *Support Vector Machines (SVMs)*

In 1992, SVM was proposed by Boser et al. (1992) for supervised classification and regression learning. Utilizing statistical learning theory, this approach enhances prediction accuracy while concurrently reducing model variation. Statistical learning theory provides a framework to analyze the issue of generating inferences and decisions based on data (Xu et al. 2020). In other words, it functions in hyperplane space in a manner analogous to its operation in the target space (Cristianini and Shawe-Taylor 2000). Consider the set $(x_1, y_1), (x_2, y_2), \ldots\ldots, (x_n, y_n)$ as training data where $x \subset R^n$ and $y \subset R^1$ denotes the n-dimensional input and one-dimensional output data, respectively. Support Vector (SV) regression aims to identify a function f(x) that deviates from the observed data by no more than a given threshold, ε, and is as flat as feasible. In other words, the error (Er) between the observed (y) and

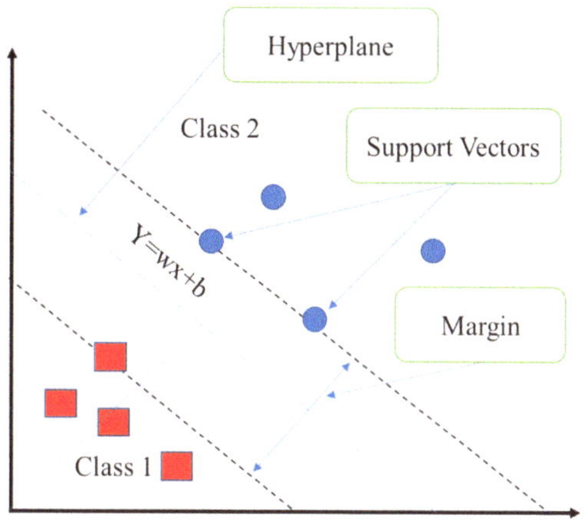

Fig. 5.10 Basic concept of SVM

simulated output (f(x)) is disregarded if it is smaller than ε; otherwise, it is equal to $|y - f(x)| - \varepsilon$ (Abyani et al. 2022).

The Lagrange multipliers approach can be employed to achieve this optimal solution. Training data x can be mapped to a higher-dimensional space using kernels such as polynomials, RBFs, and Gaussian functions. The fundamental concept of Support Vector Machines (SVM) is illustrated in Fig. 5.10.

SVMs are supervised learning models used for classification and regression tasks. In pipeline asset management, SVMs can classify pipeline segments as high-risk or low-risk based on data from sensors and monitoring systems (Chen 2024). By identifying the segments most likely to fail, SVMs help operators prioritize maintenance activities and allocate resources more efficiently. SVMs are particularly effective for analyzing high-dimensional datasets, making them well-suited for predicting pipeline failures based on multiple input variables.

SVMs are used in pipeline integrity management to classify pipeline segments based on their likelihood of failure. For example, an SVM might be trained to classify pipeline segments as high-risk or low risk based on input variables such as:

- **Wall thickness**: Data from in-line inspection (ILI) tools can be used to determine the wall thickness of the pipeline. Segments with thinner walls may be more prone to failure.
- **Corrosion rates**: Segments with higher rates of corrosion may be classified as high risk.
- **Operating conditions**: Pressure, temperature, and flow rate data from SCADA systems are also used as input variables to classify pipeline segments based on their risk level.

Once the SVM has been trained, it can be used to classify new pipeline segments, allowing operators to prioritize maintenance activities based on the segments that are most likely to fail. SVMs work by finding the hyperplane that best separates the data into different classes. In the case of pipeline integrity management, the SVM finds a hyperplane that separates high-risk pipeline segments from low-risk segments. The key features of SVMs include:

- **Support Vectors**: The support vectors are the data points that lie closest to the hyperplane. These points are the most important for defining the decision boundary between the different classes.
- **Kernel Functions**: SVMs can use different kernel functions to transform the input data into higher-dimensional spaces, making it easier to find the separating hyperplanes. Common kernel functions include linear, polynomial, and radial basis function (RBF) kernels.
- **Margin**: The margin is the distance between the hyperplane and the closest data points (support vectors). SVMs aim to maximize the margin, as this reduces the likelihood of misclassification.

SVMs offer several advantages for pipeline integrity management:

- **Effective in High-Dimensional Spaces**: SVMs are particularly effective in situations where the number of input variables is large, making them ideal for pipeline integrity management, where multiple factors influence the risk of failure.
- **Robust to Overfitting**: SVMs are less prone to overfitting than other machine learning models, especially when the number of input variables exceeds the number of training samples.
- **Flexibility with Kernel Functions**: SVMs can handle both linear and non-linear relationships between input variables and pipeline integrity outcomes, thanks to the flexibility of kernel functions.

However, like neural networks, SVMs require high-quality data for training and may be difficult to interpret in certain cases.

5.7.3 Random Forests

Random Forest (RF) is a sophisticated machine-learning technique used for regression analysis (Breiman 2001). It enhances the stability and accuracy of the model by combining the outcomes of several decision trees in an ensemble (Florez-Lopez and Ramon-Jeronimo 2015; Lin et al. 2017). This technique resembles the concurrent learning algorithm since it involves the simultaneous construction of all decision trees.

The training algorithm for random forests utilizes the general methodology of Bagging (bootstrap aggregating) to train tree learners (Liu et al. 2019; Bosch et al. 2007). To be specific, given a training set x = x1, x2, …, xn with responses y = y1, y2, …, yn, bagging is performed N times by selecting random samples with replacement from the training set and fitting decision trees to each sample, for n = 1, 2, …, N.

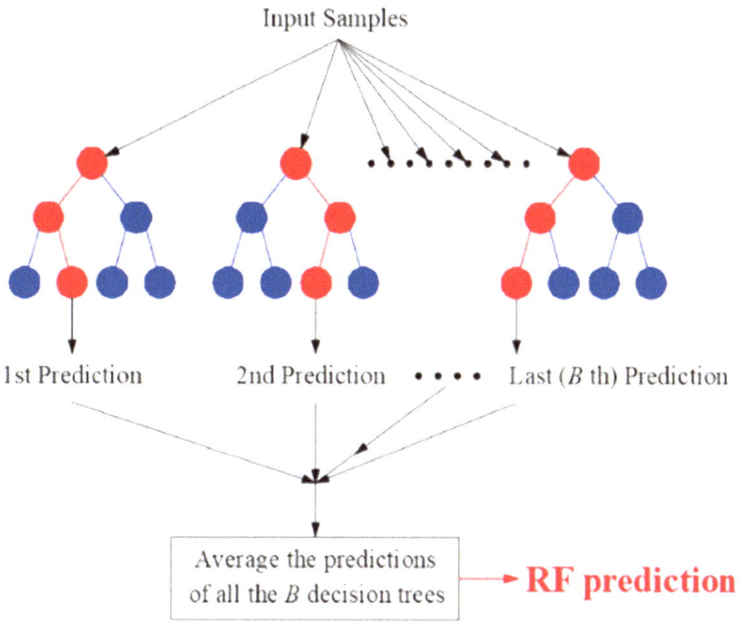

Fig. 5.11 Random forest with binary (B) decision trees

Figure 5.11 demonstrates that the variance of RF models is significantly reduced when the outcomes of all constructed trees are averaged. A fresh dataset is generated by employing the bagging predictor method and creating multiple bootstrapped replicas of the initial training data to replicate the impact of time.

Random forests are ensemble learning models that use multiple decision trees to make predictions (Putra et al. 2024). In the context of pipeline integrity management, random forests are used to predict failure probabilities and classify pipeline segments based on their risk levels (Hussain et al. 2024). Random forests are particularly effective at handling large datasets with multiple variables, making them well-suited for analyzing complex pipeline systems (Hussain et al. 2024; Latifi et al. 2024).

Random forests ensemble learning models that combine multiple decision trees to make predictions. Each tree in the forest is trained on a random subset of the data, and the final prediction is made by averaging the predictions of all the trees. Random forests are particularly well-suited for pipeline asset management because they can handle large datasets with multiple input variables and provide robust predictions even in the presence of noisy or missing data (Xiao et al. 2024).

Random forests are used in pipeline integrity management to predict failure probabilities and classify pipeline segments based on their risk level. The input data for random forests can include a wide range of variables, such as:

- **Corrosion rates**: Data from corrosion monitoring systems is used to predict which segments are most likely to experience corrosion-related failures.

- **Inspection data**: Data from in-line inspections (ILI) and non-destructive testing (NDT) methods are used to assess the pipeline's condition and predict future failures.
- **Environmental factors**: External factors, such as soil composition and weather conditions, are also incorporated into the model to provide a more comprehensive assessment of pipeline integrity.

Random forests offer several advantages for pipeline integrity management:

- **Handling of Noisy Data**: Random forests are robust to noisy or incomplete data, making them ideal for pipeline integrity management, where data quality can vary.
- **Feature Importance**: Random forests can provide insights into which input variables are most important for predicting pipeline failures. This helps operators understand which factors drive the risk of failure and allows them to take targeted preventive actions.
- **Scalability**: Random forests can handle large datasets with many input variables, making them well-suited for analyzing complex pipeline systems.

5.7.4 *Gradient Boosting*

The Gradient Boosting Regressor is an additional ensemble model consisting of a series of tree models arranged sequentially. Each subsequent model in the series learns from the mistakes made by the previous model (Konstantinov and Utkin 2021).

This regressor employs the technique of "boosting" to enhance the predictive power of a set of underperforming models, typically decision trees, resulting in a more dependable model. A gradient-boosting regression (GBR) model (Otchere et al. 2022) with M trees can be represented as

$$f_M(x_j) = \sum_{m}^{M} \gamma_m h_m(x_j)$$

where $h_m()$ represents a learner with low individual performance, γ_m is a scaling factor that incorporates the contribution of a tree into the model.

Gradient Boosting Regression (GBR) uses the gradient descent loss function to iteratively update the initial estimate using the most up-to-date estimate, hence minimizing mistakes (Begum et al. 2023). To construct a final model, all initial estimations are aggregated using appropriate weights.

5.7.5 Monto Carlo Simulation

To estimate the condition on the posterior distribution probability of failure, a crude Monte Carlo simulation usually follows as a post-processing step after the Bayesian analysis. The drawbacks of this framework are: first, the ambiguity regarding whether the final samples were properly distributed in the Markov Chain Monte Carlo (MCMC) sampling; and second the complexity of the Monte Carlo crude simulation in subsequently quantifying small probabilities of rare failure events (Betz et al. 2018). The use of the second-order polynomial dynamic linear model (DLM) to characterize the increase in the depth of corrosion defects on energy pipelines is given by Zhang and Zhou (2014) using data obtained from several ILI runs. The Markov Chain Monte Carlo simulation is used to carry out the Bayesian update of the growth model and test the later distributions of the model parameters. The analysis results show that the defect depths predicted by the proposed model are well compared to the actual depths. They are more precise than those predicted in the literature by the Gamma process- and Inverse Gaussian process-based growth models (Zhang and Zhou 2014).

To classify the parameters in the corrosion growth model based on data obtained from high-resolution ILIs, the hierarchical Bayesian approach and Markov Chain Monte Carlo (MCMC) simulation are used (Zhang and Zhou 2013). Lee et al. (2015) provide a case study of the reliability-based design and assessment (RBDA) technique for estimating the likelihood of time-dependent failure of pipelines with corrosion defects, where the separable Monte Carlo (SMC) method is used instead of the conventional crude Monte Carlo (CMC) method in the reliability calculation for corroded pipelines. In the case of multiple components with arbitrary correlation levels, the study is carried out using improved Monte-Carlo simulation methods that are very effective for quantifying device reliability (Leira et al. 2014). A reliability measurement model based on the Monte Carlo simulation for corrosion defect pipelines calculated by in-line inspection is advanced (Li 2017). The Monte Carlo simulation is conducted to estimate the precision of critical-size corrosion defects with distinct safety factors (Chen 2019). To ensure the quality of experimental findings, several fitness-for-purpose test methods were applied (Xuan et al. 2017).

Rubienstein (1986) introduce a technique of using Bayesian analysis to evaluate sufficient non-periodic inspection intervals of fatigue-sensitive structures such that their reliability during their service life remains above a defined minimum level. For the calculation of the likelihood of pipeline failure and the probability of cumulative failure due to external corrosion, the Monte Carlo approach was chosen to consider fluid pressure variations in complex flow effects about the statistical distribution of input parameters (Witek 2016). The Monte Carlo simulation model, as determined by a partial wave method and a dielectric functional approach to electron inelastic scattering with the complete Penn algorithm, is based on the Mott cross section for electron elastic scattering (Hussain et al. 2020). The mathematical model shows how the likelihood of the domino effect decreases and increases with the diameter of the target pipe (Ramírez-Camacho et al. 2015). For the burst models, the probabilistic characteristics of the model errors, including mean values, variance coefficients and probability distributions, are obtained by analyzing the ratios between the test and the expected burst pressures corresponding to the relevant model test results (Zhou and Huang 2012).

5.7.6 Dynamic Bayesian Networks (DBN)

DBNs provide a powerful tool for modeling and predicting varying growth rates of corrosion defects. These networks capture the dynamic dependencies among variables, allowing for a more nuanced representation of the evolving corrosion process. DBNs have been utilized to assess the time-dependent growth of both pitting and general corrosion in subsea pipelines (Qiu et al. 2023).

5.7.7 Dynamic Linear Model (DLM)

DLM is capable of accounting for the measurement uncertainties of defect depth and the temporal influence of defect growth. DLM in contrast to Gamma-process-based models, does not provide precise distributions to the defect depth. The Gamma-process-based models are said to be less flexible because the analysis might start without knowing the distribution type for the fault depth. According to Zhang and Zhou (2014), the technique outperforms Gamma-process-based models. Three linked equations, provided by West and Harrison (2006), make up a dynamic linear model including an observation equation, a system equation, and a system slope equation.

$$\text{Observation equation}: Y_t = u_t + v_t$$

$$\text{System equation}: u_t = u_{t-1} + \beta_{t-1} + w_{t1}$$

$$\text{System slope equation}: \beta_t = \beta_{t-1} + w_{t2}$$

where Y_t and the u_t are the measured and actual defect depth if the corrosion problem is taken as an example; β_t is the corrosion growth-rate; v_t is the measurement error, which is independent of the actual defect depth; w_{t1} represents step changes in defect depth, for example, due to the mishaps in operations and w_{t2} represents changes in corrosion growth-rate due to long term changes in operation conditions. It is often assumed that $v_t \sim Norm(0, V)$, $w_{t1} \sim Norm(0, W_{t1})$ and $w_{t2} \sim Norm(0, W_{t2})$ where V, W_{t1} and W_{t2} are the variances of the associated variables and they may vary over time.

Zhang and Zhou (2014) applied DLM to characterize growth paths of individual defects with the data collected from multiple ILIs on a gas pipeline. Adjustment to the standard definition of DLM was made to accommodate the measurement uncertainties and the non-decreasing feature of defect depth growth. In this model, an average corrosion growth rate was estimated for each independent time interval, representing defect growth as a piece-wise linear process. To estimate defect depth beyond the inspection times, extrapolation techniques were applied.

Figure 5.12 (Zhang and Zhou 2014) depicts the directed acyclic graph (DAG) for the hierarchical Bayesian updating model for DLM. In contrast to the DAG of the Gamma process-based model shown in Fig. 5.13, it should be noticed in this figure that the initiation time for each individual flaw is not considered in this model. For each of the individual faults, the parameters for this model, such as $d_{i,1}$ and $r_{i,1}$ are determined. The spatial link between various flaws cannot be represented by this method. Furthermore, there is no physical relationship associated with the use of the term $\eta_{i,j}$. For this model, the challenge to use it is how to determine those many parameters if it is used to forecast individual defects. It depends on the expert opinions in the determination of the model parameters.

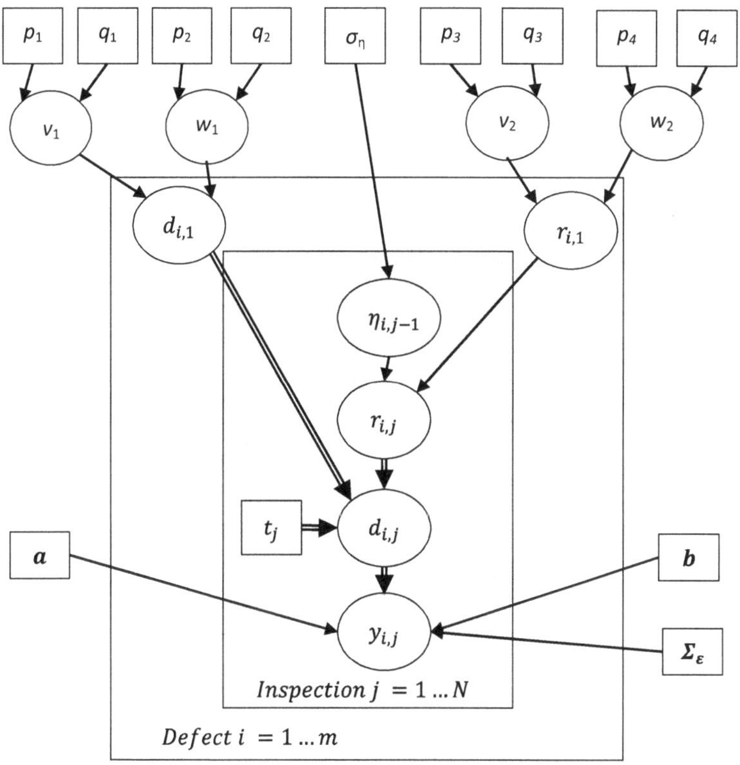

Fig. 5.12 Structure of the DLM-based degradation model (Zhang and Zhou 2014)

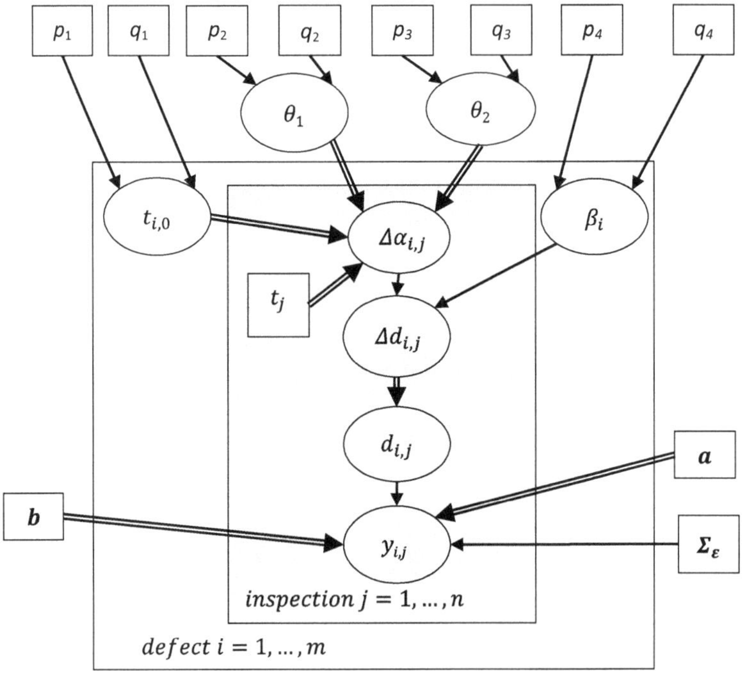

Fig. 5.13 Hierarchical Bayesian model structure (Zhang and Zhou 2013)

5.7.8 Bayesian Decision Networks (BDN)

BDNs integrate probabilistic reasoning and decision analysis, providing a comprehensive framework for incorporating uncertainty into corrosion prediction models. By leveraging Bayesian principles, these models enable a systematic representation of the relationships between different variables, allowing for informed decision-making in the face of uncertain corrosion scenarios (Khalaf et al. 2023). da Cunha Bisaggio and Netto (2015) used Bayesian reliability principles to estimate the progression of a predetermined distribution of defects acquired, for example, via an inspection campaign. Through the comparison of the anticipated chance of failure and the established reliability acceptance criteria, the operator can confidently determine the timing for defect repairs and create intervals for inspections.

5.7.9 Deterministic Linear Growth Model

Deterministic linear growth models are mathematical models that describe the growth of a quantity over time (Ma et al. 2023). They are based on two assumptions:

- The growth rate of the quantity is constant.
- The growth rate is proportional to the current size of the quantity.

The linear growth model is the simplest degradation model available and is in common use within the oil and gas industry (Heidary et al. 2018). Given a defect dimension such as: length, depth, or volume, measured at two different times, the growth-rate, r, can be estimated by

$$r = (x_2 - x_1)/(t_2 - t_1)$$

where x_i (i = 1, 2) is the relevant measurement as indicated at time t_i (i = 1, 2, and $t_2 > t_1$). The dimension of the defect at a future time, $t(t > t_2)$, can be calculated by

$$x = x_2 + r(t - t_2)$$

It is apparent that this model assumes that all factors affecting defect growth are constant and that their impact on that growth is also constant. The linear method can be useful in cases when there is only one set of measurement data. When used to estimate growth-rate using data from only one in-line inspection (ILI) run, the non-conservative error induced by assuming a defect initiation time has to be considered (Ma et al. 2023).

5.7.10 Markov Process-Based Models

When using a Markov process to simulate defect growth, it is necessary to make the assumptions that there are discrete states and that changing to a certain state only depends on the present state. The research by Valor et al. (2007) supported Provan and Rodriguez's concept for a laboratory setting by showing that all faults tend to initiate simultaneously in the laboratory setting. The diversity and intricacy of the interaction between metal and environment, however, suggest that corrosion flaws begin at different periods.

Assuming a homogeneous Poisson process for defect initiation and a homogeneous Markov process for stable defect growth, Hong (1999) was the first to include both defect initiation and defect growth processes in one model. By solving the Kolmogorov differential equation, it was possible to get an analytical answer to the model's transition probabilities. The so-called time-transformation-condensation method was used to gather analytical solutions to account for nonhomogeneous effects in both the defect creation and defect growth processes. The effectiveness of this model, however, varies depending on how many states the Markov process defines. Even though the distribution only took the greatest defect depth into account, it was discovered that it did not conform to the Gumbel distribution. Many others have since refuted it, including Provan and Rodriguez III (1989), Melchers (2004),

Valor et al. (2007, 2013). In order to forecast the development of corrosion defects, a pure-birth homogeneous Markov process model (Timashev et al. 2009) has also been applied. Based on the anticipated outcomes of the Markov model and by reducing the maintenance cost, a model to optimize inspection frequency was also created. However, like Hong (1999), the model is constrained by the assumption of a homogeneous Markov process, despite the reality that each defect often exhibits corrosion development rates that vary over time.

A model incorporating defect initiation and defect growth mechanisms was created by Valor et al. (2007). It was assumed that the defect growth process would follow a non-homogeneous Markov process, and the defect initiation process is a nonhomogeneous Poisson process. It was pre-summated that the initiation timings would follow a Weibull distribution. By solving the forward Kolmogorov equation analytically, it is simple to determine the probability of a defect whose depth is less than or equal to a given state, which acted as the parent function for the distribution of defects, each of which has a depth that is less than or equal to a specific state. This population-based distribution ended up falling into the Gumbel distribution's region of attraction for maxima. Valor et al. (2013) expanded on this by adding a second Markov process model to predict the depth growth of a single defect using additional data, like soil condition.

McCallum et al. (2014) created a stochastic model that accounts for both the defect initiation and growth phases in order to anticipate the growth of defects. Contrary to Hong (1999) and Valor et al. (2013), a new model was employed to depict the average number of defects whereas the defect start was treated as a nonhomogeneous Poisson process. They also permit transition rates that are not proportional to the state in their defective growth model.

5.7.11 Gamma Process-Based Models

The Gamma process is most suited for its stochastic modeling since pipeline defect accumulation is by definition a steady, monotonically growing process (van Noortwijk 2009). A Gamma process-based model can also convey spatial information by allowing the parameters of the Gamma process to be shared or computed independently for the individual faults in a population. It should be noted, that the assumption of a Gamma process applies to the increments in defect depth at non-overlapping time intervals rather than the corrosion development rates of individual faults. Depending on the type of Gamma processes used, such as stationary or non-stationary, it may not be possible to directly quantify the corrosion growth rates of specific faults. The accumulation of depth increments for all prior time intervals, with those increases being assumed to be independent and identically distributed (IID), determines the growth path of a defect.

van Noortwijk (2009) developed and examined a number of techniques for calculating the parameters of a Gamma process-based model, including the method of moments, the maximum likelihood approach, and Bayesian updating. Measurement uncertainties can be considered for pipeline degradation using the hierarchical Bayesian modeling (HBM) approach (Maes 2013; Pandey and Lu 2013; Zhang et al. 2014; Al-Amin et al. 2014; Dann and Maes 2015).

Measurements frequently have large inaccuracies compared to the amount of the fault. The general "bias" of the mean of subsequent measurements of the same defect and the "scatter" or repeatability of the measurement around that mean can be used to characterize this measurement error. Here, bias and scatter will be used to refer to them, respectively. The defect size may not be constant, and bias and scatter may both depend on it. It is possible to assume that the actual defect depth $d_{i,j}$, follows a Gamma process.

$$d(t) \sim Gamma(d(t)|\alpha(t), \beta)$$

where α(t) and β are the shape and rate parameters, respectively. The mean and variance of d(t) are given by α/β and α/β^2, respectively.

Different relations can be assumed between α and time t. A commonly used assumption is the power law-based function, i.e.

$$\alpha(t) = \theta_1(t - t_0)^{\theta_2}$$

where t_0 is the initiation time of the defect growth, i.e. the time elapsed from the pipe commissioning time to the time point when the defect starts to grow. It is noted that when $\theta_2 = 1$, i.e., α is linear in time, the Gamma process is said to be homogeneous. This model implies that the increments of defect depth also follow a Gamma process: indefinitely divisible, with a scale parameter $\Delta\alpha$ and rate parameter β, i.e.

$$\Delta d_{i,j}(t) \sim Gamma\left(\Delta d_{i,j}(t)|\Delta\alpha_{i,j}(t), \beta\right)$$

By further assuming Δα is specific for each defect and varies with inspections, it can be written as

$$\Delta\alpha_{i,j}(t) = \theta_1\left(t_1 - t_{i,0}\right)^{\theta_2} \quad for\ j - 1$$

$$\Delta\alpha_{i,j}(t) = \theta_1\left(\left(t_j - t_{i,0}\right)^{\theta_2} - \left(t_{j-1} - t_{i,0}\right)^{\theta_2}\right) \quad for\ j > 1$$

If posterior distributions of θ_1, θ_2 and $t_{i,0}$ are obtained then the Gamma process is uniquely determined: i.e. using distributions conditional on the measurement data as opposed to the prior distributions which represent our 'prior' belief in the absence of data. Instead of determining a posterior distribution which does not have an explicit expression for the normalization factor, the likelihood function is often derived and calculated (Zhang and Zhou 2013). It is already noted that when $\theta_2 = 1$ in $\alpha(t) = \theta_1(t - t_0)^{\theta_2}$, the Gamma process-based model is homogeneous or stationary (Zhang and Zhou 2013; Maes 2015).

The so-called directed acyclic graph (DAG), a graphical representation, was created to depict an hierarchical Bayesian model (HBM) structure (Lunn et al. 2009). Figure 5.13 depicts the DAG created for the Gamma process-based degradation model in accordance with the previously mentioned equations.

5.7.12 Stochastic Processes-Based Degradation Modeling

Several stochastic processes have represented corrosion development's initiation and steady growth stages. Alternate re-passivation and growth processes may take place between initiation and stable growth until stabilizing at a certain depth threshold (Tsukaue et al. 1994). In a stochastic model, the state of the defect for the growth phase can be viewed as either discrete or continuous, depending on how time is represented. Discrete defective state models have mainly been developed based on the Markov process. Continuous defect state models include those developed based on processes, including the Gamma process (van Noortwijk 2009) with and without measurement error (Maes 2013), the Inverse Gaussian Process (Zhang et al. 2013), Brownian motion with drift (also known as the Wiener process) (Maes 2013) and the geometric Brownian motion process (Zhang et al. 2015). A concise review is given by Ye and Xie (2015).

5.7.13 Probabilistic Models

Probabilistic models account for the uncertainty inherent in corrosion prediction by incorporating probabilistic distributions into the prediction process. These models are more versatile than deterministic models and can be used for a broader range of applications (Guan 2023). The failure criterion of a specific pipeline can be specified using the limit state equation, which is dependent on pipeline pressure.

$$G = \text{Pburst} - \text{Pop}$$

where Pburst is the burst pressure in which the pipe wall will bulge outward and reach a point of instability (Antaki 2003), and Pop is the pipeline operating pressure (Amaya-Gómez et al. 2016).

The existing models for this criterion include well-established standards such as ASME (2009), DNV (2010), CSA (2007), and the model proposed by Netto et al. (2005), refer to Fig. 5.14, which is evident that the likelihood of the pipeline failure varies considerably depending on the model employed (Amaya-Gómez et al. 2016).

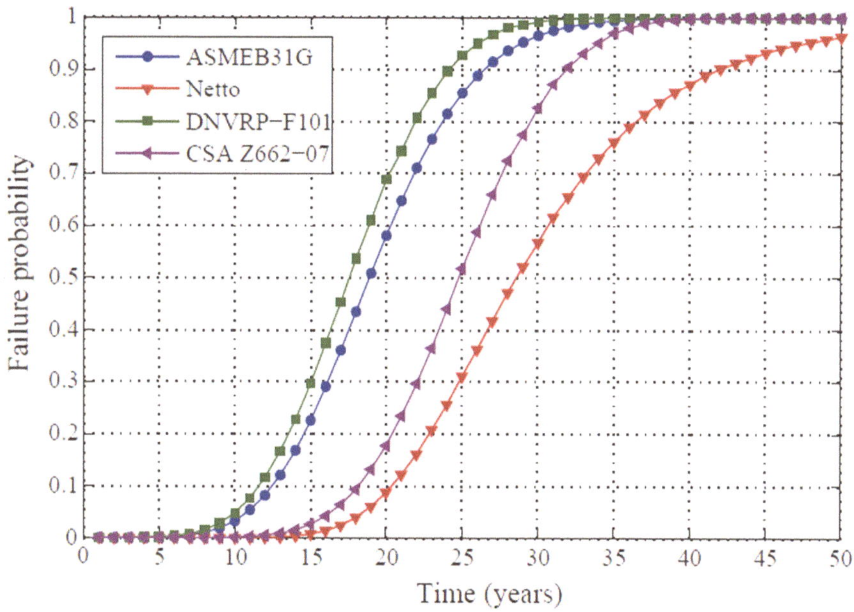

Fig. 5.14 Failure probability for individual defects (Netto et al. 2005)

5.7.14 *Population-Based HBM*

When defect reports from subsequent inspections cannot be matched, it seems to happen frequently. The individual defect-based methodologies for modeling pipeline degradation are, therefore, no longer applicable. Other HBM-based models have been created to address such circumstances (Dann and Huyse 2015). These models take population-level variations in the distribution of defect size into account. It must be expected that growth rates over the interval between measurements will follow the same distribution across all flaws. This technique may also be used to successfully monitor the integrity of a pipeline section by section if a pipeline can be divided into short pieces with homogeneous qualities in terms of the corrosion environment and pipeline properties. Since every defect measured may be utilized to develop the model, the strategy avoids the hassle of matching defect records between measurement runs. 'Population-based' approach's hierarchical structure is shown in Fig. 5.15 (Dann and Maes 2015).

5.7.15 *Statistical-Based Approach to Degradation Modeling*

According to the specification of a certain ILI tool, the measurement scatter is frequently considered to be normally distributed with a zero mean and a standard

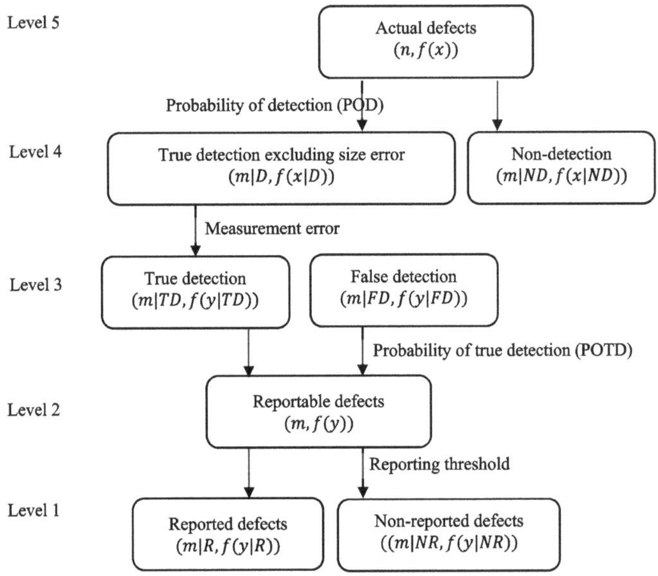

Fig. 5.15 Structure of population-based hierarch model

deviation derived from it (Maes 2013; Pandey and Lu 2013; Alamilla and Sosa 2008). But other models that take measurement scatter into account might also be considered. Weibull and Gamma distributions, in contrast to the normal distribution, do not permit negative values, according to Dann and Huyse (2015) hypothesis. As a result, they may be more accurate representations of reality. On the other hand, the defect depth is likewise constrained. Between 0 and 100% of the pipe's wall thickness should be used. A normal distribution that spans the range of minus infinite to positive infinite must be truncated to enforce compliance with this fact. Similar to the linear model, statistical approaches assume a stationary degradation growth rate, i.e., the temporal effect of the growth rate is not considered. The actual defect size $X(t)$ can be related to corrosion growth-rate, R, by

$$X(t) = Rt$$

The initial size X_0 may also be considered but here it is ignored for simplicity. The capital letters are used to denote random variables. Once R is determined, the defect size at any future time t can be determined. It is also noted that corrosion growth rate is fixed for an individual defect but follows a distribution when the whole defect population is considered in this model. In other words, the corrosion growth-rate for each defect is a sample drawn from the distribution of R. As every measurement is contaminated by noise, in this model, the measured size, y, is assumed to be related to the actual defect size, x, and the measurement error, ε, by

$$y = x + \varepsilon$$

A series of methods have been developed to estimate x when measurement error is involved, such as the method of moments, the maximum likelihood method, the regression method and the simulation method. The Maximum likelihood method was used to estimate the associated model parameters. Pandey and Lu (2013) also pointed out the danger of using increments-based methods, like that proposed by Alamilla and Sosa (2008), which could lead to erroneous distribution of corrosion growth-rate due to the inter-dependence of the depth increments.

5.7.16 Mathematical Models

Mathematical modeling plays a crucial role in various disciplines, including physics, economics, engineering, and social sciences. While models such as the Exponential, Linear, Logarithmic, Power Law, and Polynomial (Order 2) models are widely used, there remain gaps in their applications, limitations, and extensions. Oil and gas pipeline integrity management is a critical area that relies on various mathematical models to predict failures, assess risks, and optimize maintenance strategies. The Exponential, Linear, Logarithmic, Power Law, and Polynomial (Order 2) models have been applied in this domain for corrosion prediction, crack growth estimation, pressure analysis, and failure rate assessment.

Exponential Model

The exponential model serves as a foundational concept in statistics and probability, particularly in the context of understanding life data, survival analysis, and various applications across different fields. One significant characteristic of the exponential distribution is its simplicity and utility in modeling the time until an event occurs, such as failure or arrival times. This is highlighted by several studies that underscore its application in a range of fields including reliability engineering and life testing (Sule and Adegoke 2020; Ashraf et al. 2020). The exponential distribution is indeed a special case of Weibull distribution (Sule and Adegoke 2020). However, as noted by Mahdavi and Kundu (2017), alternative models such as the Weibull distribution provide more flexibility due to the introduction of additional shape parameters, enabling better fitting to various datasets with non-constant hazard rates.

Linear Model

Linear model is characterized by its simplicity and straightforward interpretation, making it particularly useful in various fields ranging from social sciences to medicine and engineering. Linear regression allows researchers to portray relationships linearly understanding how changes in independent variables influence a dependent variable.

A key advantage of linear models is its clear mathematical structure, which facilitates estimation and inference. The performance and robustness of the linear model in different scenarios, such as dose–response meta-analysis, has been well-documented. For instance, Zhang et al. (2018) emphasize the application of linear regression techniques alongside piecewise linear regression in a meta-analysis of dose–response data, which demonstrates the versatility of linear modeling in analyzing complex relationships across different datasets.

In the engineering domain, linear models have been increasingly utilized for predictive purposes. As illustrated by Perdana et al. (2025), multiple linear regression has proven effective in forecasting capacity factors in thermal power plants leading to significant improvements in operational planning and resource management. Furthermore, the study by Akpınar and Daş (2022) applied both simple and multiple linear regression to model heat transfer coefficients in solar energy systems, confirming the reliability of linear regression for engineering analyses.

Logarithmic Model

The logarithmic model plays a central role in various engineering and oil and gas applications, particularly in extraction processes. This mathematical model is beneficial for analyzing the kinetics of different extraction techniques, providing valuable insights regarding the efficiency and outcomes of various methodologies. The modeling of solid–liquid extraction processes has also been investigated, with Jokić et al. (2010) examining the application of logarithmic equations alongside other models. Their research into polyphenol extraction from soybeans emphasizes the relevance of the logarithmic model in scenarios where solvent properties and temperature fluctuations can significantly impact extraction efficiency. Such studies reinforce the necessity of accurately describing extraction kinetics in oil and gas fields to optimize production processes.

Moreover, in the context of enhanced oil recovery (EOR), Zamani and Knez (2023) discussed how empirical correlations could potentially employ logarithmic relationships to understand the effects of hydrostatic stress on extraction methods. This correlation can notably help mitigate the environmental impacts of reservoir depletion, underscoring the logarithmic model's relevance in both extraction efficiency and environmental stewardship.

In equipment maintenance and optimization, logarithmic models facilitate fault diagnosis in mechanical systems, as evidenced by Liao et al. (2021) in their exploration of gearboxes employing logarithmic penalty approaches. This approach not only enhances the processing capabilities but also ensures that extraction equipment operates at optimal levels. Application of these techniques is essential for maintaining efficiency in operations within the oil and gas sector.

As the oil and gas industry moves towards more sustainable extraction methods, methodologies that incorporate logarithmic models provide an avenue for improving the efficiency and efficacy of extraction. In applications ranging from traditional oil extractions to modern supercritical and ultrasonic methods, logarithmic modeling positions itself as a valuable tool for engineers and researchers striving to enhance production capabilities in this critical sector.

Power Law Model

The power law model is an important model in various engineering applications, particularly in the oil and gas industry. This model is characterized by a mathematical relationship where a quantity varies as a power of another quantity, often seen in fluid dynamics, material behavior, and reservoir characteristics. In the context of oil and gas, understanding and modeling the flow behavior of non-Newtonian fluids using power law representations can greatly enhance predictions of performance and efficiency in extraction processes (Gerritsen and Durlofsky 2005; Mahmood et al. 2020).

One notable application of the power law model is its use in characterizing the flow of shear-thinning fluids, which are common in petroleum engineering. While the specific study by Li et al. (2017) discusses surfactants' effects on gas holdup in shear-thinning power-law fluids, it does not propose a semi-empirical correlation based on their findings. Instead, it emphasizes the modification of flow behavior leading to enhanced gas holdup, which is crucial for optimizing gas–liquid systems in extraction operations.

The behavior of power-law fluids in microchannel flow has also been studied rigorously. Zhao et al. (2017) demonstrated that pressure-driven flows of power-law fluids can indeed be enhanced by introducing pulsatile pressure gradients, which emphasizes the nonlinear relationship between stress and strain rate. This understanding is pivotal in applications where fluid dynamics need optimization, such as in enhanced oil recovery techniques where manipulating flow characteristics can significantly impact extraction efficiency.

Polynomial (Order 2) Model

The polynomial (order 2) model is a widely utilized mathematical model that aids in modeling complex relationships between variables through quadratic equations. This model is particularly effective in cases where data exhibits non-linear patterns that cannot be adequately explained by simple linear relationships. The second-order polynomial is characterized by its ability to capture curvature in the data, making it suitable for various applications across disciplines, including agriculture, engineering, and statistics.

One prominent application of second-order polynomial models is in the field of dairy science, where such models are employed to analyze the genetic parameters for milk production traits. Research by Aspilcueta-Borquis et al. (2012) demonstrated that lactation curves can be accurately modeled using polynomial functions, with findings indicating that the models suitably described the variation of milk, fat, and protein production during lactation. This demonstrates the model's capability to capture variations over time, which is crucial for understanding biological processes.

In addition to agricultural applications, polynomial models have been employed across engineering domains, particularly in the fields of materials science and thermodynamics. Aworanti et al. (2012) illustrated the application of second-degree polynomial equations in correlating the variations of densities and viscosities of binary and ternary blends of biodiesel and petroleum diesel as a function of temperature, successfully demonstrating that polynomial regression effectively predicts the interactions of complex fluid properties under changing thermal conditions.

The significance of polynomial (order 2) models can also be observed in statistical modeling, where they serve as a valuable tool for approximating relationships in various datasets. Nandi's (2020) work demonstrates that polynomial representations can provide meaningful connections with autoregressive time series in data modeling, allowing researchers to capture detailed dynamics in temporal datasets while leveraging established statistical principles. Similarly, research by Kozłowski et al. (2010) confirmed that second-degree polynomial models were the closest fit to empirical growth data of vendace fish populations, reinforcing the practical effectiveness of polynomial modeling in biological research.

References

Abdullahi MA, Aliyu S, Isa A (2019) Development of hybrid software architectural model for big data analytics in the oil and gas industry. In: 4th YUMSCIAIC, July 2019

Abyani M et al (2022) Predicting failure pressure of the corroded offshore pipelines using an efficient finite element based algorithm and machine learning techniques. Ocean Eng 254:111382

Adegboye MA, Fung W-K, Karnik A (2019) Recent advances in pipeline monitoring and oil leakage detection technologies: principles and approaches. Sensors 19(11):2548

Ahmad W et al (2018) Formal reliability analysis of oil and gas pipelines. Proc Inst Mech Eng, Part O: J Risk Reliab 232(3):320–334

Ahmad MW et al (2022) Intelligent framework for automated failure prediction, detection, and classification of mission critical autonomous flights. ISA Trans 129:355–371

Ahmed SK, Kabir G (2021) An integrated approach for failure analysis of natural gas transmission pipeline. CivilEng 2(1):87–119

Aiello G et al (2020) The employment of unmanned aerial vehicles for analyzing and mitigating disaster risks in industrial sites. IEEE Trans Eng Manage 67(3):519–530

Akpınar EK, Daş M (2022) Modeling of a solar air collector heat transfer coefficient with regression algorithms. Firat Univ J Exp Comput Eng 1(1):14–23

Akram NA et al (2014) Active incremental Support Vector Machine for oil and gas pipeline defects prediction system using long range ultrasonic transducers. Ultrasonics 54(6):1534–1544

Alamilla JL, Sosa E (2008) Stochastic modelling of corrosion damage propagation in active sites from field inspection data. Corros Sci 50(7):1811–1819

Al-Amin M et al (2014) Hierarchical Bayesian corrosion growth model based on in-line inspection data. J Pressure Vessel Technol 136(4):041401–041401

AL-Shareefi HKM (2009) Neutral network corrosion control by impressed cathodic protection. Thesis for Master of Science in Mechanical Engineering, University of Technology, Baghdad, Iraq

Amaya-Gómez R, Sánchez-Silva M, Muñoz F (2016) Pattern recognition techniques implementation on data from In-Line Inspection (ILI). J Loss Prev Process Ind 44:735–747

Antaki GA (2003) Piping and pipeline engineering: design, construction, maintenance, integrity, and repair. CRC Press

Ashraf JH et al (2020) A new generalized exponential distribution: properties and applications. Int J Anal Appl 18(5):799–818

Aspilcueta-Borquis RR et al (2012) Genetic parameters for test-day yield of milk, fat and protein in buffaloes estimated by random regression models. J Dairy Res 79(3):272–279

Aworanti OA, Agarry SE, Ajani AO (2012) A laboratory study of the effect of temperature on densities and viscosities of binary and ternary blends of soybean oil, soy biodiesel and petroleum diesel oil. Adv Chem Eng Sci 2(04):444
Baduge SK et al (2022) Artificial intelligence and smart vision for building and construction 4.0: machine and deep learning methods and applications. Autom Constr 141:104440
Barani R, Lakshmi VJ (2013) Oil well monitoring and control based on wireless sensor networks using Atmega 2560 controller. Int J Comp Sci Commun Netw 3(6):341
Begum M et al (2023) Software defects identification: results using machine learning and explainable artificial intelligence techniques. IEEE Access 11:132750–132765
Betz W et al (2018) Bayesian inference with reliability methods without knowing the maximum of the likelihood function. Probab Eng Mech 53:14–22
Bhutada G (2023) How big is the market for crude oil? June 2023
Bosch A, Zisserman A, Munoz X (2007) Image classification using random forests and ferns. In: 2007 IEEE 11th international conference on computer vision. IEEE
Boser BE, Guyon IM, Vapnik VN (1992) A training algorithm for optimal margin classifiers. In: Proceedings of the fifth annual workshop on Computational learning theory
Breiman L (2001) Random forests. Machine Learn 45:5–32
Brewer D (2023) The role of AI in the oil and gas value chain—midstream and downstream segment [cited 2023 July]; Available from: https://www.citationcompliance.com/blog/role-of-ai-in-the-oil-and-gas-value-chain-midstream-downstream-segment
Bubbico R (2018) A statistical analysis of causes and consequences of the release of hazardous materials from pipelines. The influence of layout. J Loss Prevent Process Industries 56:458–466
Buddhika T et al (2021) Living on the edge: data transmission, storage, and analytics in continuous sensing environments. ACM Trans Internet of Things 2(3):1–31
Canadian Energy Pipeline Association (CEPA) (2007) Stress corrosion cracking: recommended practices. SCC Working Group, Canada. 2nd Edition, December, 2007
Chaari M et al (2020) An integrated genetic-algorithm/artificial-neural-network approach for steady-state modeling of two-phase pressure drop in pipes. SPE Prod Oper 35(03):628–640
Chen J et al (2019) A study on the safety factor for corrosion assessment of oil and gas pipeline through in-line inspection. In: Key engineering materials. Trans Tech Publ
Chen Y (2024) Risk and reliability analysis for oil and gas pipelines using data-driven methods. Thesis for Doctor of Philosophy, Department of Mechanical Engineering, University of Alberta, Canada
Cobanoglu C et al (2022) A systematic review of big data: research approaches and future prospects. J Smart Tourism 2(1):21–31
The Business Research Company (2023) Crude oil global market report, Jan 2023
Cristianini N, Shawe-Taylor J (2000) An introduction to support vector machines and other kernel-based learning methods. Cambridge University Press
da Cunha Bisaggio H, Netto TA (2015). Predictive analyses of the integrity of corroded pipelines based on concepts of structural reliability and Bayesian inference. Marine Struct 41:180–199
Dann MR, Huyse L (2015) Bayesian approach to estimate corrosion growth from a limited set of matched features. In: Proceedings of the 12th International Conference on Applications of Statistics and Probability in Civil Engineering, ICASP12. Vancouver, Canada, July 12-15, 2015
Dann MR, Maes MA (2015) Population-based approach to estimate corrosion growth in pipelines. In: Proceedings of International conference on applications of statistics and probability in civil engineering, ICASP12. Vancouver, Canada
Davies A (2012) Handbook of condition monitoring: techniques and methodology. Springer Science & Business Media
El-Abbasy MS et al (2014) Artificial neural network models for predicting condition of offshore oil and gas pipelines. Autom Constr 45:50–65
El-Abbasy MS et al (2015) A condition assessment model for oil and gas pipelines using integrated simulation and analytic network process. Struct Infrastruct Eng 11(3):263–281

Elijah O et al (2021) A survey on Industry 4.0 for the oil and gas industry: upstream sector. IEEE Access 9:144438–144468
El-Shenawy A, Shehadeh M (2014) Prognosis the erosion-corrosion rates for slurry seawater flow in steel pipeline using neural system. In: Advanced materials research. Trans Tech Publ
Eze J, Nwagboso C, Georgakis P (2017) Framework for integrated oil pipeline monitoring and incident mitigation systems. Robot Comp Integr Manuf 47:44–52
Fan M-W, Ao C-C, Wang X-R (2019) Comprehensive method of natural gas pipeline efficiency evaluation based on energy and big data analysis. Energy 188:116069
Florez-Lopez R, Ramon-Jeronimo JM (2015) Enhancing accuracy and interpretability of ensemble strategies in credit risk assessment. A correlated-adjusted decision forest proposal. Expert Syst Appl 42(13):5737–5753
Franconi NG et al (2014) Wireless communication in oil and gas wells. Energ Technol 2(12):996–1005
Gerritsen MG, Durlofsky LJ (2005) Modeling fluid flow in oil reservoirs. Annu Rev Fluid Mech 37(1):211–238
Gu J et al (2019) The application of the Big Data algorithm for pipeline lifetime analysis. In: 2019 Chinese automation congress (CAC). IEEE
Guan X (2023) Probabilistic modeling of threshold stress intensity factor for fatigue endurance reliability prediction. Probab Eng Mech 72:103417
Gupta R et al (2021) A review of sensing technologies for non-destructive evaluation of structural composite materials. J Compos Sci 5(12):319
Gupta M et al (2022) Advances in applications of non-destructive testing (NDT): a review. Adv Mater Process Technol 8(2):2286–2307
Gupta M, Ahmad M (2021) Advances in applications of non-destructive testing (NDT): a review. Int Res J Eng IT Scient Res 7(3):76–86
Habeeb RAA et al (2019) Real-time big data processing for anomaly detection: a survey. Int J Inf Manage 45:289–307
Hancke GP, de Carvalho e Silva B, Hancke GP Jr (2012) The role of advanced sensing in smart cities. Sensors 13(1):393–425
Hassani H, Silva ES (2018) Big Data: a big opportunity for the petroleum and petrochemical industry. OPEC Energy Review 42(1):74–89
He G et al (2019) A framework of smart pipeline system and its application on multiproduct pipeline leakage handling. Energy 188:116031
Heggab A, El-Nemr A, El Aghoury IM (2023) Numerical sensitivity analysis of corroded pipes and burst pressure prediction using finite element modeling. Int J Press Vessels Pip 202:104906
Heidary R et al (2018) A review of data-driven oil and gas pipeline pitting corrosion growth models applicable for prognostic and health management. Int J Prognostics Health Manage 9(1):009
Holdaway KR (2014) Harness oil and gas big data with analytics: optimize exploration and production with data-driven models. John Wiley & Sons
Hong HP (1999) Application of the stochastic process to pitting corrosion. Corrosion 55(1):10–16
Hussain A et al (2020) Monte Carlo simulation study of electron yields from compound semiconductor materials. J Appl Phys 128(1):015305
Hussain M et al (2021) Application of big data analytics to energy pipeline corrosion management. Corros Manage 2021:28–29
Hussain M et al (2024) Review of prediction of stress corrosion cracking in gas pipelines using machine learning. Machines 12(1):42
Hussain M, Zhang T, Seema M (2023) Adoption of big data analytics for energy pipeline condition assessment. Int J Pressure Vessels Piping: 206:105061
Hussain M et al (2024) Energy pipeline degradation condition assessment using predictive analytics–challenges, issues, and future directions. J Pipeline Sci Eng 4(3):100178
Ikegwu AC et al (2022) Big data analytics for data-driven industry: a review of data sources, tools, challenges, solutions, and research directions. Clust Comput 25(5):3343–3387

Isermann R (2011) Fault-diagnosis applications: model-based condition monitoring: actuators, drives, machinery, plants, sensors, and fault-tolerant systems. Springer Science & Business Media

Javaid M et al (2022) Exploring contributions of drones towards Industry 4.0. Industrial Robot: Int J Robot Res Appl 49(3):476–490

Jensen MH, Nielsen PA, Persson JS (2023) From Big Data technologies to Big Data benefits. Computer 56(6):52–61

Jokić S et al (2010) Modelling of the process of solid-liquid extraction of total polyphenols from soybeans. Czech J Food Sci 28(3):206–212

Khalaf AH et al (2023) Emerging AI technologies for corrosion monitoring in oil and gas industry: a comprehensive review. Eng Failure Anal: 155:107735

Khan F, Yarveisy R, Abbassi R (2021a) Cross-country pipeline inspection data analysis and testing of probabilistic degradation models. J Pipeline Sci Eng 1(3):308–320

Khan F, Yarveisy R, Abbassi R (2021b) Risk-based pipeline integrity management: a road map for the resilient pipelines. J Pipeline Sci Eng 1(1):74–87

Kong W et al (2020) Model building and simulation for intelligent early warning of long-distance oil & gas storage and transportation pipelines based on the probabilistic neural network. IOP Conf Ser: Earth Environ Sci

Konstantinov AV, Utkin LV (2021) Interpretable machine learning with an ensemble of gradient boosting machines. Knowl-Based Syst 222:106993

Kozłowski K et al (2010) Age and growth of vendace, Coregonus albula (L.), from Lake Wigry (northeast Poland). Fisheries & Aquatic Life 18(4):239–245

Kumar A, Shankar R, Thakur LS (2018) A big data driven sustainable manufacturing framework for condition-based maintenance prediction. J Comput Sci 27:428–439

Lam C, Zhou W (2016) Statistical analyses of incidents on onshore gas transmission pipelines based on PHMSA database. Int J Press Vessels Pip 145:29–40

Lampropoulos G, Siakas K, Anastasiadis T (2019) Internet of things in the context of industry 4.0: overview. International Journal of Entrepreneurial Knowledge 7(1):4–19

Latifi M et al (2024) Efficacy of tree-based models for pipe failure prediction and condition assessment: a comprehensive review. J Water Resour Plan Manag 150(7):03124001

Layouni M, Hamdi MS, Tahar S (2017) Detection and sizing of metal-loss defects in oil and gas pipelines using pattern-adapted wavelets and machine learning. Appl Soft Comput 52:247–261

Lee J-H, Jo Y-D, Kim LH (2015) Reliability assessment for corroded pipelines by separable Monte Carlo method. J Korean Inst Gas 19(5):81–86

Leira BJ, Næss A, Næss OEB (2014). Reliability of corroded pipelines by enhanced Monte Carlo simulation. In: ASME 2014 33rd international conference on ocean, offshore and arctic engineering. American Society of Mechanical Engineers Digital Collection

Leis B et al (2017) Minimize model uncertainty in current corrosion assessment criteria. In: Proceedings of the 21th PRCI-APGA-EPRG joint technical meeting on pipeline research

Li S, Huang S, Fan J (2017) Effect of surfactants on gas holdup in shear-thinning fluids. Int J Chem Eng 2017(1):9062649

Li MF et al (2017) Reliability assessment of gas pipelines with corrosion defects based on in-line inspection. In: Applied mechanics and materials. Trans Tech Publ

Liao Y et al (2021) Enhanced sparse regularization based on logarithm penalty and its application to gearbox compound fault diagnosis. IEEE Trans Instrum Meas 70:1–12

Liao K, Cao B, Liu Z (2011) An effective internal corrosion rate prediction model for the wet natural gas gathering pipeline. In: 2011 International conference on computational and information sciences (ICCIS)

Lin L et al (2017) Random forests-based extreme learning machine ensemble for multi-regime time series prediction. Expert Syst Appl 83:164–176

Liu H et al (2019) Matching pipeline in-line inspection data for corrosion characterization. NDT and E Int 101:44–52

Liu P et al (2023) A CNN-based transfer learning method for leakage detection of pipeline under multiple working conditions with AE signals. Process Saf Environ Prot 170:1161–1172

Lunn D et al (2009) The BUGS project: evolution, critique and future directions. Stat Med 28(25):3049–3067

Lv Z et al (2017) Next-generation big data analytics: state of the art, challenges, and future research topics. IEEE Trans Industr Inf 13(4):1891–1899

Ma H et al (2023) Advances in corrosion growth modeling for oil and gas pipelines: a review. Process Saf Environ Prot 171:71–86

Maes MA (2013) Corrosion growth models and ILI-based estimation procedures for reliability-based and deterministic pipeline integrity assessments. Phase 1: Formulation of probabilistic corrosion growth models and corrosion growth rate estimation, in contract PR-328-133600. Blade Energy Partners

Maes MA (2015) Corrosion growth models and ILI-based estimation procedures for reliability-based and deterministic pipeline integrity assessments. Phase II: implementation of probabilistic corrosion growth models and corrosion growth rate estimation, in contract PR-328-133600. Blade Energy Partners

Mahdavi A, Kundu D (2017) A new method for generating distributions with an application to exponential distribution. Commun Statistics-Theory Methods 46(13):6543–6557

Mahmood R et al (2020) A comparative analysis of flow features of Newtonian and power law material: a new configuration. J Market Res 9(2):1978–1987

Mantzoukas K (2020) Runtime monitoring of security SLAs for big data pipelines: design implementation and evaluation of a framework for monitoring security SLAs in big data pipelines with the assistance of run-time code instrumentation. City Research Online, University of London

McCallum K et al (2014) Localized corrosion risk assessment using Markov analysis. Corrosion 70(11):1114–1127

McMahon P, Zhang T, Dwight R (2020) Requirements for Big Data adoption for railway asset management. IEEE Access 8:15543–15564

Melchers RE (2004) Pitting corrosion of mild steel in marine immersion environment—part 2: variability of maximum pit depth. Corrosion 60(10):937–944

Mohammadpoor M, Torabi F (2020) Big Data analytics in oil and gas industry: an emerging trend. Petroleum 6(4):321–328

Mora RG et al (2016) Pipeline integrity management systems: a practical approach, vol 374. ASME Press, New York

Muñoz-Zavala AE et al (2024) A literature review on some trends in artificial neural networks for modeling and simulation with time series. Algorithms 17(2):76

Nandi AK (2020) Data modeling with polynomial representations and autoregressive time-series representations, and their connections. IEEE Access 8:110412–110424

Nasurudeen AN, Ayesha A, Karthikeyan P (2023) Industry 4.0: features adopting digital technologies in the oil and gas industry. In: Sharma V, Balusamy B, Sabharwal M, Ouaissa M (Eds.) Sustainable Digital Technologies. CRC Press

Netto T, Ferraz U, Estefen S (2005) The effect of corrosion defects on the burst pressure of pipelines. J Constr Steel Res 61(8):1185–1204

Nguyen T, Gosine RG, Warrian P (2020) A systematic review of big data analytics for oil and gas industry 4.0. IEEE Access 8:61183–61201

Noshi CI, Assem AI, Schubert JJ (2018). The role of big data analytics in exploration and production: a review of benefits and applications. In: SPE international heavy oil conference and exhibition. SPE

Otchere DA et al (2022) Application of gradient boosting regression model for the evaluation of feature selection techniques in improving reservoir characterisation predictions. J Petrol Sci Eng 208:109244

Pandey MD, Lu D (2013) Estimation of parameters of degradation growth rate distribution from noisy measurement data. Struct Saf 43:60–69

Papadakis GA (1999) Major hazard pipelines: a comparative study of onshore transmission accidents. J Loss Prev Process Ind 12(1):91–107

Provan JW, Rodriguez ES III (1989) Part I: development of a Markov description of pitting corrosion. Corrosion 45(3):178–192

Putra PH et al (2024) Random forest and decision tree algorithms for car price prediction. Jurnal Matematika Dan Ilmu Pengetahuan Alam LLDikti Wilayah 1 (JUMPA) 4(1):81–89

Qiu S et al (2023) Deep learning techniques in intelligent fault diagnosis and prognosis for industrial systems: a review. Sensors 23(3):1305

Rachman A, Zhang T, Ratnayake RC (2021) Applications of machine learning in pipeline integrity management: a state-of-the-art review. Int J Press Vessels Pip 193:104471

Raj A, Sharma V, Shanu AK (2022) Comparative analysis of security and privacy technique for federated learning in IOT based devices. In: 2022 3rd international conference on computation, automation and knowledge management (ICCAKM). IEEE

Ramírez-Camacho JG et al (2015) Analysis of domino effect in pipelines. J Hazard Mater 298:210–220

Rawat N (2014) Big Data analytics in oil and gas industry. Int J Scien Eng Res 5(5):2601–2611

Rouhanizadeh B, Kermanshachi S (2019) Predictive model development to perform condition assessment on pipeline networks. Computing in civil engineering 2019: smart cities, sustainability, and resilience. American Society of Civil Engineers Reston, VA, pp 24–32

Sarker S et al (2022) A comprehensive review on Big Data for industries: challenges and opportunities. IEEE Access 11:744–769

Seghier MEAB et al (2021) Advanced intelligence frameworks for predicting maximum pitting corrosion depth in oil and gas pipelines. Process Saf Environ Prot 147:818–833

Selmy HA, Mohamed HK, Medhat W (2024) Big data analytics deep learning techniques and applications: a survey. Inf Syst 120:102318

Senouci A et al (2013) A model for predicting failure of oil pipelines. Struct Infrastruct Eng 10(3):375–387

Seo Y et al (2021) Development of AI-based diagnostic model for the prediction of hydrate in gas pipeline. Energies 14(8):2313

Shah V et al (2022) Big Data analytics in oil and gas industry. In: Emerging technologies for sustainable and smart energy. CRC Press Boca Raton and London, pp 37–55

Shahriar A, Sadiq R, Tesfamariam S (2012) Risk analysis for oil and gas pipelines: a sustainability assessment approach using fuzzy based bow-tie analysis. J Loss Prev Process Ind 25(3):505–523

Shaik NB et al (2022) Recurrent neural network-based model for estimating the life condition of a dry gas pipeline. Process Saf Environ Prot 164:639–650

Shipley RJ, Miller BA, Parrington RJ (2022) Introduction to failure analysis and prevention. J Fail Anal Prev 22(1):9–41

Singh M, Hetlevik S (2017) Data–information–knowledge hierarchy based decision support system for risk based inspection analysis. Int J Syst Assurance Eng Manage 8(2):1588–1595

Sinha SK, Pandey MD (2002) Probabilistic neural network for reliability assessment of oil and gas pipelines. Comp Aided Civil Infrastruct Eng 17(5):320–329

Skrynkovskyy R et al (2018) Big Data approach application for steel pipelines in the conditions of corrosion fatigue. J Eng Sci 5(2):E27–E32

Soomro AA et al (2022) Integrity assessment of corroded oil and gas pipelines using machine learning: a systematic review. Eng Fail Anal 131:105810

Spandonidis C, Theodoropoulos P, Giannopoulos F (2022) A combined semi-supervised deep learning method for oil leak detection in pipelines using IIoT at the edge. Sensors 22(11):4105

Sule BO, Adegoke TM (2020) Bayesian approach in estimation of shape parameter of an exponential inverse exponential distribution. Asian J Prob Statistics 9(1):13–27

Swetapadma A, Yadav A (2016) Data-mining-based fault during power swing identification in power transmission system. IET Sci Meas Technol 10(2):130–139

Telford S, Mazhar MI, Howard I (2011) Condition based maintenance (CBM) in the oil and gas industry: an overview of methods and techniques. In: Proceedings of the 2011 international conference on industrial engineering and operations management, Kuala Lumpur, Malaysia
Timashev SA et al (2009) Markov description of corrosion defects growth and its application to reliability based inspection and maintenance of pipelines. In: Proceedings of the 7th International Pipeline Conference, Volume 4:IPC2008-64546, p 525–533, September 29–October 3, 2008, Calgary, Alberta, Canada
Tsukaue Y et al (1994) Initiation behavior of pitting in stainless steels by accumulation of triiodide ions in water droplets. Corrosion 50(10):755–760
Valencia Parra Á (2019) Analysis of Big Data architectures and pipelines: challenges and opportunities. Thesis for Máster Universitario en Ingeniería Informática, Departamento de Lenguajes y Sistemas Informáticos, Universidad de Sevilla, Spain
Valor A et al (2007) Stochastic modeling of pitting corrosion: a new model for initiation and growth of multiple corrosion pits. Corros Sci 49(2):559–579
Valor A et al (2013) Markov chain models for the stochastic modeling of pitting corrosion. Math Probl Eng 2013:108386
van Noortwijk JM (2009) A survey of the application of gamma processes in maintenance. Reliab Eng Syst Saf 94(1):2–21
Wan J et al (2011) Hierarchical leak detection and localization method in natural gas pipeline monitoring sensor networks. Sensors 12(1):189–214
Wang HY, He S (2010) Predict soil corrosion rate of pipeline steel using ANFIS. In: 2010 International conference on computational and information sciences (ICCIS)
West M, Harrison J (2006) Bayesian forecasting and dynamic models. Springer Science & Business Media
Wikipedia (2015) List of countries by total length of pipelines [Cited 25 June 2023]
Witek M (2016) Gas transmission pipeline failure probability estimation and defect repairs activities based on in-line inspection data. Eng Fail Anal 70:255–272
Wu H et al (2019) One-dimensional CNN-based intelligent recognition of vibrations in pipeline monitoring with DAS. J Lightwave Technol 37(17):4359–4366
Xiao R et al (2024) Predicting failure pressure of corroded gas pipelines: a data-driven approach using machine learning. Process Saf Environ Prot 184:1424–1441
Xu Y et al (2020) Efficient prediction of wind and wave induced long-term extreme load effects of floating suspension bridges using artificial neural networks and support vector machines. Ocean Eng 217:107888
Xuan WB et al (2017) Research on full-scale hydrostatic burst testing of different pipeline girth weld defects. Applied Mechanics and Materials 853:351–355. Trans Tech Publ
Ye Z-S, Xie M (2015) Stochastic modelling and analysis of degradation for highly reliable products. Appl Stoch Model Bus Ind 31(1):16–32
Yi P, Lizhi X, Yuanzhong Z (2010) Remote real-time monitoring system for oil and gas well based on wireless sensor networks. In: 2010 International conference on mechanic automation and control engineering. IEEE
Yin H et al (2021) An integrated framework for criticality evaluation of oil and gas pipelines based on fuzzy logic inference and machine learning. J Nat Gas Sci Eng 96:104264
Zahedi P et al (2018) Random forest regression prediction of solid particle Erosion in elbows. Powder Technol 338:983–992
Zamani MAM, Knez D (2023) Experimental investigation on the relationship between biot's coefficient and hydrostatic stress for enhanced oil recovery projects. Energies 16(13):4999
Zhang S, Zhou W (2013) System reliability of corroding pipelines considering stochastic process-based models for defect growth and internal pressure. Int J Press Vessels Pip 111–112:120–130
Zhang S, Zhou W (2014) Bayesian dynamic linear model for growth of corrosion defects on energy pipelines. Reliab Eng Syst Saf 128:24–31
Zhang S, Zhou W, Qin H (2013) Inverse Gaussian process-based corrosion growth model for energy pipelines considering the sizing error in inspection data. Corros Sci 73:309–320

Zhang L et al (2015) Incorporating the CALPHAD sublattice approach of ordering into the phase-field model with finite interface dissipation. Acta Mater 88:156–169

Zhang C et al (2018) Introduction to methodology of dose–response meta-analysis for binary outcome: with application on software. J Evid Based Med 11(2):125–129

Zhang H et al (2020) A modified method for the safety factor parameter: the use of big data to improve petroleum pipeline reliability assessment. Reliab Eng Syst Saf 198:106892

Zhang S et al (2014) Time-dependent corrosion growth modeling using multiple in-line inspection data. J Pressure Vessel Technol 136(4):7

Zhao C, Zhang W, Yang C (2017) Dynamic electroosmotic flows of power-law fluids in rectangular microchannels. Micromachines 8(2):34

Zheng Y et al (2011) Prediction of gathering pipeline internal corrosion rate based on improved wavelet neural network model. In: ICPTT 2011: sustainable solutions for water, sewer, gas, and oil pipelines, pp 90–99

Zhou W, Huang G (2012) Model error assessments of burst capacity models for corroded pipelines. Int J Press Vessels Pip 99:1–8

Perdana E, Sulardjaka S, Warsito B (2025) Prediction modeling of capacity factor of Rembang Coal-Fired Steam Power Plant based on machine learning to improve the accuracy of primary energy planning. Int J Artificial Int Res 9(1):1–15

Chapter 6
Using Big Data Analytics in PIMS

6.1 Introduction

The integration of Big Data Analytics (BDA) into Pipeline Integrity Management Systems (PIMS) has revolutionized how the oil and gas industry approaches pipeline monitoring, maintenance, and risk management.

Traditionally, pipeline operators relied on periodic inspections and reactive maintenance strategies, which often left pipelines vulnerable to unanticipated failures. Big Data Analytics offers a more proactive and predictive approach, allowing operators to monitor pipelines in real time, identify potential risks before they lead to failures, and optimize maintenance schedules to reduce operational costs.

The Oil and Gas (O&G) industry participants anticipate changes in the O&G value chain due to technological adoption (Wanasinghe et al. 2021). The worldwide Big Data market's growth rate from 2012 to 2027 is illustrated in Fig. 6.1 (Wikibon. 2018). In 2018, the Big Data market is anticipated to expand by 20 percent, reaching 42 billion dollars. According to Wikibon (2018), the global Big Data industry is expected to grow at a Compound Annual Growth Rate (CAGR) of 10.48%, from \$42 billion in 2018 to \$103 billion in 2027. Global software sales are expected to total \$628 billion, with \$302 billion coming from applications.

Big Data Analytics has emerged as a powerful tool for the condition assessment and real-time monitoring of energy pipelines. By leveraging the massive volumes of data generated by sensors, meters, and other monitoring devices installed on these pipelines. Big Data Analytics is all about managing and analyzing large amount of data in order to improve asset performance, predict anomaly, understand the current condition of the asset, reduce maintenance costs, better decision-making, and reduce workplace risks (Dijck 2014). Implementing Big Data Analytics enables proactive maintenance interventions, reduces downtime, enhances safety, and optimizes maintenance costs. This helps pipeline owners and operators to take proactive actions on time, preventing failures, and ensuring the safe and efficient operation of the system. In such instances, Big Data Analytics helps in better ways

M. Hussain and T. Zhang, *Big Data Analytics in Energy Pipeline Integrity Management*, Lecture Notes in Energy 104, https://doi.org/10.1007/978-981-96-8019-1_6

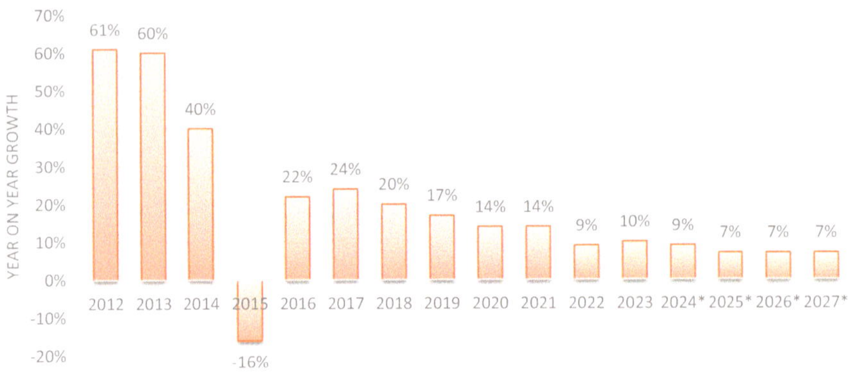

Fig. 6.1 Global Big Data market year-on-year growth from 2012 to 2027

to analyze the large data effectively that is collected by the system from different sources such as weather conditions, geospatial information, operating parameters, and historical maintenance record. By analyzing diverse datasets, pipeline companies will get a holistic overview of the pipeline's current condition and identify correlations between different variables, enabling more accurate assessments and informed decision-making process.

Hassani and Silva (2018) presented the 5V's of Big Data such as volume, velocity, variety, veracity, and value as illustrated in Fig. 6.2 and Big Data 5Vs of Big Data in the energy sector in Table 6.1 (Gohil and Shah 2022). This is a process of Big Data Analytics which extract a significantly large volume of a wide variety of data. Any Big Data system is expected to deliver one or more of them for energy pipeline asset management through a cost-effective mechanism and its architecture to extract information from a very large volume of data (Mohammadpoor and Torabi 2020). However, the successful exploitation of Big Data requires the alignment between business services and IT services so that the necessary infrastructure can be established (Rawat 2014).

Big Data Analytics will become more crucial as this development progresses in maintaining the dependability and integrity of energy infrastructure. Therefore, Big Data Analytics has received the most research attention trying to better tune

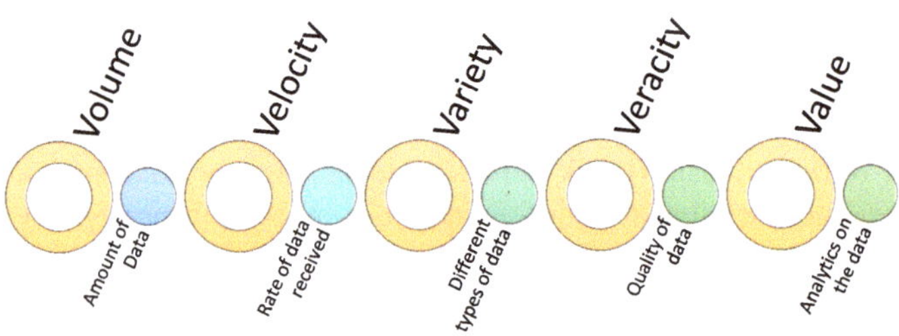

Fig. 6.2 The 5V's of Big Data

Table 6.1 Big Data 5Vs in the energy sector

Industry type	5V's	Data type
Big Data's 5Vs in production and transportation	Volume	• Production operations' sensor data • Transportation operations' sensor data • Automated SCADA • Archived company data and processing
	Velocity	• Real-time data streaming of data from operation-specific sensors • Location-based varied data stream for Big Data Analytics • Evaluation-based varied data stream (flow, pressure, temperature, etc.) for Big Data Analytics
	Variety	• Structured (sensory data) • Unstructured (images, maps, audio, video) • Semi-structured (operational data analysis reports)
	Veracity	• Operation-specific data cleaning and segregation • Amalgamation of all data sources for a quality-rich dataset • Combination of selected geological locations' data for specific analysis
	Value	• Reduced safety hazards and environmental impacts • Optimum and enhanced recovery of hydrocarbons • Economic viability determination • Operation-based optimization
Big Data's 5Vs in energy (natural gas)	Volume	• Extraction operations' sensor data • Treatment operations' sensor data • Storage and transportation operations' sensor data • Distribution operations' sensor data • Automated SCADA • Archived company data and processing
	Velocity	• Real-time data streaming of data from activity-specific sensors • Equipment-based varied data stream for Big Data Analytics • Evaluation-based varied data stream (flow, pressure, temperature, etc.) for Big Data Analytics
	Variety	• Structured (sensory data) • Unstructured (images, maps, audio, and video) • Semi-structured (operational data analysis reports)
	Veracity	• Activity-specific data cleaning and segregation • Amalgamation of all data sources for a quality-rich dataset
	Value	• Reduced safety hazards and environmental impacts • Optimum and enhanced natural gas generation operations • Economic viability determination and operational optimization

data acquisition and processing, to get a clearer outcome. The precise prediction of corrosion and other such defects in oil and gas pipelines holds significant importance when it comes to making decisions about pipe material, predicting remaining useful life, planning maintenance, and more (Xu et al. 2022).

Figure 6.3 illustrates (Sivarajah et al. 2017; Hussain et al. 2023) the classification of Big Data Analytics techniques and its brief details on how and where these methods can be used. These techniques can assist plant owners and operators in pipeline design and operation as well as control decision-making. The evolution of data-driven approaches across various process safety domains, spanning from fault detection to risk analysis, has been significantly influenced by the accessibility of extensively historical and real-time data (Senouci et al. 2014; Bhadriraju et al. 2021; Chen et al. 2021).

6.2 Big Data in the Oil and Gas Industry

The oil and gas industry generates massive volumes of data from a variety of sources, including SCADA systems, in-line inspection (ILI) tools, sensor networks, and environmental monitoring systems. This data provides valuable insights into pipeline conditions, operational parameters, and external environmental factors that influence pipeline integrity. However, the sheer volume and complexity of this data can make it difficult for operators to process and analyze using traditional methods.

Big Data Analytics is now one of the most important aspects of the oil and gas industry's digital transformation (Sivarajah et al. 2017). With the help of Big Data technology, the energy sector can control the massive volumes of data, for instance, Table 6.2 shows data that was reviewed by Subramaniyan in 2017 (Subramaniyan 2017) and also presented in Lu et al. (2019).

In the oil and gas sector, Big Data plays a crucial role in improving operational efficiency, optimizing exploration and production activities, reducing costs, and enhancing safety. Table 6.3 shows some key areas where Big Data is making a significant impact.

It is anticipated that Big Data Analytics will be beneficial for efficient integrity management of energy pipelines. Mohammad and Torabi (2020) gives an overall overview of the Big Data Analytics in oil and gas sector. Ockree et al. (2018), employed Big Data Analytics to create AI-based production-type curves using a complex data processing. Big Data was used by Cadei et al. (2018) to create prediction software that predicted operational disruptions and hazard incidents during oil and gas operations. The methodology used was tested on a real-world dataset and was able to predict operational upsets with great accuracy. According to Pettinger (2014), safety predictive analytics can be created using the information obtained from safety inspections. Big Data Analytics was reviewed by Tarrahi and Shadravan (2016), using Bureau of Labour Statistics (BLS) statistics, to manage risk and increase safety in the oil and gas industry. They reviewed the potential benefits of leveraging Big Data Analytics for improved risk assessment, incident prevention, and overall, Health,

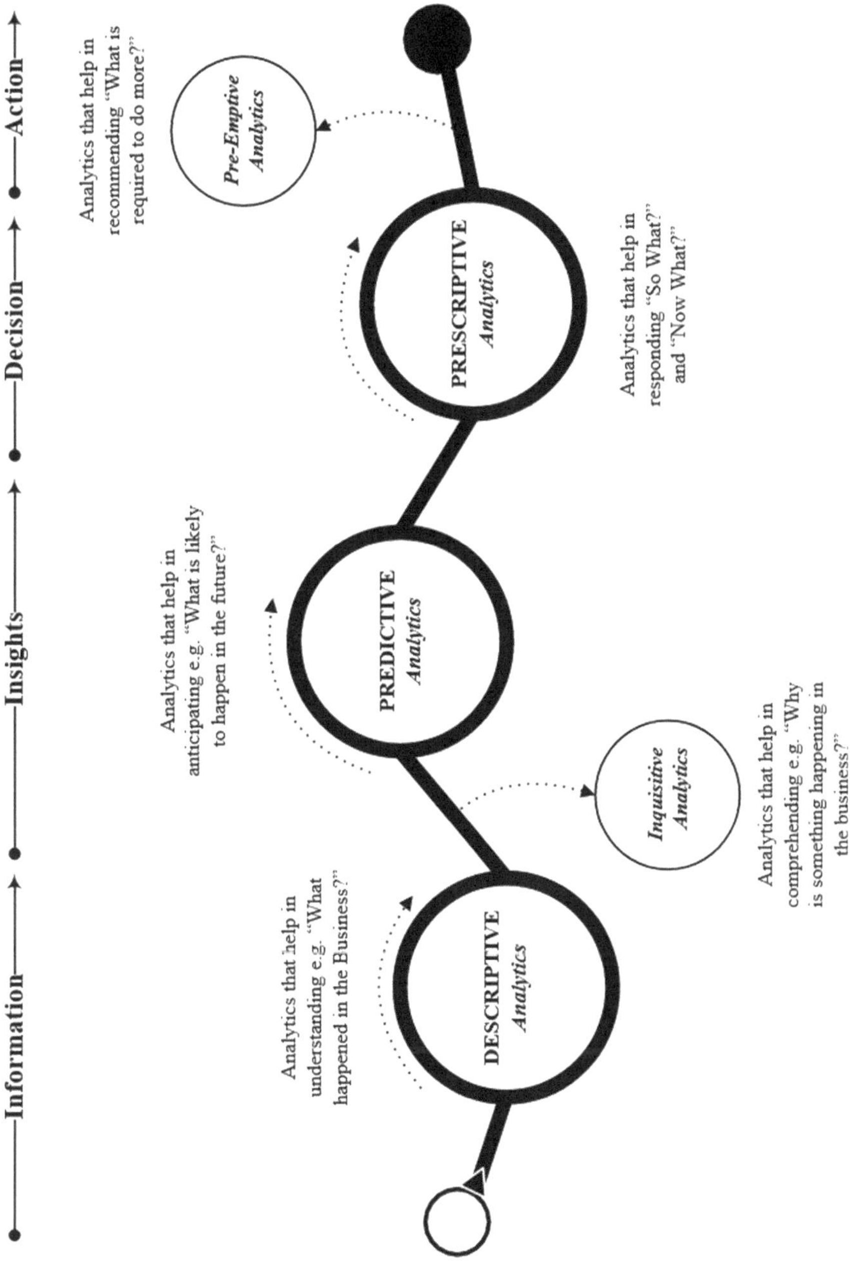

Fig. 6.3 Categorization of Big Data Analytical methods

Table 6.2 The amount of data captured by various sectors of the oil and gas industry (Lu et al. 2019)

Data collection section	Amount of data
Pipeline inspection	1.5 TB (600 km)
Plant operation	8 GB (annually)
Vibration	7.5 GB (per customer annually)
Plant process	4–6 GB (daily)
Submersible pump monitoring	0.4 GB/well (daily)
Drilling	0.3 GB/well (daily)
Wireline	5 GB/well (daily)
Seismic	100 GB–2 TB/survey

Table 6.3 Impact of Big Data in oil and gas

Key areas	Big Data impact
Exploration and reservoir analysis	Big Data Analytics allows oil and gas companies to process and analyze seismic data, well logs, and other geospatial information to identify potential drilling locations and assess the viability of reservoirs. This helps in minimizing risks, improving success rates, and maximizing the recovery of hydrocarbon resources
Predictive maintenance	By analyzing large volumes of operational data, including sensor readings and maintenance records, Big Data Analytics can predict equipment failures before they occur. This allows oil and gas companies to schedule maintenance activities proactively, minimize unplanned downtime, and reduce maintenance costs
Asset performance management	Big Data Analytics allows oil and gas companies to analyze historical and real-time data on asset performance, including drilling rigs, pipelines, and refineries. By identifying patterns and trends, companies can optimize maintenance schedules, improve asset reliability, and extend the lifespan of critical infrastructure
Predictive analytics and decision support	Big Data Analytics provides valuable insights for decision-making in the oil and gas sector. By leveraging advanced analytics techniques, such as machine learning and artificial intelligence, companies can predict reservoir behaviour, optimize drilling techniques, assess investment opportunities, and make informed business decisions
Supply chain management	Big Data Analytics helps optimize the complex supply chains in the oil and gas sector. By analyzing data related to inventory levels, transportation logistics, demand forecasts, and market trends, companies can improve procurement processes, minimize inventory holding costs, and ensure timely delivery of products
Safety and environmental monitoring	Big Data Analytics can analyze data from various sources, such as sensors, drones, and satellite imagery, to monitor and detect potential safety hazards and environmental risks. This includes identifying leaks, assessing air quality, monitoring pipeline integrity, and ensuring compliance with regulatory standards
Production optimization	Big Data Analytics enables real-time monitoring and analysis of production processes, equipment performance, and sensor data from oil wells and refineries. This helps in identifying bottlenecks, predicting equipment failures, and optimizing production rates to enhance efficiency and reduce downtime

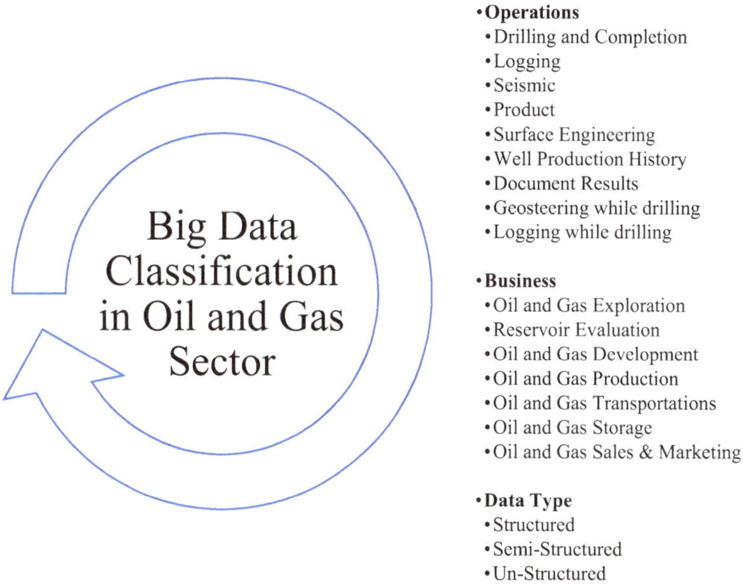

Fig. 6.4 Classification of Big Data in the oil and gas sector

Safety, and Environment (HSE) performance. As technology continues to evolve, the effective utilization of Big Data will become increasingly essential for staying competitive in the dynamic and data-intensive oil and gas industry.

6.2.1 Classification of Big Data in the Oil and Gas Sector

In the oil and gas sector, Big Data can be classified into various categories based on different aspects and applications, Fig. 6.4 shows a classification of Big Data in the oil and gas sector.

6.2.2 Big Data Analytics for Energy Pipeline

Big Data plays a pivotal role in handling the vast amounts of data produced by sensors, monitoring systems, and other sources associated with pipeline operations. Figure 6.5 shows that the Big Data for the energy pipeline is extracted from single or multiple components of data, namely, design data, operational data, attribute data, inspection data, and maintenance data (Hassani and Silva 2018; Hussain et al. 2023; Nguyen et al. 2020). These categories are depicted in Table 6.4.

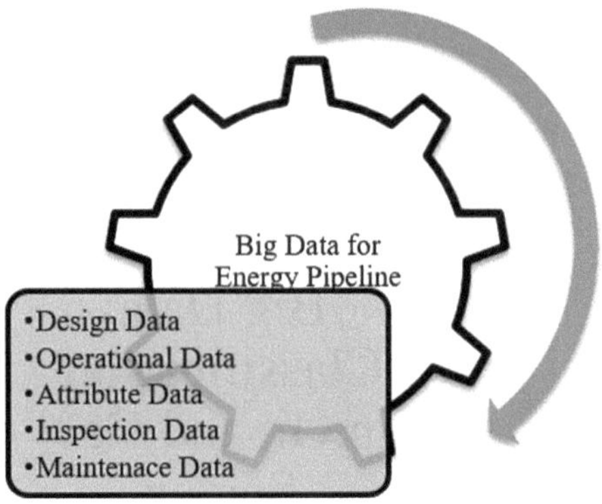

Fig. 6.5 Data utilized in pipeline asset management

According to Feldman (2025), there is a projected exponential growth in the global revenue generated by artificial intelligence products and services, anticipated to surge from $643.7 million in 2016 to an astonishing $36.8 billion by 2025, as depicted in

Table 6.4 Co-related data

Category	Data
Design data	• Year of construction • Joining method • Hydro test • Materials type • Materials specification • Field coating methods • Coating type • Bending method • Soil, backfill • Cathodic protection (CP) installed • Quality control document • Quality assurance document • Final delivery inspection reports
Operational data	• Gas quality • Flow rate • Normal maximum and minimum operating pressure • Leak/failure history • Coating condition • CP system performance • Pipe wall temperature • Pipe inspection reports • OD/ID corrosion monitoring • Pressure fluctuations • Regulator/relief performance • Encroachments • Repairs • Vandalism • External forces

(continued)

Table 6.4 (continued)

Category	Data
Attribute data	• Pipe wall thickness • Diameter • Seam type and joint factor • Manufacturer • Manufacturing date • Material properties • Equipment properties
Inspection data	• Pressure tests • In-line inspections • Geometry tool inspections • Bell hole inspections • CP inspection (CIS) • Coating condition inspections (DCVG) • Condition monitoring data • Non-destructive testing (NDT) data • Audits and reviews
Maintenance	• Maintenance repair cost • Downtime • Production losses • Repair cost • Man-days loss

Fig. 6.6. This extraordinary growth amounts to a staggering 57-fold increase over the specified timeframe, positioning artificial intelligence as the most rapidly expanding sector within the IT industry, regardless of its scale.

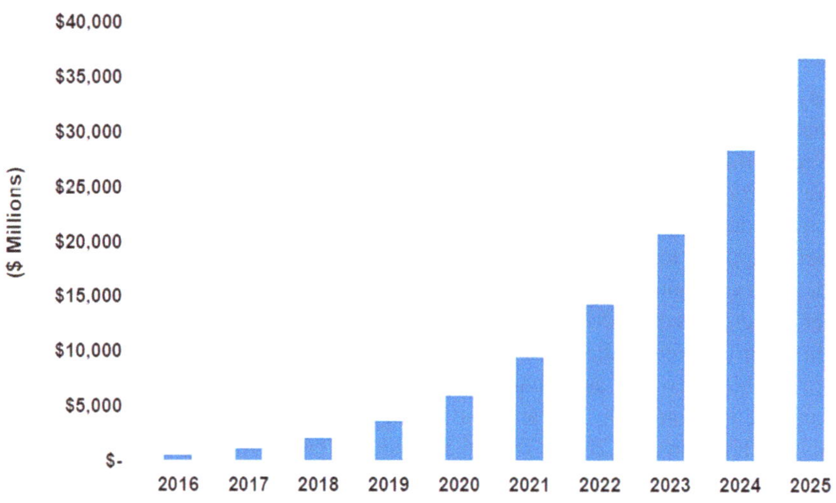

Fig. 6.6 Artificial intelligence revenue (2016–2025)

Table 6.5 Application of Big Data Analytics in pipeline monitoring

Application	Application details
Data collection and integration	• Big Data Analytics in pipeline monitoring starts with the collection and integration of diverse data sources, including sensor data, operational data, maintenance records, and external data such as weather conditions and geographical information • Integration techniques, such as data fusion and data harmonization, enable a unified view of the pipeline system, facilitating comprehensive analysis
Real-time monitoring and anomaly detection	• Big Data Analytics allows for real-time monitoring of pipeline parameters, such as pressure, temperature, flow rates, and vibration levels, providing early detection of abnormal conditions • Advanced anomaly detection algorithms analyze streaming data to identify deviations from normal operating patterns, enabling proactive maintenance and mitigating potential risks
Predictive maintenance and asset management	• By analyzing historical and real-time data, Big Data Analytics can facilitate predictive maintenance models that forecast potential failures or degradation of pipeline components • Predictive analytics algorithms and machine learning techniques help optimize maintenance schedules, reduce downtime, and extend the lifespan of pipeline assets
Data-driven decision making	• Big Data Analytics enables data-driven decision making by providing actionable insights and visualizations to support risk assessment, operational planning, and asset integrity management • Visualization tools, dashboards, and interactive interfaces allow stakeholders to easily interpret complex data and make informed decisions
Continuous improvement and optimization	• Big Data Analytics facilitates the continuous improvement of pipeline monitoring systems through the analysis of historical data, identifying patterns, trends, and areas for optimization • By applying machine learning algorithms and statistical models, pipeline operators can optimize operational parameters, enhance energy efficiency, and reduce costs

By leveraging advanced analytical techniques, researchers and industry professionals can extract valuable insights from this data to enhance the safety, efficiency, and reliability of pipeline infrastructure. The use of improved monitoring systems, more sensor data collection, larger storage volumes, quick data processing, and improved decision-making processes are a few examples of how Big Data Analytics is being applied to understanding the current condition of a gas pipeline in order to develop strategies to extend asset life and prevent the pipeline from corrosion or other defects (Gu et al. 2019).

In order to forecast the current status of assets including pipelines, and develop more flexible and robust asset management systems, oil and gas companies are accelerating the implementation of artificial intelligence with machine learning for data analytics (Hussain et al. 2022). Table 6.5 provides a brief overview of the application of Big Data Analytics in pipeline monitoring and assessment.

Some key applications and benefits of Big Data Analytics in energy pipeline condition and integrity assessment are listed in Table 6.6.

Table 6.6 Applications and benefits of Big Data Analytics in energy pipeline condition

Application/Benefits	Brief details
Predictive maintenance	By analyzing historical and real-time data from sensors and other sources, Big Data Analytics can identify patterns and anomalies that indicate potential equipment failures or maintenance needs. This enables the implementation of predictive maintenance strategies, where maintenance activities are scheduled based on the actual condition of the assets rather than fixed intervals. This approach reduces downtime, improves asset reliability, and minimizes maintenance costs
Condition monitoring	Big Data Analytics allows for continuous monitoring of asset conditions by processing large volumes of sensor data. By analyzing trends and patterns, operators can detect early signs of asset deterioration or performance degradation. This enables timely intervention and prevents potential failures, ensuring the overall health and longevity of the pipeline infrastructure
Real-time monitoring and alerts	Real-time analytics on streaming data from sensors and internet of things (IoT) devices can provide immediate insights into pipeline operations. By setting up automated alerts and notifications, operators can quickly respond to abnormal conditions, such as pressure drops, leaks, or equipment malfunctions. Real-time monitoring improves operational efficiency, enhances safety, and minimizes the risk of incidents
Risk assessment and mitigation	Big Data Analytics enables comprehensive risk assessments by analyzing a wide range of data sources, including historical pipeline data, environmental conditions, geological information, and asset metadata. By identifying potential risks, such as corrosion, ground movement, or external threats, operators can implement targeted mitigation measures to minimize the probability and impact of incidents
Optimization of asset performance	By integrating various data sources, including operational data, maintenance records, and external factors like market conditions or demand forecasts, Big Data Analytics facilitates the optimization of asset performance. Advanced algorithms can identify opportunities for operational improvements, such as optimizing flow rates, reducing energy consumption, or maximizing throughput, leading to cost savings and increased efficiency
Regulatory compliance and reporting	Big Data Analytics can help streamline regulatory compliance by capturing, analyzing, and reporting relevant data to regulatory authorities. This ensures adherence to safety, environmental, and operational standards, reducing the risk of penalties or legal issues

6.3 Big Data Analytics Techniques for Pipeline Integrity Management

The application of Big Data Analytics in PIMS relies on several advanced techniques that allow operators to analyze large datasets, detect patterns, and make data-driven decisions. Some of the most commonly used techniques include machine learning, data mining, predictive analytics, and anomaly detection.

6.3.1 Machine Learning for Predictive Analytics

Machine learning algorithms are widely used in Big Data Analytics to develop predictive models that forecast pipeline failures and optimize maintenance schedules (Hussain et al. 2023; Al-Sabaeei et al. 2023). These algorithms analyze historical data on pipeline performance, such as corrosion rates, wall thickness measurements, and pressure fluctuations, to identify patterns that indicate a higher likelihood of failure (Al-Sabaeei et al. 2023).

- **Supervised Learning**: In supervised learning, machine learning models are trained on labeled data, meaning that the input data is associated with known outcomes (Tariq 2021). For example, a supervised learning algorithm might be trained on historical data that includes instances of pipeline failures. The model then learns to predict future failures based on patterns in the input data.
- **Unsupervised Learning**: Unsupervised learning algorithms are used when there is no labeled data available. Instead, these algorithms identify hidden patterns or clusters in the data without prior knowledge of the outcomes. Unsupervised learning is particularly useful for anomaly detection in pipeline integrity management, as it can identify unusual patterns in real-time data that may indicate a potential problem (Tariq 2021; Dong et al. 2020).
- **Reinforcement Learning**: Reinforcement learning is a type of machine learning in which algorithms learn to make decisions by interacting with their environment (Dong et al. 2020). In the context of pipeline integrity management, reinforcement learning algorithms can optimize maintenance schedules by learning from feedback on past maintenance decisions.

6.3.2 Data Mining and Pattern Recognition

Data mining involves extracting meaningful patterns and trends from large datasets. In PIMS, data mining techniques are used to analyze historical data on pipeline performance and identify factors that contribute to pipeline failures. For example, by analyzing data on corrosion rates, operators can identify the environmental conditions or operating parameters that accelerate the degradation of pipeline materials.

- **Association Rule Mining**: Association rule mining is used to discover relationships between different variables in the data. For example, it might reveal that pipelines operating in certain geographic regions are more likely to experience corrosion-related failures due to environmental factors such as soil composition or temperature fluctuations.
- **Clustering**: Clustering algorithms group similar data points together based on their characteristics. In pipeline integrity management, clustering can be used to group pipeline segments with similar risk profiles, allowing operators to prioritize inspections and maintenance activities for high-risk segments.

6.3.3 Real-Time Anomaly Detection

Anomaly detection algorithms are used to identify unusual patterns in real-time data from SCADA systems and other monitoring tools (Babayeju et al. 2024). These algorithms can detect sudden changes in pressure, flow rate, or temperature that may indicate a leak, rupture, or other integrity issue.

- **Threshold-Based Detection**: Threshold-based detection involves setting predefined limits for key pipeline parameters, such as pressure or temperature. If the value of a parameter exceeds the threshold, the system generates an alert. While threshold-based detection is simple to implement, it may not be sensitive enough to detect subtle anomalies.
- **Statistical Anomaly Detection**: Statistical anomaly detection algorithms analyze the distribution of data points and flag any values that deviate significantly from the expected range. For example, if the pressure in a pipeline segment deviates significantly from the historical average, the system would flag this as a potential anomaly.
- **Machine Learning-Based Anomaly Detection**: More advanced anomaly detection systems use machine learning models to identify patterns in the data that indicate potential issues. These models are trained on historical data and can detect more complex anomalies that may not be obvious using traditional methods.

6.3.4 Big Data Analytics Techniques

Many latest technologies are in use to detect defects and enhance the integrity of pipelines. Over the years, from these inspection and maintenance activities, huge amounts of data have been gathered. These vast volumes of data provide us with great value, but it's also a great challenge to understand and deal with this data. The conventional way of managing data, including data storage, processing, and management, is unable to meet the current needs of industrial activities (McMahon et al. 2020; Baccarelli et al. 2017; Wang et al. 2016). Hence, well-developed systems that can deal with large data volumes require the appearance of a new 'Big Data Analytics' concept. According to a market research report in 2018 by Research and Markets, the pipeline integrity management market is expected to exceed $ 10.4 billion by 2023 (Brulé 2015).

Energy companies need tools that incorporate and synthesize disparate data sources into a single whole to support real-time decision-making. The ability to process Big Data allows for the extraction of information from the relationships that emerge when all these sources are processed together. However, to realize this benefit, these businesses must have access to the necessary technology, resources, and expertise. Table 6.7 showcases various Big Data Analytics techniques used in pipeline monitoring and condition assessment.

Table 6.7 Big Data Analytics techniques

Technique	Description
Data integration	Integrating diverse data sources, such as sensor data, operational data, and external data
Data fusion	Combining multiple data sources to create a unified view of the pipeline system
Data harmonization	Standardizing and aligning data from different sources to ensure consistency and compatibility
Real-time monitoring	Monitoring pipeline parameters in real time, such as pressure, temperature, and flow rates
Anomaly detection	Identifying abnormal patterns or deviations from normal operating conditions
Predictive maintenance	Using historical and real-time data to forecast potential failures and optimize maintenance
Machine learning	Employing algorithms that learn from data to make predictions or take automated actions
Statistical analysis	Analyzing data using statistical methods to uncover trends, patterns, and correlations
Visualization	Presenting data in a visual format, such as graphs, charts, and dashboards
Data mining	Extracting meaningful insights and patterns from large datasets
Text mining	Analyzing and extracting valuable information from unstructured text data
Pattern recognition	Identifying recurring patterns or sequences in data
Predictive analytics	Applying statistical models and algorithms to make predictions and forecast future outcomes
Clustering	Grouping similar data points together based on their characteristics or attributes
Association rule mining	Discovering relationships and dependencies between variables or events in the data
Sentiment analysis	Evaluating and determining the sentiment or opinion expressed in textual data
Deep learning	Utilizing neural networks with multiple layers to learn complex patterns and make predictions
Natural language processing	Analyzing and understanding human language to extract information and meaning

6.3.5 *Big Data Analytics Tools*

It is critical to select tools and software that will assure the greatest results when performing high-quality data analysis (Brandt 2014). Table 6.8 depicted some of the most common types of data analysis tools.

6.3.6 *Data Clean-Up Methods*

To develop well advanced analytical system, data cleansing is one of the crucial steps. Data cleaning is basically a process to ensure that the data used for analysis and modeling is accurate, consistent, and reliable. Few common methods for data cleaning is listed in Table 6.9 and some common tools for cleaning are listed in Table 6.10.

The above methods and tools for data cleaning basically depends on the type and nature of the collected data, the extent of cleaning required, and the goal of the analysis.

6.3.6.1 Challenges with Pipeline Data

Pipeline data comes from multiple sources, including SCADA systems, in-line inspection (ILI) tools, and remote sensors, each of which can introduce unique challenges. The primary issues that affect data quality include:

- **Missing Data**: Pipeline monitoring systems may occasionally fail to capture data due to equipment malfunctions, communication errors, or environmental conditions, leading to gaps in the dataset.
- **Inconsistent Data**: Different systems may use different formats or units of measurement, making it difficult to combine data from multiple sources. For example, one system may record pressure in PSI, while another records it in bar.
- **Outliers and Noise**: Data from sensors may include erroneous readings or outliers caused by sensor malfunctions or environmental interference. These outliers can distort the results of data analysis if not properly addressed.

6.3.6.2 Data Cleaning Techniques

Several data cleaning techniques can be applied to ensure that the data used for analysis is accurate and reliable. These methods include:

- **Handling Missing Data**: There are several techniques for addressing missing data, depending on the severity of the issue:

Table 6.8 Big Data Analytics tools

Analytical tools	Tools brief description
Business intelligence (BI) tools	BI tools allow you to process large amounts of data from a variety of sources in any format. It provides a full-service solution, including cutting-edge data analysis, key performance indicators (KPI) visualization, live dashboards, and reporting, as well as artificial intelligence technology to predict trends and reduce risk (Brandt 2014; Zikopoulos and Eaton 2011)
Statistical analysis	These tools are often created for data scientists, statisticians, market researchers, and mathematicians to do complex statistical analyses using methods such as regression analysis, predictive analysis, and statistical modeling (Zikopoulos and Eaton 2011)
SQL consoles	Structured Query Language (SQL) is a programming language that is commonly used in relational databases to manage structured data. These technologies include a visual tool for database modeling and monitoring, comprehensive SQL optimization, administrative tools, and visual performance dashboards to track key performance indicators (Zikopoulos and Eaton 2011)
Data visualization	Data visualization tools are used to depict data in the form of charts, graphs, and maps, allowing analysts to spot patterns and trends (Zikopoulos and Eaton 2011)
Apache Hadoop	Originally developed in Java (Pence 2014), Hadoop makes use of a sizable computer cluster for distributed processing (Anuradha 2015). Hadoop tracks unstructured data from a variety of sources, including social media, geolocation, and sensor and machine data. It is intended specifically for the storing of large amounts of data and the execution of concurrent processes. Hadoop can also be utilized to improve security and compliance (Anuradha 2015; Ghazi and Gangodkar 2015)
MongoDB	MongoDB is a technology that can regulate and handle unstructured and chaotic data such as documents, multi-media, and social media. MongoDB is incompatible with applications that demand durability, isolation, consistency, and atomicity, such as database-level transactions and the design of a bank's core financial system (Trifu and Ivan 2014). Additionally, MangoDB offers a dynamic and adaptable structure that may be tailored to meet the needs of different users (Győrödi et al. 2015; Mounir et al. 2018; Warden 2011; Kudo et al. 2016)
Cassandra	It is more effective when more time can be spent studying a complex system that provides a lot of power and flexibility. Cassandra is utilized in real-time analytics, streaming data, and fraud and criminal detection (Manoj 2014)
Spark	An open-source framework for handling enormous amounts of data, Apache Spark is made for speed and sophisticated analyses. You may simply construct apps in Scala, Java, or Python with Spark (2022). Large-scale datasets that are structured, semi-structured, or unstructured can all be handled using Spark, an incredibly powerful technology. In contrast, Spark saves data in memory, reducing the number of read and write cycles (Ahmed et al. 2020). Due to its broad range of capabilities, many research fields, including pattern mining and machine learning, have adopted Spark as a processing engine to handle Big Data difficulties (Hicham and Anis 2021)

Table 6.9 Data cleaning methods

Data cleaning methods		Brief details
Outlier detection and handling	Statistical methods	Statistical methods for data cleaning involve using various statistical techniques to identify and handle anomalies, errors, and inconsistencies in a dataset. These methods rely on the underlying distribution and statistical properties of the data to detect unusual or erroneous observations
	Visualization	Using scatter plots, box plots, and other visualizations to identify data points that deviate from the norm or defined operating parameters
	Transformations	Applying transformations like logarithmic scaling to mitigate the impact of outliers
Missing data handling	Imputation	Replacing missing values with estimated or calculated values
	Deletion	Removing rows or columns with missing data, if they are not critical for analysis
	Predictive modeling	Using other variables to predict missing values based on patterns in the data
Inconsistent data handling	Standardization	Converting data to a consistent format, such as converting dates to a common format
	Normalization	Data normalization is a data cleaning technique used to transform numerical data into a standard scale, making it easier to compare and analyze across different variables
	Regular expressions	Cleaning text data using pattern matching and substitution
Duplicate data removal	Exact matching	Identifying and removing exact duplicates
	Fuzzy matching	Detecting approximate duplicates using similarity measures
Data validation	Cross-field validation	Checking relationships between fields to identify inconsistent or incorrect data
	Domain-specific validation	Ensuring that data adheres to the constraints and rules of the specific domain

- **Mean/Median Imputation**: For continuous variables, missing values can be replaced with the mean or median value of the dataset.
- **Interpolation**: For time-series data, missing values can be estimated based on surrounding data points using linear or polynomial interpolation methods.
- **Multiple Imputation**: In cases where the missing data is extensive, multiple imputation techniques generate several possible datasets with imputed values and combine them to create more accurate predictions.

Table 6.10 Data cleaning tools

Cleaning tools	Brief details
Pandas	A popular Python library for data manipulation and analysis, offering tools for cleaning, pre-processing, and transforming data
Microsoft Excel	Widely used for basic data cleaning tasks, such as removing duplicates, filtering data, and basic transformations
OpenRefine (formerly Google Refine)	A powerful open-source tool for data cleaning and transformation, particularly useful for cleaning messy, large datasets
Trifacta Wrangler	A user-friendly data preparation tool that assists in cleaning, transforming, and structuring data for analysis
R	A programming language commonly used for statistical analysis and data manipulation, with packages for data cleaning
SQL	Used for data cleaning tasks involving database querying and manipulation
DataRobot	An automated machine learning platform that also provides data preparation tools for cleaning and transforming data
Open Data Kit (ODK)	Primarily for mobile data collection, it includes features for data validation and cleaning

- **Outlier Detection and Removal**: Outliers can distort the results of predictive models, so it's important to identify and remove them. Techniques such as Z-score analysis, interquartile range (IQR) methods, and density-based clustering can help detect outliers in the data.
- **Data Normalization and Standardization**: To address inconsistencies in data formats or units of measurement, data normalization or standardization is used to ensure that all data is expressed in comparable terms. For example, temperature data from different systems may be converted to a standard unit (e.g., Celsius or Fahrenheit) to facilitate analysis.
- **Data Deduplication**: Duplicate data points can occur when data from multiple sources is merged, particularly if data is collected at different frequencies. Deduplication techniques identify and remove redundant data points to ensure that each observation is unique.

6.3.6.3 Data Integration

After the data has been cleaned, the next step is to integrate it into a unified dataset that can be used for analysis. This involves combining data from multiple sources, such as SCADA systems, ILI tools, and external environmental sensors. Data integration is crucial for developing a comprehensive view of pipeline conditions, allowing

Table 6.11 Evaluation criteria for Big Data Analytics techniques

Evaluation criteria	Description
Accuracy	The degree of precision and correctness in predicting pipeline conditions or identifying anomalies
Reliability	The consistency and stability of the results produced by the analytics techniques over multiple evaluations
Scalability	The ability of the technique to handle large volumes of pipeline data and scale with increasing data sizes
Computational efficiency	The speed and efficiency of the technique in processing and analyzing data within acceptable time frames
Interpretability	The ease of understanding and interpreting the results generated by the analytics technique
Flexibility	The adaptability and versatility of the technique to different types of pipeline data and varying monitoring requirements
Robustness	The ability of the technique to handle noisy or incomplete data and still provide reliable insights and predictions
Resource requirements	The hardware, software, and infrastructure resources needed for implementing and running the analytics technique
Integration capability	The extent to which the technique can be integrated with existing pipeline monitoring systems or data management platforms
Cost-effectiveness	The balance between the benefits provided by the technique and the associated costs in terms of implementation and usage

operators to analyze the relationships between different factors that influence pipeline integrity.

6.3.7 Evaluation Criteria

Brulé (2015) stated that the oil and gas sector spends over half of its time managing datasets. For this purpose, Big Data is rapidly becoming one of the most important aspects of the oil and gas industry's digital transformation (Gohil and Shah 2022). Anand (2013) discussed the benefits of Big Data and how it can now uncover a lot of hidden information from the large amount of data that is available in the oil and gas sector. Table 6.11 provides an evaluation criterion for Big Data Analytics techniques in pipeline condition assessment.

6.3.8 Big Data Architectures

A Big Data framework is shaped to handle the intake, processing, and analysis of data that surpasses the capacity or complexity thresholds of conventional database

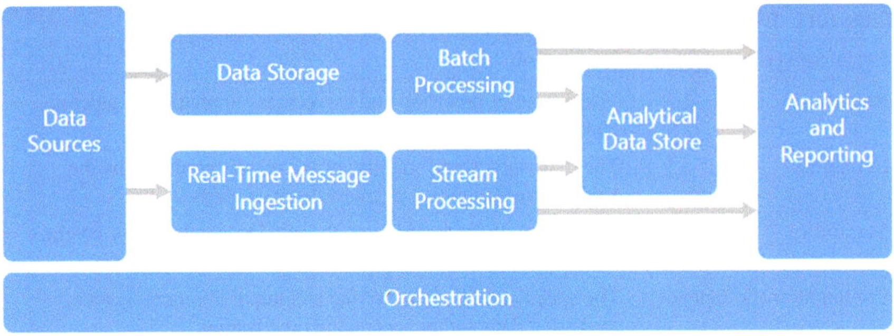

Fig. 6.7 Big Data architecture (Azure) (Microsoft 2018)

systems. Big Data Analytics architectures are carefully designed frameworks that facilitate the processing, storage, and analysis of massive volumes of data, often in real-time (Mohamed et al. 2020; Wang et al. 2020).

Figure 6.7 shows Azure Data Factory (ADF) pipeline. ADF is a cloud based ETL (extract, transform, load) and data integration service that allows you to create and manage data pipelines. The pipeline in the diagram takes data from various sources, processes it, and then stores it in a data store. Most Big Data architectures include some or all of the following components:

- Data sources
- Data storage
- Batch processing
- Real-time message ingestion
- Stream processing
- Analytical data store
- Analysis and reporting
- Orchestration.

These architectures are instrumental in addressing the complex challenges faced by the industry, ranging from reservoir management to operational efficiency. These architectures play a pivotal role in managing the vast amount of data generated by sensors, meters, and other monitoring devices deployed throughout the pipeline infrastructure. By leveraging distributed computing frameworks and real-time analytics, the industry can gain valuable insights into the condition of pipelines, enabling proactive maintenance and minimizing the risk of disruptions.

Figure 6.8 shows a high-level overview of a cloud-based data processing architecture (Microsoft 2018). It shows how data from various devices is ingested into the cloud, processed, and then used to generate insights and actions.

Operational efficiency is another area where Big Data Analytics architectures bring significant benefits to the oil and gas sector (Lu et al. 2019). By integrating data from sensors and Internet of Things (IoT) devices deployed throughout oilfields, companies can monitor equipment health, predict maintenance needs, and optimize

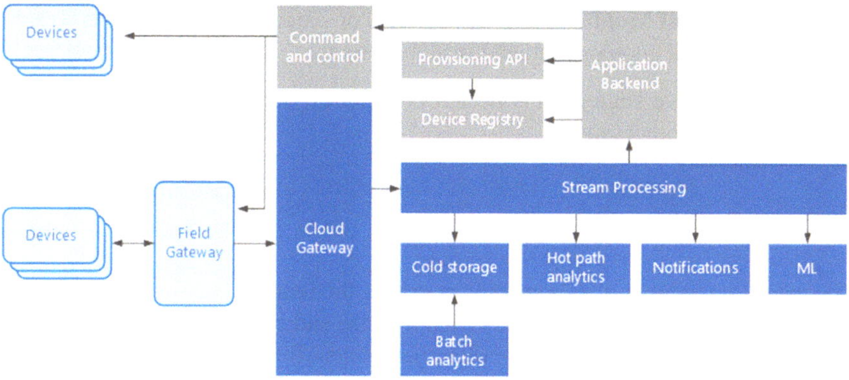

Fig. 6.8 IoT architecture (Microsoft 2018)

overall operational performance (Wanasinghe et al. 2020). This predictive maintenance approach helps prevent costly downtime and extends the lifespan of critical assets, ultimately improving the bottom line for oil and gas organizations (Mqadi 2021).

One common feature of these architectures is their scalability, allowing organizations to handle increasing amounts of data as their needs evolve. At the core of Big Data Analytics architectures are distributed computing frameworks such as Apache Hadoop and Apache Spark (Salloum et al. 2016). These frameworks enable the parallel processing of data across a cluster of computers, making it feasible to manage and analyze data on a massive scale (Zaharia 2016). Additionally, storage solutions like Hadoop Distributed File System (HDFS) provide fault tolerance and high availability, ensuring that data remains accessible even in the face of hardware failures (Zhou et al. 2019).

To effectively implement Big Data Analytics in PIMS, pipeline operators need a robust and scalable Big Data architecture that can handle the collection, storage, processing, and analysis of large datasets. These architectures are designed to ensure that data flows smoothly from the point of collection to the analytics platforms that generate insights.

6.3.8.1 Data Collection Layer

The data collection layer is responsible for gathering data from all relevant sources, including:

- **SCADA systems**: Real-time operational data on pipeline conditions, such as pressure, flow rate, and temperature.
- **In-Line Inspection (ILI) Tools**: High-resolution data on pipeline wall thickness, corrosion, and structural integrity.

- **Environmental Monitoring Systems**: Data on external factors such as weather conditions, soil composition, and seismic activity.

Data from these sources is collected continuously and transmitted to the data storage layer for processing.

6.3.8.2 Data Storage Layer

The data storage layer is responsible for storing the large volumes of data collected from pipeline monitoring systems. Given the size and complexity of pipeline data, traditional relational databases are often insufficient for storing and managing this data. Instead, distributed storage systems are used to handle the scale and variety of pipeline data.

- **Hadoop Distributed File System (HDFS)**: HDFS is commonly used for storing large datasets in a distributed manner, allowing data to be spread across multiple servers to improve storage capacity and retrieval speed.
- **Cloud-Based Storage Solutions**: Many pipeline operators are moving to cloud-based storage solutions, such as Amazon S3 or Microsoft Azure Blob Storage, which offer flexible storage options and can scale to accommodate growing data volumes.

6.3.8.3 Data Processing Layer

The data processing layer is responsible for transforming raw data into a format that can be used for analysis. This layer includes:

- **Batch Processing**: Large datasets are processed in batches, allowing operators to run complex analytics queries on historical data. Batch processing systems, such as Apache Hadoop and Apache Spark, are commonly used for this purpose.
- **Real-Time Processing**: For real-time monitoring and anomaly detection, stream processing frameworks like Apache Kafka and Apache Flink are used to process data as it is generated. Real-time processing allows operators to detect potential issues as they arise and take immediate action.

6.3.8.4 Analytics and Decision Support Layer

The analytics and decision support layer is where advanced analytics models are applied to the data to generate actionable insights. This layer includes:

- **Machine Learning Models**: Predictive models that forecast pipeline failures based on historical and real-time data.
- **Anomaly Detection Systems**: Algorithms that identify deviations from normal operating conditions and flag potential issues.

- **Visualization Tools**: Dashboards and reporting tools that provide operators with real-time insights into pipeline conditions, risk levels, and recommended maintenance activities.

6.3.8.5 Security and Governance Layer

The **security and governance layer** ensures that data is protected and that all analytics activities comply with regulatory requirements. This layer includes:

- **Data Encryption**: All data transmitted between layers and stored in the system is encrypted to protect it from unauthorized access.
- **Access Controls**: Role-based access controls (RBAC) ensure that only authorized personnel have access to sensitive pipeline data and analytics tools.
- **Audit Logs**: Comprehensive audit logs track all actions taken within the system, ensuring that any changes or decisions made based on the data are fully documented and traceable.

6.4 Why Big Data Framework for PIMS?

Frameworks offer a systematic and organized structure to understand the overall process and pathway. The primary goal of the Big Data Framework is to offer a framework for oil and gas sectors seeking to capitalize on Big Data application. To attain enduring accomplishment, Big Data necessitates more than simply the combination of proficient individuals and technology; it demands a well-defined framework and aptitude to understand the existing condition of pipelines and predict any anomaly or degradation like corrosion.

The development of the Big Data Framework was motivated by the fact that, despite the evident advantages and commercial justifications of Big Data, oil and gas companies encounter difficulties in effectively implementing a successful Big Data practice within their organization due to data availability, required skill set, data format etc. The Big Data Framework offers a comprehensive structure that considers all the organizational competencies necessary for a successful Big Data implementation. From the inception of a Big Data strategy to the requisite technical tools and competencies that an organization must possess to get the best from the data that they collected over the years.

The primary advantages of implementing a Big Data architecture in energy pipelines encompass:

- The Big Data Framework offers a systematic framework to initiate or enhance their Big Data capabilities
- The Big Data Framework encompasses all the organizational aspects that must be considered in a Big Data organization

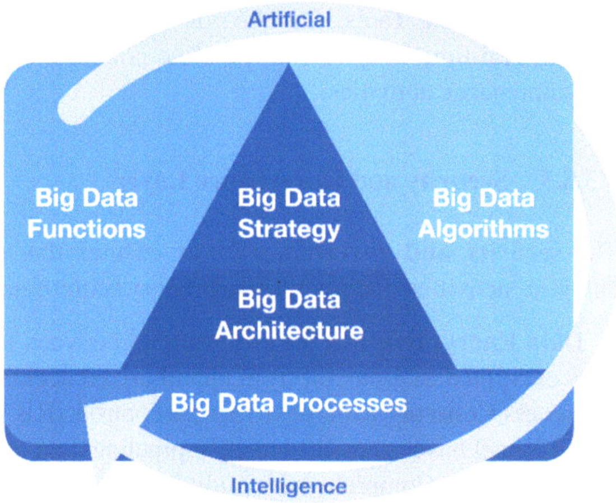

Fig. 6.9 Big Data framework

- The Big Data Framework offers a universal reference model that may be applied to various departmental functions or geographical boundaries of pipeline segment, based on process and environment
- Predict any anomaly or degradation in pipeline system, assist pipeline owners to avoid unwanted downtimes.

The Big Data Framework delineates fundamental and quantifiable competencies within each of its six categories (Fig. 6.9), enabling the pipeline operator to progress gradually.

To get the best from Big Data depend on human interactions and activities. Pipeline or oil and gas companies cannot achieve success solely through the use of advanced computers, but they also require the necessary knowledge and skills. The primary objective of the Big Data Framework is to enhance the understanding of individuals who have an interest in Big Data. The Big Data framework is a methodical approach that encompasses six fundamental skills, Fig. 6.9 illustrates the Big Data Framework.

6.5 Workflow to Develop and Implement Big Data Analytics in PIMS

The successful implementation of **Big Data Analytics (BDA)** in **Pipeline Integrity Management Systems (PIMS)** requires a well-structured workflow that guides operators through each stage of the process, from defining goals to deploying analytics models and monitoring performance. This workflow ensures that all stakeholders are aligned, that data is collected and processed efficiently, and that insights are effectively translated into actionable decisions.

6.5.1 Define Scope and Goals

The first step in implementing Big Data Analytics in PIMS is to define the **scope** of the project and establish clear **goals**. This involves identifying the specific challenges that the organization aims to address using Big Data and defining the desired outcomes.

Key Considerations

- **Identify Key Challenges**: For example, an operator might be dealing with frequent pipeline failures due to corrosion or mechanical damage. The goal of the Big Data Analytics initiative could be to predict these failures more accurately and schedule preventive maintenance before issues arise.
- **Set Measurable Goals**: It's essential to establish clear, measurable goals that can be tracked throughout the project. These might include reducing unplanned downtime by a certain percentage, improving failure prediction accuracy, or optimizing maintenance schedules to reduce costs.
- **Stakeholder Alignment**: Ensure that all stakeholders, including data scientists, engineers, and decision-makers, are aligned with the project's goals and understand their roles in the implementation process.

6.5.2 Data Acquisition and Integration

The next step is to acquire and integrate data from all relevant sources, ensuring that the data used for analysis is accurate, comprehensive, and up to date. In pipeline operations, data typically comes from a variety of systems, including **in-line inspection (ILI) tools**, **SCADA systems**, and **remote sensors**.

Key Tasks

- **Identify Data Sources**: This includes operational data from pipeline monitoring systems, historical inspection records, environmental data, and any external data that may influence pipeline integrity, such as weather conditions or seismic activity.
- **Ensure Data Quality**: Implement data cleaning techniques to address issues such as missing data, outliers, and inconsistencies. Ensuring data quality at this stage is critical for the success of the analytics models.
- **Data Integration**: Use data integration tools to combine data from multiple sources into a unified dataset. This ensures that all data is available for analysis in a consistent format, enabling more accurate and comprehensive insights.

6.5.3 Data Analysis and Modeling

Once the data has been acquired and cleaned, the next step is to analyze the data and develop predictive models that will support pipeline integrity management. This step involves using advanced **analytics techniques** and **machine learning models** to generate insights from the data.

Key Tasks

- **Exploratory Data Analysis (EDA)**: Perform an initial analysis of the dataset to identify patterns, trends, and correlations. This step helps data scientists understand the structure of the data and select the most appropriate analytics techniques (Martinez et al. 2017).
- **Select Machine Learning Models**: Based on the characteristics of the data and the goals of the project, select machine learning models that will be used to predict pipeline failures and optimize maintenance schedules (Liu and Bao 2022). Common models include **neural networks**, **random forests**, and **support vector machines (SVMs)**.
- **Train and Validate Models**: Train the selected models on historical data and validate their performance using a separate test dataset. It's important to assess the accuracy and reliability of the models before deploying them in real-world applications. Techniques such as **cross-validation** and **hyperparameter tuning** can be used to improve model performance.
- **Predictive Maintenance**: Use predictive models to forecast when and where pipeline failures are most likely to occur. These predictions can be used to schedule preventive maintenance activities, reducing the likelihood of unplanned downtime and costly repairs.

6.5.4 Deployment and Monitoring

Once the models have been developed and validated, the next step is to deploy them in the field and monitor their performance. This step involves integrating the predictive models into the organization's existing IT infrastructure and using them to make data-driven decisions about pipeline operations.

Key Tasks

- **Integrate Models with Operational Systems**: Deploy the models within the organization's operational systems, such as SCADA systems, maintenance planning tools, and decision support systems (DSS). This allows real-time data from the pipeline to be continuously fed into the models, ensuring that predictions are based on the latest information.
- **Monitor Model Performance**: Continuously monitor the performance of the predictive models to ensure they are providing accurate and reliable predictions.

Metrics such as **precision**, **recall**, and **F1-score** can be used to assess the accuracy of the models.

- **Real-Time Anomaly Detection**: In addition to predictive maintenance, real-time anomaly detection systems can be used to identify sudden changes in pipeline conditions that may indicate potential failures. For example, a sudden drop in pressure or an unexpected increase in temperature may trigger an alert, allowing operators to take immediate action.

6.5.5 Governance and Security

As with any large-scale data initiative, it's important to ensure that data governance and security are maintained throughout the implementation of Big Data Analytics in PIMS. This involves establishing policies and procedures for managing data, protecting sensitive information, and ensuring compliance with industry regulations.

Key Considerations

- **Data Privacy**: Ensure that data privacy is maintained, particularly if sensitive information about pipeline operations or third-party contractors is being collected and analyzed. Implement **encryption** and **access control** measures to protect data from unauthorized access.
- **Regulatory Compliance**: Ensure that all data collection, analysis, and reporting activities comply with industry regulations, such as **PHMSA** and **API 1160** guidelines for pipeline safety. Compliance with these regulations is essential for minimizing legal risks and maintaining a high standard of pipeline integrity.
- **Audit and Documentation**: Maintain detailed records of all analytics activities, including the data used, the models developed, and the decisions made based on the insights generated by Big Data Analytics. These records are essential for regulatory audits and for ensuring that the organization can demonstrate the effectiveness of its pipeline integrity management practices.

6.5.6 Risk Management and Contingency Planning

Even with predictive models and real-time monitoring in place, it's important to have a comprehensive **risk management** and **contingency planning** strategy to address unexpected failures or issues that may arise during pipeline operations.

Key Considerations

- **Risk Assessment**: Continuously assess the risks associated with pipeline operations, using data from predictive models and real-time monitoring systems. Develop **risk matrices** to prioritize maintenance activities and allocate resources to the most critical areas of the pipeline network.

- **Contingency Plans**: Develop contingency plans for responding to unexpected failures or emergencies. These plans should outline the steps to be taken in the event of a pipeline failure, including communication protocols, repair procedures, and environmental protection measures.

6.6 Proposed Framework for PIMS Using Big Data Analytics

In today's modern advanced maintenance era, characterized by the proliferation of digital information, organizations across various sectors are inundated with vast amounts of data. The ability to utilize the potential of this collected real-time data has become an important aspect in gaining a competitive edge and making informed decisions. Big Data Analytics (BDA) emerges as a transformative paradigm, offering the tools and methodologies to extract meaningful insights from massive datasets. By putting various parts of vast sensory datasets to use with the aid of Big Data Analysis, Big Data has the potential to increase the lifespan of expensive equipment used by asset owners. With Big Data support, businesses may proactively decide when to maintain the appropriate equipment to increase both the equipment's performance and lifespan and to ensure that as few operational hiccups or halts as possible are caused by maintenance. Using IoT data analysis, Shell was able to save $1 million in Nigeria (2016).

The prediction capabilities that Big Data carries with it because of technology like AI and accompanying predictive models are one of the main advantages of employing it on energy pipelines. This is very beneficial for almost every aspect of pipelines because it allows for the most precise understanding possible by feeding enormous amounts of live operations data to the AI model(s), which then provides an accurate picture of the current condition of the pipeline for each operation. Using such patterns, if there are any variations in pipeline operating parameters, a model can detect the operating window and will calculate the impact on the overall pipeline's remaining life or materials deterioration.

Figure 6.10 presents a proposed conceptual framework with layers implemented. This framework represents the Big Data value chain (horizontal axis) and business value chain (vertical axis). In the value chain, the value can be achieved by accessing relevant data streams through data collection, analysis, processing, and visualization which lead to business insights for energy pipeline asset management. In the business value chain dimension, the Big Data value can be achieved by providing infrastructure, platforms, and relevant application tools. This provides energy pipeline asset management with a layered process to develop a holistic approach to policy development through managing resources, processes, competencies, and technologies. Most importantly, this framework supports managing the most important part and right assets or associated component or a segment in the first place, by knowing when and how to get rid of them. Therefore, to reap the advantages of asset measurement, Big Data Analytics can be used to align business direction with the best whole-life

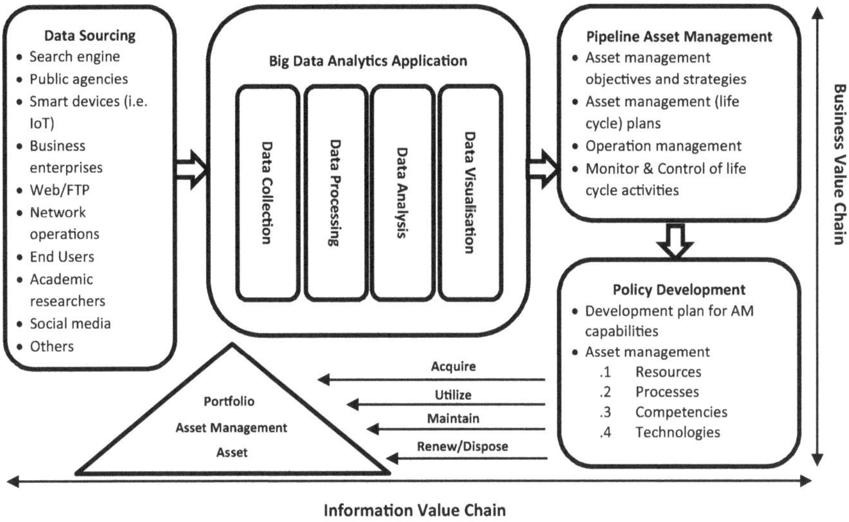

Fig. 6.10 Conceptual framework of Big Data adoption for pipeline asset management

cycle through the combination of acquiring, utilizing, maintaining, and renewing assets which turn re-organization into rewards, and assets deliver the best possible value. The process will be used to further manage assets more effectively, and better coordinate and performance over the whole life cycle.

This conceptual framework presents the flow of the process on how Big Data Analytics applications interact with pipeline integrity management system. Figure 6.11 shows digital transformation review for pipeline integrity using Big Data Analytics. It shows various data, collected from different sources either from online system (Real-time data) or offline (Field inspection and maintenance history). Using artificial intelligence (AI) and machine learning to analyze this real-time or field inspection data to identify potential problems or predict any anomaly. For example, the AI could analyze inspection data to identify cracks in welds or other structural problems. These advanced analytics help pipeline owners to make better and informed decisions about how to maintain and repair pipeline and associated infrastructure. The following sections provides conceptual framework of Big Data adoption for pipeline asset integrity management step by step and recommendations are provided at the end of this section.

6.7 The Data Sourcing

In any **Big Data Analytics (BDA)** initiative, **data sourcing** is a critical first step. For **Pipeline Integrity Management Systems (PIMS)**, data must be gathered from multiple internal and external sources to provide a comprehensive view of pipeline

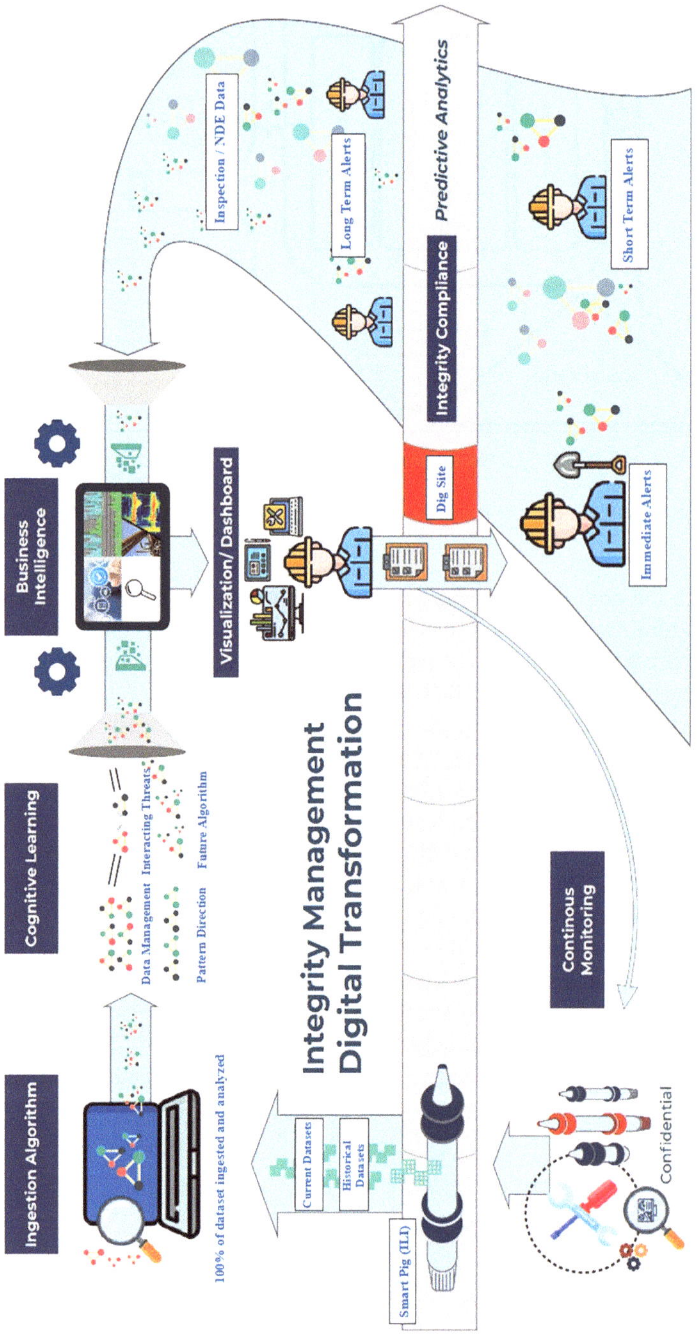

Fig. 6.11 Digital transformation view for pipeline integrity management

conditions, operational risks, and potential failure points. The goal is to capture all relevant data streams to create a unified dataset that supports predictive modeling and real-time decision-making.

6.7.1 Types of Data in Pipeline Integrity Management

The data required for pipeline integrity management typically falls into several categories, each of which plays a crucial role in assessing and predicting pipeline conditions.

- **Operational Data**: This includes real-time data from **SCADA systems**, which monitor key operational parameters such as pressure, flow rate, and temperature. SCADA data is critical for identifying deviations from normal operating conditions that could indicate potential integrity issues.
- **Inspection Data**: Data from **in-line inspection (ILI) tools** provides detailed information on the physical condition of the pipeline, including wall thickness, corrosion rates, and the presence of cracks or mechanical damage. ILI data is often collected periodically, providing a historical record of pipeline integrity.
- **Environmental Data**: External factors such as weather patterns, soil composition, and seismic activity can influence pipeline integrity. For example, pipelines in regions with high seismic activity may be more prone to mechanical damage, while pipelines in coastal areas may experience faster corrosion rates due to saltwater exposure.
- **Maintenance and Repair Data**: Historical records of past repairs, maintenance activities, and inspections are essential for understanding how previous interventions have affected pipeline integrity. This data can be used to identify patterns and trends that may indicate future maintenance needs.

6.7.2 Sources of Data

Data can be sourced from a variety of systems and tools, including:

- **SCADA Systems**: SCADA systems provide continuous, real-time data on pipeline operations. This data is often critical for detecting anomalies and responding quickly to potential integrity issues.
- **In-Line Inspection Tools (Smart Pigs)**: ILI tools generate high-resolution data on the internal condition of the pipeline. These tools are typically used during periodic inspections to detect corrosion, cracks, and other forms of degradation.
- **External Sensors**: Environmental sensors, such as weather stations, soil monitoring systems, and seismic sensors, provide valuable data on external conditions that may impact pipeline integrity.

- **Third-Party Data Providers**: In some cases, external data may be sourced from third-party providers, such as meteorological agencies or environmental monitoring firms. This data can be integrated into the analytics platform to provide a more comprehensive view of the factors affecting pipeline integrity.

6.7.3 The Data Sourcing Mechanism

In the energy sector, operators face issues and challenges to keep pipelines safe, reliable, and in operations in order to avoid catastrophic situations and also it ensures to keep it sustainable and environmentally friendly. In this purpose, data related to failure mechanism, maintenance, repairs, downtimes, design, construction, and other which related to the industrial facilities and operation have become most important and significant to the industry. The data sourcing of the Big Data Analytics are the benchmarks which helps to measuring every aspect of both internal and external environment turns data analytics into the competitive advantages and opportunities for new decision-making process in a business management and environmental aspects. In this respect, a number of technologies, models, tools, and techniques are being used to collect data from inside and outside of the pipeline to control stations. In this regard, the data collection will be via automatic system as manual data collection process is much slower process as compared to automated data collection procedures. Further, manual data collection can lead to delays especially when the data contains critical information and is used in decision-making process. Therefore, automatic mechanism is mainly recommended for data collection as delays that associated with manual data collection negatively impact on the ability to respond to the critical changes occurs into pipeline integrity management system especially to changes in demand or production issues. At the moment the data collection is performed manually in the energy pipeline sector, it needs to be updated. As mentioned above, manual systems tend to result in delays higher costs, and significant time consumption. Human errors are likely causing the compliance issues. However, manual system for data collection for Big Data Analytics will result in less human errors, increased efficiency, maximized productivity and performance, greater accuracy, real-time access and optimized costs (as this is discussed in recommendation for effective digital data management section).

Different sensing tools and networks are being utilized to support pipeline monitoring, controlling and maintenance. One of the most important data sourcing for pipeline includes: ETL (extract, transform, and load) and ELT (extract, load and transform) pipelines. Figure 6.12 illustrates a complete data sourcing mechanism in this conceptual framework. In this process, different sources can be used to collect data which extract raw data and transform and load to the data warehouse. The main purpose of this is instead of converting significantly large raw data, first it moves to data warehouse or Data Lake and the further assess and analyze and prepare data into meaningful form. For example, sources can be IoT devices, user applications, sensors networks, search engines, and others with real-time or streaming analytics

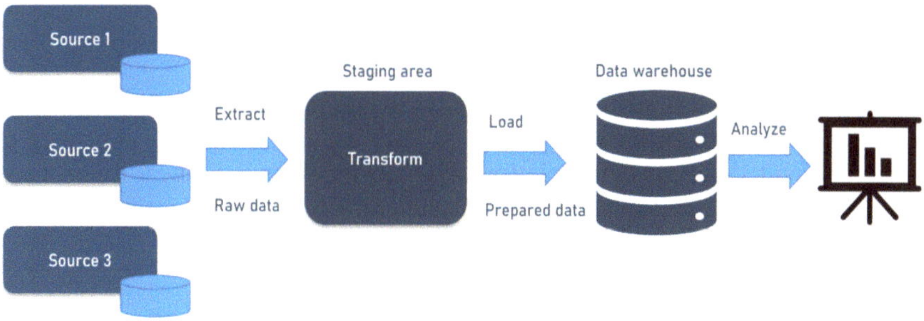

Fig. 6.12 The data source for pipeline integrity management process

to support constant flow of data. This mechanism allows to provide solutions or interventions for smart monitoring and infrastructure performance.

6.7.4 Recommendations for Effective Digital Data Management

In term of data sourcing in the Big Data Analytics, it is a pivotal procedure (Fig. 6.12) which can help energy sector to retrieve right and accurate information which they need to use in strategic decision-making process and also use in their development of portfolio management and design marketing and management strategies. Using digital or automated system for data collection and data management will improve not only minimize human errors but also increase accuracy, efficiency, performance, optimizing costs and maximizing productivity.

There are different sources of retrieving data for the Big Data Analytics applications which includes: search engines, business enterprises, public engines, network operations, web or FTPs, academic researchers, end users and many more which facilitate in data collection procedures. However, there are challenges as well with some of the sources of data which can be in the form of legal and security issues. As there is no effective legal framework for digital data management which can create issues because of the lack of governing mechanism with varying jurisdiction in different regions or states. Similarly, the security of sensitive information needs to be remained protected and unreached. It is recommended that a comprehensive legal framework is required for digital data management in terms of setting of policies, procedures, and guidelines that the energy sector should use to manage its data under the internationally defined standards.

6.8 Data Processing and Analytics

Once the data has been sourced and stored, it must be processed and prepared for analysis. This step involves transforming raw data into a format that can be used by predictive models and analytics platforms. Data processing is critical for ensuring that the insights generated by Big Data Analytics are accurate and actionable.

6.8.1 Data Preprocessing Techniques

Data preprocessing is the process of cleaning, transforming, and organizing the raw data to make it suitable for analysis. Key preprocessing steps include:

- **Data Cleaning**: As mentioned in previous sections, data cleaning involves handling missing values, detecting and removing outliers, and resolving inconsistencies in the data. This step ensures that the data is accurate and free from errors.
- **Data Transformation**: In some cases, data may need to be transformed into a different format or scale to facilitate analysis. For example, temperature data from different sources may be converted into a standard unit (e.g., Celsius) to ensure consistency.
- **Feature Engineering**: Feature engineering involves creating new variables or features from the existing data that can help improve the performance of machine learning models. For example, combining pressure and temperature data to create a "stress" variable that represents the overall strain on the pipeline.

6.8.2 Data Analytics and Modeling

After the data has been pre-processed, the next step is to apply advanced **analytics techniques** and **predictive models** to generate insights. This step involves selecting the appropriate machine learning models and running them on the processed data to identify patterns, predict failures, and optimize maintenance schedules.

- **Predictive Modeling**: Machine learning models, such as **random forests**, **neural networks**, and **support vector machines (SVMs)**, are used to predict pipeline failures based on historical and real-time data. These models provide operators with forecasts of when and where failures are most likely to occur, allowing for proactive maintenance.
- **Anomaly Detection**: In addition to predictive modeling, anomaly detection algorithms are applied to real-time data from SCADA systems to identify deviations from normal operating conditions. For example, a sudden drop in pressure or an unexpected spike in temperature may indicate a potential integrity issue.

6.9 Data Storage in Big Data Analytics for PIMS

Efficient and scalable **data storage** is essential for managing the large volumes of data generated by pipeline monitoring systems. In the context of **Pipeline Integrity Management Systems (PIMS)**, data is collected continuously from a variety of sources, including **SCADA systems**, **in-line inspection (ILI) tools**, **environmental sensors**, and **maintenance records**. As the dataset grows, traditional storage systems may become inadequate, making it necessary to adopt more advanced storage architectures that can handle large-scale data processing and retrieval.

6.9.1 Distributed Storage Systems

Given the vast amount of data generated in real time, **distributed storage systems** are widely used in Big Data Analytics to store and manage pipeline data. These systems allow data to be stored across multiple servers or cloud-based platforms, ensuring that storage capacity can be scaled as needed.

- **Hadoop Distributed File System (HDFS)**: **HDFS** is a popular distributed storage system used in Big Data environments. It allows data to be distributed across a network of servers, enabling high-volume storage and fast retrieval of large datasets. HDFS is particularly well-suited for batch processing tasks, such as analyzing historical pipeline data or running complex machine learning models on large datasets.
- **Cloud-Based Storage**: Many pipeline operators are transitioning to cloud-based storage solutions, such as **Amazon Web Services (AWS)**, **Microsoft Azure**, or **Google Cloud Platform**. These platforms offer flexible and scalable storage options, allowing operators to store vast amounts of data without the need for on-premise infrastructure. Cloud-based storage also provides the advantage of easy access to data from remote locations, making it ideal for geographically distributed pipeline operations.

6.9.2 Data Lake Architectures

A **data lake** is a centralized repository that allows pipeline operators to store structured, semi-structured, and unstructured data in its native format. This approach enables the integration of diverse data sources, including SCADA logs, ILI reports, and environmental data, into a single storage environment.

- **Advantages of Data Lakes**:
  - **Flexibility**: Data lakes can store large volumes of data without the need to preprocess or transform the data before storage, making them ideal for handling diverse data types.

  - **Cost-Effectiveness**: Since data lakes store data in its raw format, they are often more cost-effective than traditional databases, especially for long-term storage.
  - **Data Availability**: With all data stored in one place, operators can easily access both historical and real-time data for analysis.
- **Challenges**: The main challenge with data lakes is managing the **quality** and **organization** of the data. Without proper governance, data lakes can become disorganized, leading to the so-called **data swamp** problem, where data is difficult to retrieve or analyze effectively.

6.9.3 Hybrid Storage Solutions

For some pipeline operators, a **hybrid approach** that combines on-premise storage with cloud-based solutions may be the most effective strategy. This approach allows operators to store sensitive or frequently accessed data on local servers while leveraging cloud storage for archiving historical data or processing large datasets.

- **On-Premise Storage**: Operators can store real-time SCADA data and critical operational information on secure, on-premise servers to ensure fast access and compliance with data security policies.
- **Cloud Integration**: Historical data and large-scale analysis tasks can be offloaded to cloud platforms, where scalable storage and computing resources are available for processing large datasets.

6.10 Data Visualization Techniques for PIMS

Effective **data visualization** is essential for translating the insights generated by Big Data Analytics into actionable decisions. Visualization tools help operators understand complex data patterns, track the performance of pipelines, and identify high-risk areas that require immediate attention (Kharakhash 2023; Miller 2017).

6.10.1 Dashboards and Monitoring Interfaces

Dashboards provide a **centralized view** of pipeline operations, displaying key performance indicators (KPIs), risk assessments, and real-time data from monitoring systems. These dashboards allow operators to track pipeline integrity at a glance and respond to emerging issues quickly.

- **Real-Time Dashboards**: Real-time dashboards display live data from SCADA systems, sensors, and ILI tools, allowing operators to monitor pipeline conditions continuously. These dashboards often include features such as **heat maps**, **line**

graphs, and **gauge indicators** that highlight deviations from normal operating conditions.

- **Customizable Interfaces**: Many data visualization tools allow operators to customize their dashboards based on specific operational needs. For example, a pipeline operator might configure a dashboard to display real-time pressure readings alongside historical failure rates for a specific pipeline segment.

6.10.2 Risk Heat Maps

Risk heat maps are widely used in pipeline integrity management to visualize the relative risk levels across different pipeline segments. These maps use color coding to represent areas of higher or lower risk, allowing operators to focus their attention on the segments that are most likely to experience failures.

- **Corrosion Risk**: For example, a heat map might display segments of a pipeline with varying levels of corrosion risk, with red indicating high-risk areas and green indicating low-risk areas. This allows operators to prioritize maintenance activities and inspections in high-risk zones.
- **Failure Probability**: Heat maps can also be used to visualize the probability of failure based on data from predictive models. By overlaying predictive model outputs onto a geographic map of the pipeline, operators can see where failures are most likely to occur and take preventive measures.

6.10.3 Anomaly Detection Visualizations

Visualization tools can be used to highlight **anomalies** detected by Big Data Analytics models. For example, an anomaly detection system might flag a sudden drop in pressure or an unexpected increase in temperature as a potential integrity issue.

- **Time-Series Graphs**: Time-series graphs allow operators to track changes in pipeline conditions over time, making it easier to identify when an anomaly occurred and how long it has persisted. These graphs are particularly useful for detecting slow-developing issues, such as gradual increases in corrosion or pressure fluctuations.
- **Alert Systems**: Visualization tools can also include **alert systems** that notify operators when an anomaly is detected. These alerts are often displayed on dashboards and may include visual cues, such as flashing icons or color changes, to indicate the severity of the issue.

References

Ahmed N et al (2020) A comprehensive performance analysis of Apache Hadoop and Apache Spark for large scale data sets using HiBench. J Big Data 7(1):1–18

Al-Sabaeei AM et al (2023) Prediction of oil and gas pipeline failures through machine learning approaches: a systematic review. Energy Rep 10:1313–1338

Anand P (2013) Big Data is a big deal. J Petrol Technol 65(04):18–21

Anuradha J (2015) A brief introduction on Big Data 5Vs characteristics and Hadoop technology. Procedia Comp Sci 48:319–324

Babayeju OA et al (2024) Advancements in predictive maintenance for aging oil and gas infrastructure. World J Adv Res Rev 22(3):252–266

Baccarelli E et al (2017) Fog of everything: energy-efficient networked computing architectures, research challenges, and a case study. IEEE Access 5:9882–9910

Bhadriraju B, Kwon JS-I, Khan F (2021) Risk-based fault prediction of chemical processes using operable adaptive sparse identification of systems (OASIS). Comput Chem Eng 152:107378

Brandt S (2014) Data Analysis: Statistical and computational methods for scientists and engineers. Springer

Brulé M (2015) The data reservoir: how big data technologies advance data management and analytics in E&P. In: SPE digital energy conference and exhibition. OnePetro

Cadei L et al (2018) Big data advanced analytics to forecast operational upsets in upstream production system. In: Abu Dhabi international petroleum exhibition & conference. OnePetro

Chen K et al (2021) Operational risk analysis of blowout scenario in offshore drilling operation. Process Saf Environ Prot 149:422–431

Dong H et al (eds) (2020) Deep reinforcement learning. Springer

Feldman M (2016) Market for artificial intelligence projected to hit $36 billion by 2025. Top500.org

Ghazi MR, Gangodkar D (2015) Hadoop, MapReduce and HDFS: a developers perspective. Procedia Comp Sci 48:45–50

Gohil J, Shah M (2022) Application of Big Data in petroleum streams. CRC Press

Gu J et al (2019) The application of the Big Data algorithm for pipeline lifetime analysis. In: 2019 Chinese automation congress (CAC). IEEE

Győrödi C et al (2015) A comparative study: MongoDB vs. MySQL. In: 2015 13th International conference on engineering of modern electric systems (EMES). IEEE

Hassani H, Silva ES (2018) Big Data: a big opportunity for the petroleum and petrochemical industry. OPEC Energy Review 42(1):74–89

Hicham R, Anis BM (2021) Processes meet Big Data: scaling process discovery algorithms in Big Data environment. J King Saud Univ Comp Inf Sci 33(10):8478–8489

Hussain M, Zhang T, Naseer MS (2022) Impact of Covid-19 and needs of digital transformation to protect assets from corrosion:31–33

Hussain M, Zhang T, Seema M (2023) Adoption of big data analytics for energy pipeline condition assessment. Int J Pressure Vessels Piping 206:105061

Kharakhash O (2023) Data visualization: transforming complex data into actionable insights. Autom Technol Bus Processes 15(2):4–12

Kudo T et al (2016) A proposal of transaction processing method for MongoDB. Procedia Comp Sci 96:801–810

Liu Y, Bao Y (2022) Review on automated condition assessment of pipelines with machine learning. Adv Eng Inform 53:101687

Lu H et al (2019) Oil and gas 4.0 era: a systematic review and outlook. Comp Industry 111:68–90

Manoj V (2014) Comparative study of NOSQL document, column store databases and evaluation of Cassandra. Int J Database Manage Syst 6(4):11

Martinez WL, Martinez AR, Solka J (2017) Exploratory data analysis with MATLAB. Chapman and Hall/CRC

McMahon P, Zhang T, Dwight R (2020) Requirements for Big Data adoption for railway asset management. IEEE Access 8:15543–15564

Microsoft (2018) Big data architecture style. https://learn.microsoft.com/en-us/azure/architecture/guide/architecture-styles/big-data [cited 2 May 2024]
Miller JD (2017) Big data visualization. Packt Publishing Ltd
Mohamed A et al (2020) The state of the art and taxonomy of big data analytics: view from new big data framework. Artif Intell Rev 53:989–1037
Mohammadpoor M, Torabi F (2020) Big Data analytics in oil and gas industry: an emerging trend. Petroleum 6(4):321–328
Mounir N et al (2018) Integrating Big Data: simulation, predictive analytics, real time monitoring, and data warehousing in a single cloud application. In: Offshore technology conference. OnePetro
Mqadi N (2021) Application of IoT (Internet of Things) in the monitoring and control of offshore oil & gas production platforms. A research report submitted to the Faculty of Engineering and the Built Environment, University of the Witwatersrand, in partial fulfillment of the requirements for the degree of Master of Science in Engineering, 2021.
Nguyen T, Gosine RG, Warrian P (2020) A systematic review of big data analytics for oil and gas industry 4.0. IEEE Access 8:61183–61201
Nigeria S (2016) Shell saves $1 million using IoT (2016) Internet of Business: IoT, AI, data and edge computing in the connected world
Ockree M et al (2018) Integrating big data analytics into development planning optimization. In: SPE/AAPG eastern regional meeting. OnePetro
Pence HE (2014) What is big data and why is it important? J Educ Technol Syst 43(2):159–171
Pettinger CB (2014) Leading indicators, culture and big data: Using your data to eliminate death. In: ASSE professional development conference and exposition. OnePetro
Rawat N (2014) Big Data analytics in oil and gas industry. Int J Scien Eng Res:2601–2611
Salloum S et al (2016) Big data analytics on Apache Spark. Int J Data Sci Anal 1:145–164
Senouci A, El-Abbasy MS, Zayed T (2014) Fuzzy-based model for predicting failure of oil pipelines. J Infrastruct Syst 20(4):04014018
Sivarajah U et al (2017) Critical analysis of Big Data challenges and analytical methods. J Bus Res 70:263–286
Spark A (2022) Cluster mode overview. https://spark.apache.org/docs/latest/cluster-overview.html [cited 20 March 2025]
Subramaniyan A (2017) Putting analytics to work for you: data and models
Tariq Z et al (2021) A systematic review of data science and machine learning applications to the oil and gas industry. J Petrol Explor Prod Technol:1–36
Tarrahi M, Shadravan A (2016) Advanced big data analytics improves HSE management. In: SPE Bergen one day seminar. OnePetro
Trifu MR, Ivan ML (2014) Big Data: present and future. Database Syst J 5(1):32–41
Van Dijck J (2014) Datafication, dataism and dataveillance: Big Data between scientific paradigm and ideology. Surveill Soc 12(2):197–208
Wanasinghe TR et al (2020) The internet of things in the oil and gas industry: a systematic review. IEEE Internet Things J 7(9):8654–8673
Wanasinghe TR et al (2021) Human centric digital transformation and operator 4.0 for the oil and gas industry. IEEE Access 9:113270–113291
Wang H et al (2016) Towards felicitous decision making: an overview on challenges and trends of Big Data. Inf Sci 367:747–765
Wang J et al (2020) Big data service architecture: a survey. J Internet Technol 21(2):393–405
Warden P (2011) Big data glossary. O'Reilly Media, Inc
Wikibon (2018) Global big data market year-on-year growth from 2012 to 2027 [cited 25 June 2023] https://www.statista.com/statistics/255968/global-big-data-market-growth-forecast/
Xu L et al (2022) The research progress and prospect of data mining methods on corrosion prediction of oil and gas pipelines. Eng Failure Anal 144(6):106951
Zaharia M (2016) An architecture for fast and general data processing on large clusters. The Association for Computing Machinery and Morgan & Claypool Publishers

Zhou J et al (2019) A highly reliable metadata service for large-scale distributed file systems. IEEE Trans Parallel Distrib Syst 31(2):374–392

Zikopoulos P, Eaton C (2011) Understanding big data: analytics for enterprise class Hadoop and streaming data. McGraw-Hill Osborne Media

Chapter 7
Data Quality Issues in Model Testing

Data quality issues can significantly impact the accuracy and reliability of the assessment process (Handfield et al. 2020; Correia et al. 2023). An often-encountered obstacle is the presence of inadequate or missing data, which may lack information regarding the pipeline's history, maintenance records, or operational conditions. Lacking a comprehensive dataset poses difficulties in making the right decisions regarding the condition of the pipeline and successfully prioritizing repair or replacement activities (Koroteev and Tekic 2021).

Data inconsistency is a prevalent data quality problem that arises during the assessment of pipeline conditions. Inconsistencies may arise due to disparities in data formats, units, or measurement standards. An instance of this occurs when multiple departments or contractors employ different units of measurement for pipeline diameters or corrosion levels, leading to ambiguity and mistakes in the evaluation process. Inconsistencies in the data can compromise the dependability of the analysis, impeding the capacity to offer a conclusive and accurate evaluation of the pipeline's integrity.

7.1 Introduction

Data quality issues can significantly impact the accuracy and reliability of the assessment process (Handfield et al. 2020; Correia et al. 2023). An often-encountered obstacle is the presence of inadequate or missing data, which may lack information regarding the pipeline's history, maintenance records, or operational conditions. Lacking a comprehensive dataset poses difficulties in making the right decisions regarding the condition of the pipeline and successfully prioritizing repair or replacement activities (Koroteev and Tekic 2021).

Data inconsistency is a prevalent data quality problem that arises during the assessment of pipeline conditions. Inconsistencies may arise due to disparities in

M. Hussain and T. Zhang, *Big Data Analytics in Energy Pipeline Integrity Management*, Lecture Notes in Energy 104, https://doi.org/10.1007/978-981-96-8019-1_7

data formats, units, or measurement standards. An instance of this occurs when multiple departments or contractors employ different units of measurement for pipeline diameters or corrosion levels, leading to ambiguity and mistakes in the evaluation process. Inconsistencies in the data can compromise the dependability of the analysis, impeding the capacity to offer a conclusive and accurate evaluation of the pipeline's integrity.

The existence of inaccurate data poses a significant risk when assessing the integrity of pipelines. Errors in measurement, calibration, or recording might lead to inaccurate evaluations of the pipeline's reliability. Erroneous data can prompt erroneous maintenance judgments, potentially leading to unavoidable expenditures or, more significantly, overlooking genuine risks to the integrity and dependability of the pipeline. Addressing data quality issues in pipeline condition assessment is crucial for ensuring the reliability of results and making informed decisions that improve the overall integrity and safety of the pipeline infrastructure.

7.2 Background

7.2.1 Data Quality in Energy Pipelines Integrity Management

The accuracy and reliability of the data collected directly influence decision-making processes aimed at maintaining the safety and operational efficiency of these critical infrastructures (Ouyang 2014). The veracity of this data is paramount, as it forms the foundation for identifying potential threats to pipeline integrity (Zhang et al. 2020; Asha et al. 2023). Pipeline condition assessment programs typically collect and manage a wide range of data, including:

- Pipeline asset data (e.g., type, material, diameter, age, location)
- Operating data (e.g., pressure, temperature, flow rate)
- Inspection data (e.g., corrosion, cracks, dents)
- Maintenance and repair data.

The data used in this research is ILI data. The precision of the ILI data impacts its decisions on maintenance, repair, and risk avoidance measures to be implemented (Zhou et al. 2017). However, quality issues affecting ILI data include missing numbers, differences, and outliers. The observed problems concerning the quality of ILI data are summarized in Table 7.1. These deficiencies can lead to serious negligence in the safety and reliability of pipelines regarding asset integrity.

For enhancing the accuracy of the pipeline assessment, this work proposes to apply machine learning models to mitigate the data quality issues. Improved data quality increases the confidence of evaluating the status of pipeline potential defects and allows for a more effective approach toward pipeline integrity. This lies in the

Table 7.1 Significance of addressing data quality gaps in ILI data

Challenge	Impact	Benefits of improved data quality
Missing values	Underestimation or overestimation of defect severity (Meeker and Hong 2014)	By imputation of missing data, the improved data quality can help achieve more accurate and reliable assessments of potential pipeline anomalies
Inconsistent measurements	Reduced confidence in overall data	By improving the ILI data measurement mapping accuracy, the data quality will be significantly improved, and enhanced accuracy of measurement data
Outliers	May cause misdiagnosis of defects or unnecessary repairs	By identifying the outliers, it can help achieve optimized inspection and maintenance strategies
Overall poor data quality	Inefficient resource allocation (Tabrizchi and Kuchaki Rafsanjani 2020)	By taking measures to improve data quality, better data-driven decision-making for pipeline management could be achieved

fact that it can avoid pipeline failures, reduce risks, and optimize resource utilization for maintenance practices to ensure safer and more efficient use of pipelines.

7.2.2 *Importance of Data Quality for Predictive Models*

The success of predictive models in pipeline integrity management depends on the quality of the data they are trained on. High-quality data allows models to learn the underlying patterns and relationships between variables, leading to more accurate predictions (Ewuzie et al. 2022). Conversely, poor data quality can lead to biased models, overfitting, and incorrect predictions, which may result in:

- **Falsc Alarms**: Models may flag non critical issues as high-risk, leading to unnecessary maintenance activities and increased operational costs.
- **Missed Failures**: Inaccurate models may fail to detect critical pipeline integrity issues, potentially leading to pipeline ruptures, leaks, or other catastrophic failures.
- **Inefficient Maintenance Schedules**: Poor data quality can cause models to generate suboptimal maintenance schedules, resulting in increased downtime or premature maintenance (Ewuzie et al. 2022).

7.2.2.1 Common Data Quality Issues in Pipeline Operations

Several common data quality issues affect pipeline integrity management. These include:

- **Missing Data**: Missing data is one of the most frequent issues in pipeline monitoring systems. Gaps in the data can occur due to sensor malfunctions, communication failures, or incomplete inspection records. Missing data can introduce bias into the model, making it difficult to train accurate predictive models.
- **Noise and Outliers**: Sensor readings may contain noise or outliers caused by equipment malfunctions or environmental factors. These outliers can distort the results of data analysis, leading to inaccurate predictions or false alarms.
- **Inconsistent Data**: Inconsistent data refers to data that does not follow a uniform format or measurement system. For example, pressure may be measured in different units (PSI vs. bar) across different systems, leading to discrepancies when the data is combined.
- **Duplicated Data**: Data duplication can occur when data from multiple sources is merged. Duplicates can skew model performance by inflating the importance of certain data points, resulting in inaccurate predictions.

7.2.2.2 Impact of Data Quality on Machine Learning Model Testing

The impact of data quality on machine learning model testing cannot be overstated. When testing machine learning models for pipeline integrity management, poor data quality can lead to several issues:

- **Overfitting**: Models may learn to fit noise or outliers in the data, leading to overfitting. Overfitted models perform well on training data but fail to generalize to new data, resulting in poor performance in real-world applications (Gudivada et al. 2017).
- **Bias in Predictions**: Missing or incorrect data can introduce bias into the model, causing it to make inaccurate predictions. For example, if data from high-risk pipeline segments is missing, the model may underestimate the overall risk of failure (Ewuzie et al. 2022).
- **Reduced Predictive Accuracy**: Inconsistent or incomplete data can reduce the model's ability to accurately predict pipeline failures. Poor data quality increases the risk of both false positives (predicting failures that do not occur) and false negatives (failing to predict actual failures) (McDonald 2021).

7.3 Anomaly/Defect Matching Using ILI Data

7.3.1 Measurement of an Anomaly/Defect by an ILI Tool

In essence, a pipeline can be seen as a three-dimensional longitudinal system, and it is reasonable to accurately employ models to simulate pipeline corrosion in three dimensions. Corrosion occurs exclusively within the pipe wall, which can be visualized as a two-dimensional manifold, as depicted in Fig. 7.1.

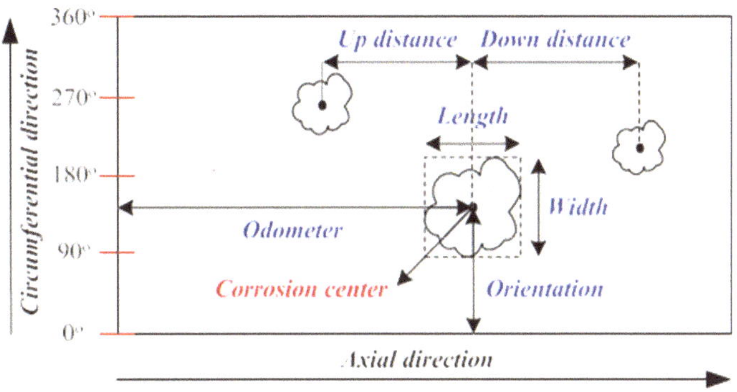

Fig. 7.1 Schematic of a corrosion feature on a 2D plane from one segment (Liu et al. 2019)

The wall can be regarded as a 2D steel sheet curved into a cylindrical shape and fused along the lengthwise direction to form a pipe joint. Therefore, the corrosions are depicted on a two-dimensional flat surface, as illustrated in Fig. 7.1 (Liu et al. 2019).

The ILI data contains pre-processed and selected intrinsic features, which are listed below:

- **Odometer**: The Odometer measures the distance between the center point of this corrosion and the head or tail joint.
- **Orientation**: Orientation refers to the angle position in the circumferential direction, which is represented using a 0°–360° angle analogy as illustrated in Fig. 7.1.
- **Length**: It is the horizontal span of a corrosion defect.
- **Width**: It is the vertical span of a corrosion defect.
- **Depth percentage**: It is the proportion of the depth of a corrosion defect to the wall thickness, the depth is measured from the flat surface of the pipe to the culet (the bottom of the corrosion defect), in the radial direction of the pipe.

These five attributes are the outcomes derived from the original ILI data (Desjardins 2001; Huyse and van Roodselaar 2010; Li et al. 2017). Based on several manual matching methods, we posit that these variables are closely correlated with the location of the corrosion defect in both the reference and current ILIs. Two additional variables, "up corrosion distance" and "down corrosion distance," are derived from the raw ILI data using mathematical procedures to enhance the comprehensiveness and effectiveness of model training features.

The positions of anomalies determined from the ILI measurement correspond to the precise coordinates of the centre of the features as shown in Fig. 7.2. In feature matching procedures, a coordinate system consisting of two characteristics, namely "absolute odometer" and "orientation," was utilized to match center points. The matching process was conducted using the theory of picture pattern recognition.

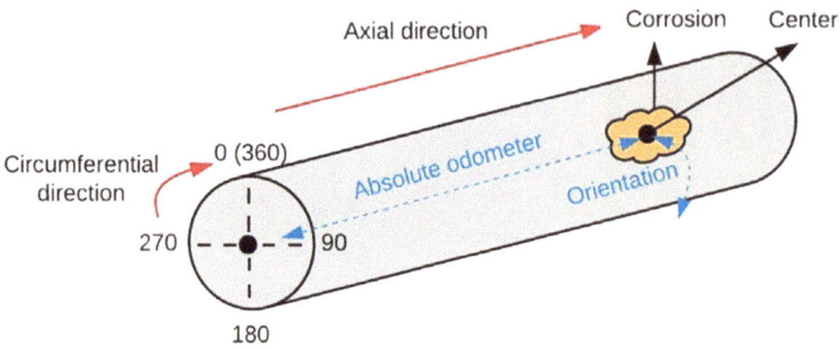

Fig. 7.2 Location of a corrosion defect on a pipeline (Dann and Dann 2017)

Through the examination of raw ILI data, many characteristics are identified, including absolute odometer and orientation, as well as additional variables such as the dimensions of a corrosion defect length and width. However, not all these aspects might be deemed effective. Feature selection is a process that involves analyzing raw data to identify and select useful and potentially relevant characteristics to reduce the training burden before inputting the data into a model (Liu et al. 2019).

7.3.2 Anomaly/defect Matching Using Machine Learning Techniques

Liu et al. (2019) tested and validated a model using machine learning techniques to match defects in the ILI data. Their overall process of training a classification model is outlined in Fig. 7.3. This diagram illustrates the stages employed in developing a machine-learning model. The experimental data were compiled using 15% of the data from the first pair of ILIs, along with the data from the other two pairs. After identifying and validating the model, the experimental data were inputted into the trained machine-learning model for classification and subsequent cluster trimming (Liu et al. 2019).

The obtained model was able to match some anomalies measured in two ILI runs, see Figs. 7.4 and 7.5. The dimensions and the extent of corrosion are illustrated in Figs. 7.4 and 7.5, depicting a comparison between reference (ILI 1) and current (ILI 2) ILI data for a specific pipeline segment. The corrosion features, characterized by their length, width, and depth percentage (e.g., 0.10 = 10% of the pipe wall thickness) are illustrated in Fig. 7.5. The fraction next to each rectangle represents the depth percentage of the respective corrosion. The total number of identified corrosion features is 17 in ILI1 and 45 in ILI2. The developed matching model was implemented to align the corrosion features from ILI2 with those in ILI1. Figure 7.5 presents the matched corrosion defects using the trained ensemble classification model, where corrosion features (i.e., corrosion defects) assigned the same integer value denote a successful match. A comparison between Figs. 7.4 and 7.5 reveals a

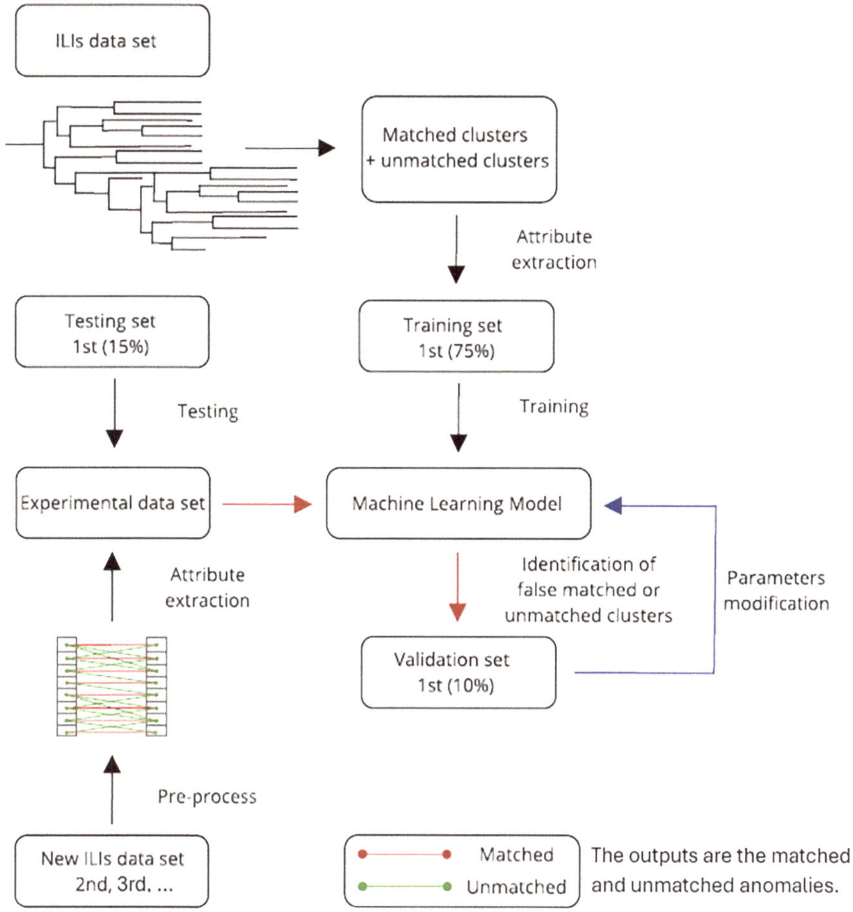

Fig. 7.3 A flowchart showing model training process (Liu et al. 2019)

clear pattern of location matching. Several corrosion features detected in ILI1 align well with those identified in ILI2. However, a significant number of features remain unmatched, likely due to repairs conducted on the reference pipeline (ILI1) and the formation of new corrosion features in the current pipeline (ILI2) over time.

Figure 7.5 displays the corresponding feature areas identified by the trained ensemble classification model. The first number (like 1, 2, ..., 7) next to each rectangle indicates the matched anomaly ID. The comparison between Figs. 7.4 and 7.5 reveals a noticeable correspondence in position. The corrosion features identified in ILI 1 correspond closely to the findings of ILI 2. The discrepancy in the number of corrosions may be attributed to the fact that the corrosions in the reference pipeline have been remedied, while new corrosions have emerged in the present pipeline over time (Liu et al. 2019). Continuous implementation of pipeline integrity is required to prevent avoidable interruptions, failures, and property damage (Bjørnøy et al. 2004;

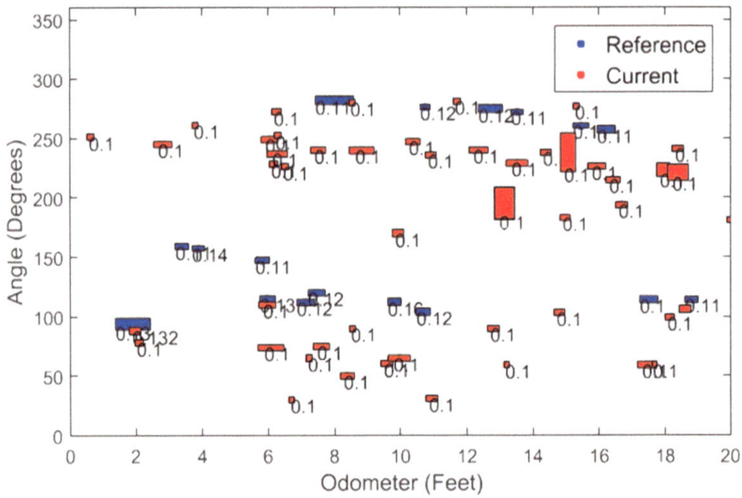

Fig. 7.4 Unmatched results (Liu et al. 2019)

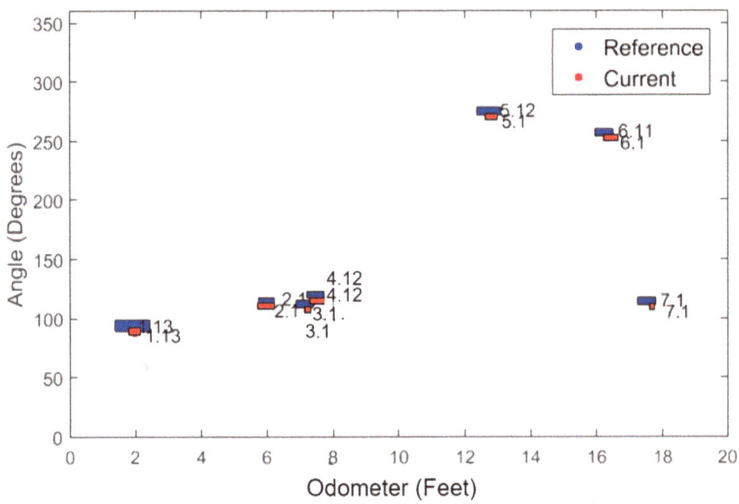

Fig. 7.5 Matched results (Liu et al. 2019)

Zhou et al. 2009). One aspect of the analytical integrity assessment is estimating the growth of corrosion features, which may reveal the remaining safe lifespan of each corrosion feature.

7.4 Brief Overview of Machine Learning Models

The increasing use of machine-generated data in several fields, including energy pipelines, requires a strong framework to ensure data quality, especially in detecting outliers and identifying false data values in raw datasets collected from field inspection (Faroukhi et al. 2020). Machine learning models are powerful in identifying anomalies and outliers that could affect the accuracy of datasets (Salem et al. 2022; Figueroa Barraza et al. 2022; Ragab et al. 2021; Rachman et al. 2021).

The models were trained using a range of methodologies, including unsupervised learning, supervised learning, statistical methods, ensemble approaches, hybrid techniques, and deep learning (Usama et al. 2019). Understanding the strengths, limitations, and applications of these models is crucial for developing successful tactics to tackle data quality issues and improve the reliability of raw datasets (Karkouch et al. 2016). Table 7.2 shows a few machine learning models to detect outliers and fake data in raw data.

7.4.1 Selection Criteria for Machine Learning Models

Choosing the most suitable machine learning model for identifying and eliminating outliers in pipeline ILI data is essential to achieve optimal outcomes to maintain pipeline integrity and reliability. A few key criteria for consideration are discussed below.

7.4.1.1 Data Characteristics

- **Data size and dimensionality**: Working on a large ILI dataset, there are several features to be considered like defect size, length, depth, pipeline segment, location, orientation, etc. Keep in mind selecting a model that is efficient for specific data size and dimensionality.
- **Nature of data**: Consider whether the raw data is numerical, categorical, sequential, or a combination of all of them. Model selection needs to consider if a model is better suited to handle specific data types.
- **Distribution of outliers**: Examine the distribution of data points and assess whether it follows a normal distribution or exhibits skewed patterns, as this can impact the performance of certain models.

7.4.1.2 Training Data Size

- Assess the availability of a substantial amount of training data. Deep learning models often require large datasets for effective training, while simpler models may perform well with smaller datasets (Brigato and Iocchi 2020).

Table 7.2 Machine learning models for outlier and false data detection

Model	Description	Advantages	Disadvantages
Isolation forest	Isolates outliers by randomly partitioning the data	Effective for high-dimensional data, resistant to masking effects	Black box model, limited interpretability
Local outlier factor (LOF)	Computes the local density of data points to identify outliers	Good at detecting local outliers, works well with noisy data	Sensitive to parameter tuning, computationally expensive
Anomaly detection algorithms	Learns the "normal" data and identifies deviations	Handles high-dimensional data, efficient with limited labeled data	It can be computationally expensive, and performance depends on data distribution
Autoencoders	Neural network architecture for unsupervised learning is useful for capturing data representations	Can handle complex data patterns; effective in feature learning	Requires careful tuning; may not scale well for large datasets
K-means	Clustering-based unsupervised algorithm that partitions data into clusters (Zhong and Ghosh 2003)	Simple and computationally efficient	Sensitive to initial cluster centroids; struggles with varying cluster densities
Support vector machines (SVM)	Supervised learning algorithm that separates data into classes using hyperplanes (Osisanwo et al. 2017)	Robust in high-dimensional spaces; effective for binary classification	Sensitive to the choice of kernel and parameter tuning
Random forests	Ensemble of decision trees used for both supervised learning and anomaly detection (Biswas and Samanta 2021)	Robust and versatile; handles high-dimensional data well	May overfit on noisy data; computational overhead in training large forests
Neural network	A network of interconnected neurons is used for learning and inference (Alobaidi et al. 2022)	• Can model complex, non-linear relationships • Robust to noise and missing data	• Requires large amounts of data for training • Vulnerable to overfitting
Regression analysis	Predicts the value of a dependent variable based on the values of one or more independent variables (Hariharan et al. 2017)	• Can handle both continuous and categorical data • Various algorithms available (e.g., linear regression, logistic regression)	Sensitive to outliers and influential observations

(continued)

Table 7.2 (continued)

Model	Description	Advantages	Disadvantages
Bayesian reference	Utilizes Bayesian inference to model uncertainty and update beliefs based on observed data (Soomro et al. 2022; Hassan et al. 2022)	• Allows for uncertainty quantification • Suitable for small datasets	• Requires careful selection of prior distributions
K-nearest neighbors	Instance-based method	Works well with small to medium-sized datasets	Performance can degrade with high-dimensional data
Gaussian mixture model	Probabilistic model	Identifies outliers based on probability	Requires estimation of multiple parameters

7.4.1.3 Detection Goals

- **Sensitivity**: Consider the sensitivity of the chosen model to outliers. Some models, like Z-score and Mahalanobis Distance, are sensitive to outliers, while others, such as Isolation Forests, are designed to be robust against them (Baesens et al. 2021).
- **Specificity**: How crucial is it to minimize false positives and accurately identify only true outliers? If specificity is paramount, models like interquartile range (IQR) or isolation forest might be suitable (Staberg 2020).

7.4.1.4 Computational Sources

- **Computational Complexity**: Consider the computational resources available. Deep learning models can be computationally intensive, while simpler models may be more computationally efficient.

7.4.1.5 Performance Evaluation

- **Evaluation metrics**: Define appropriate metrics for evaluating model performance, considering aspects such as precision, recall, F1-score, and area under the Receiver Operating Characteristic (ROC) curve depending on the specific goals of the analysis (Giglioni et al. 2021).
- **Baseline comparison**: Compare the performance of the chosen model with simpler baselines like statistical methods or domain-specific rules to assess its added value.

By carefully considering these criteria and conducting a thorough evaluation, pipeline operators can select the most effective machine learning model for addressing data quality gaps and enhancing the accuracy and reliability of pipeline condition assessments.

7.5 Guideline of ML Model Selection for Detection of Outliers in ILI Dataset

Outliers' assessment is a critical thing to do in pipeline ILI data to determine if there is any data that needs to be exploited to make the right decisions on pipeline condition assessment. While applying machine learning on the ILI dataset, it was observed that all models for outlier detection are not equally efficient. Therefore, the selection of the appropriate machine learning method must be based on the nature of the ILI data and the objectives of outlier identification. To choose an appropriate machine learning model for outlier detection in pipeline in-line inspection (ILI) dataset, this work develops a guideline as shown in Fig. 7.6 based on contents and model types.

7.6 Data Collection and Analysis

Several probes can be used to collect pipeline condition inspection data using in-line inspection. Each type has its strengths and weaknesses. Table 7.3 shows some of the most common probes used for pipeline assessment.

Current approaches to matching an anomaly in the context of anomaly detection and analysis can be generally categorized into two main types:

- Signal comparison and
- Anomaly size and location (data derived from the signal) comparison.

Figure 7.7 illustrates the signal comparison approach, wherein raw signal data, particularly magnetic data (Spencer et al. 2010; Moreno et al. 2012), is visually represented, where the data shown is collected from two different inspection runs. By comparing the two graphs, i.e., the left and right images in Fig. 7.7, changes in defect size become apparent, enabling the observation of variations in the inspected material. While automatic detection is feasible, as demonstrated by Spencer et al. (2010), its practical application is typically constrained to scenarios where the defect density is low in a given area.

The signal comparison method is favored for its enhanced accuracy compared to the second approach, as it circumvents the uncertainties inherent in algorithms that convert signals into defect sizes. While the signal comparison and defect size and location comparison methods each have its own merits, achieving unequivocal identification of a particular defect from one inspection to another remains a challenging task. This limitation underscores the complexity and dynamic nature of defect analysis, prompting ongoing efforts to enhance matching techniques for more accurate and reliable results. Researchers and industry professionals continue to explore innovative approaches to addressing these challenges and improve the overall effectiveness of defect detection. Achieving a definitive identification of a specific defect from one inspection to another is not directly feasible. It requires running defect-matching algorithms.

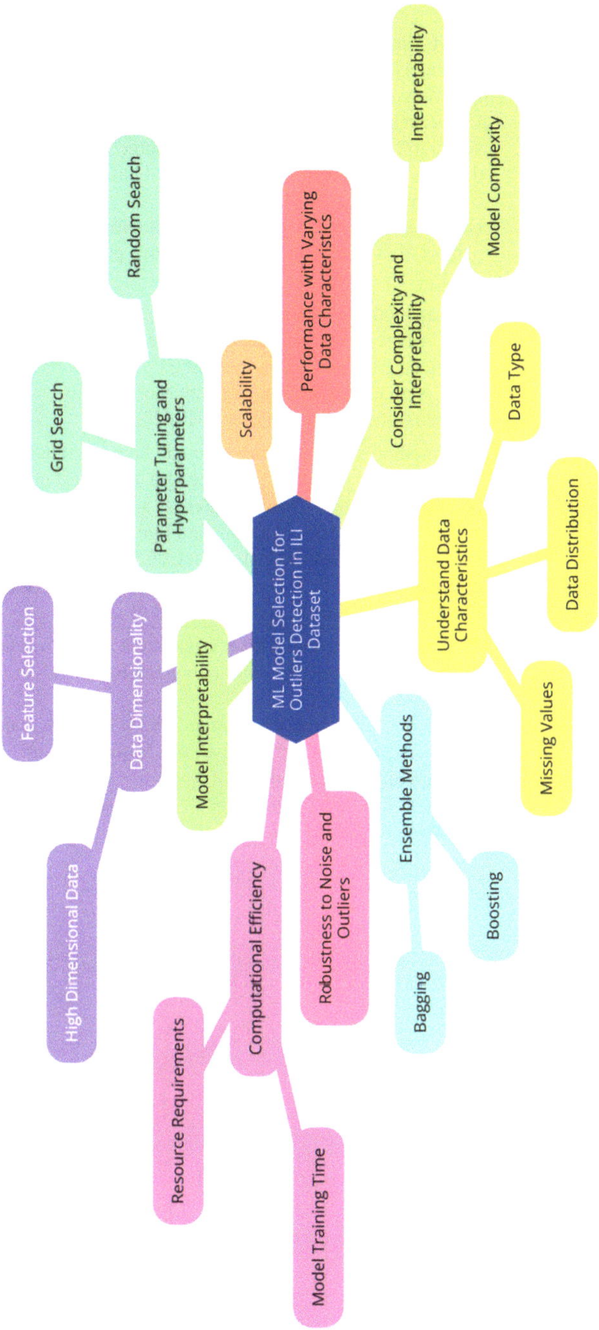

Fig. 7.6 Guideline of ML model selection for detection of outliers in ILI dataset

Table 7.3 ILI probes and details

ILI probes	Details
Ultrasonic testing (UT)	• This is the most widely used method for ILI due to its ability to accurately measure wall thickness and detect a variety of defects, including corrosion, pitting, and cracks (Vanaei et al. 2017) • UT probes use sound waves to measure the time it takes for a signal to travel from the probe to the pipe wall and back. The thickness of the wall can be calculated based on this travel time (Brockhaus et al. 2014) • Different types of UT probes are available, such as thickness gauges, flaw detectors, and guided wave probes (Olisa et al. 2021)
Magnetic flux leakage (MFL)	• This method uses magnets to create a magnetic field around the pipe. Any wall loss will disrupt the field, and this disruption can be detected by sensors on the probe • MFL is good at detecting large areas of wall loss, but it can be less sensitive to smaller defects (Ru 2023)
Eddy current testing (ECT)	• This method uses an electromagnetic field to induce eddy currents in the pipe wall. The strength and distribution of these currents are affected by the thickness and conductivity of the wall (Yin et al. 2018; Coramik and Ege 2017) • ECT is good at detecting small defects, such as pitting and cracks, but it can be less effective on thicker pipes or pipes made of certain materials (García-Martín et al. 2011)
Electromagnetic acoustic transducer (EMAT)	• This method uses sound waves generated by electromagnetic fields to inspect the pipe wall (Ling and Rahim 2017; Ho et al. 2020) • EMAT does not require a couplant like UT, which makes it useful for pipelines carrying gases or liquids that are incompatible with a couplant • EMAT is still under development, but it has the potential to be a valuable tool for ILI (Ma et al. 2021)

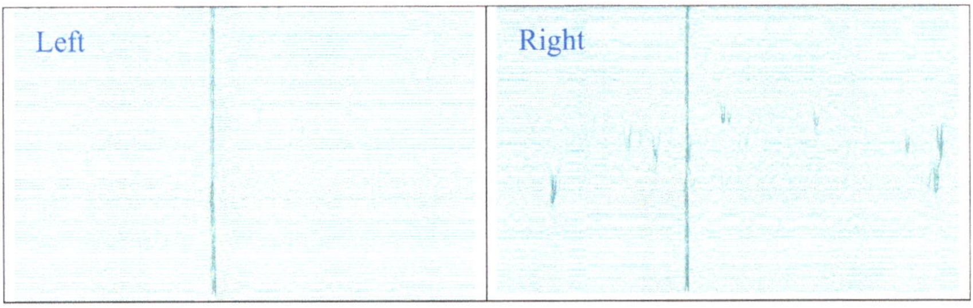

Fig. 7.7 Defect matching by comparing the raw magnetic signals

In this work, the ILI data was collected from a pipeline company in the Middle East for a pipeline commissioned in 1978. The data includes a wide range of features crucial for pipeline integrity assessment, such as pipeline segment distance (km), wall losses (%), anomaly length (mm), width (mm), corrosion growth rate and year of data collection. The ILI data was exported from ILI database to MS Excel spreadsheet with a clear identification of pipeline location, length, and wall losses. The ILI probes utilized to collect data for this selected oil pipeline are UT thickness measurement and MFL signal, where UT thickness was used for any wall losses and MFL for cracks. We analyzed wall losses to show data variations and their impact on pipeline condition assessment.

Out of a total dataset comprising 42,000 samples, a subset of 500 data samples was randomly selected from various segments of the pipeline for further analysis. The selection process ensured a representative distribution of samples across different pipeline sections to enhance the robustness of the findings. The objective of this analysis was to investigate variations in in-line inspection (ILI) measurements over an extended period of 22 years, thereby identifying potential discrepancies, trends, and influencing factors.

As seen in Fig. 7.8, the raw ILI data exhibits significant variations across different measurement intervals. The first in-line inspection (ILI 1) was conducted in the year 2000, with subsequent inspections performed in 2005 (ILI 2), 2010 (ILI 3), 2015 (ILI 4), and 2022 (ILI 5). Notably, the data collected during the initial inspection in 2000 recorded relatively higher levels of wall loss in the pipeline. However, an unexpected trend was observed in the subsequent years, where the recorded wall loss appeared to be lower in 2005, followed by additional variations in the later inspection years.

Such variations could happen for a few reasons, these include, but are not limited to:

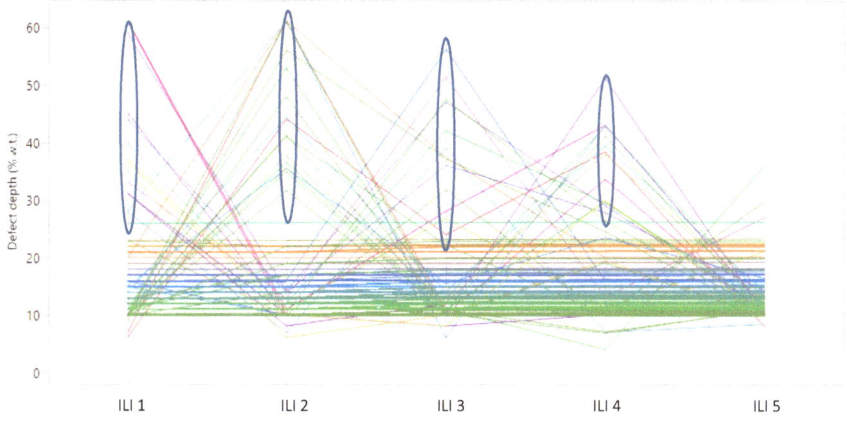

Fig. 7.8 Growth paths of individual anomalies determined by ILI measurements

1. **Probe Selection and Sensor Type**: Variability in probe selection, including differences in the type, model, and sensitivity of the sensors used in each inspection cycle, could result in discrepancies in the measured wall loss values.
2. **Probe Orientation and Positioning**: The orientation and positioning of the probe within the pipeline may impact measurement accuracy, particularly in areas with complex geometries or varying surface conditions.
3. **Human Influence and Data Collection Methods**: Differences in the personnel responsible for data collection may introduce subjectivity or procedural inconsistencies, affecting the accuracy of the recorded measurements.
4. **Probe Calibration and Instrumentation Drift**: Calibration errors or variations in sensor calibration procedures over the years could contribute to fluctuations in the recorded data.
5. **Environmental and Operational Factors**: Changes in pipeline operating conditions, such as variations in pressure, temperature, and flow rate, may influence corrosion rates and, consequently, the measured wall loss values.
6. **Data Transmission and Processing Errors**: The method of data transfer from inspection devices to analysis software could introduce errors, especially if different processing techniques or software versions were used across different inspection periods.
7. **Material and Structural Changes**: Variations in material properties, potential repairs, and modifications to the pipeline structure over time may also influence the measured degradation levels.

7.7 Statistical Analysis of ILI Data

ILI 1 data is analyzed by using the distribution method to identify outliers. This method identified a total of 12 outliers (red circle) in ILI 1 raw data (Fig. 7.9). The distribution method is a statistical approach that involves analyzing the distribution of the data within a dataset to identify any outliers. The ILI data can be visualized using histograms, Q-Q plots, and box plots to identify data points that deviate significantly from the expected value.

From Box plot, it is observed that the data points with values above 28.15 may be considered as outliers in raw ILI 1 data. Quantiles details for ILI 1 data are shown in Table 7.4, where the 97.5th percentile is 29.580 which represents a defect depth in percentage of the pipe wall thickness (% w.t.). Therefore, any values above 29.580 (percentage) are considered outliers. Table 7.5 shows summary statistics details for ILI 1 raw data.

The analysis results for ILI 2 data are illustrated in Fig. 7.10 with outliers indicated within the red circle. Table 7.6 presents detailed quantiles information for the ILI 2 inspection data, where data points exceeding 26.16 are considered outliers. The 97.5th percentile is 36.01 and there is a total of 13 data points with values larger than 36.01. Specifically, the ILI 2 raw data reveal the presence of 17 outliers. Table 7.7 shows summary statistics details for ILI 2 raw data.

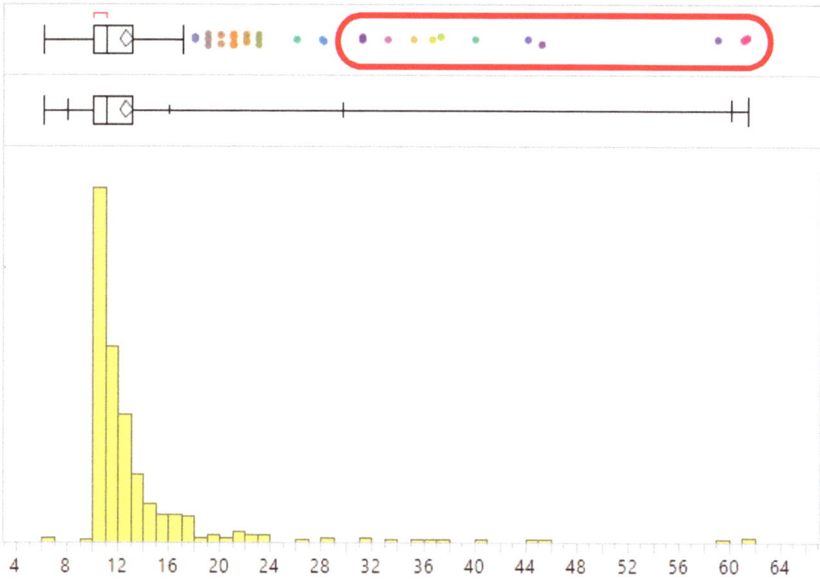

Fig. 7.9 ILI 1 data outliers

Table 7.4 Quantiles details for ILI 1 data

Quantile rank		
100.0%	Maximum	61.25
99.5%		59.995
97.5%		29.580
90.0%		16.00
75.0%	Quartile	13.00
50.0%	Median	11.00
25.0%	Quartile	10.00
10.0%		10.00
2.5%		10.00
0.5%		8.046
0.0%	Minimum	6.12

Analysis results for ILI 3 data are presented in Fig. 7.11, where outliers are highlighted within the red circle. A total of 10 outliers are observed in ILI 3 dataset. Table 7.8 shows detailed information on the quantiles for ILI 3 inspection data. The 97.5th percentile is 26.54. If this value is set as a threshold, there are 12 outliers identified in ILI 3 data. Table 7.9 shows summary statistics details for ILI 3 raw data.

Figure 7.12 shows analysis results for ILI 4 data, with outliers shown inside the red circle where 9 outliers are identified. Table 7.10 gives quantiles details for ILI 4

Table 7.5 Summary statistics

Summary statistics	
Mean	12.620
Std. Dev.	5.7651
Std. Err. mean	0.257824
Upper 95% mean	13.1267
Lower 95% mean	12.1136
N	500

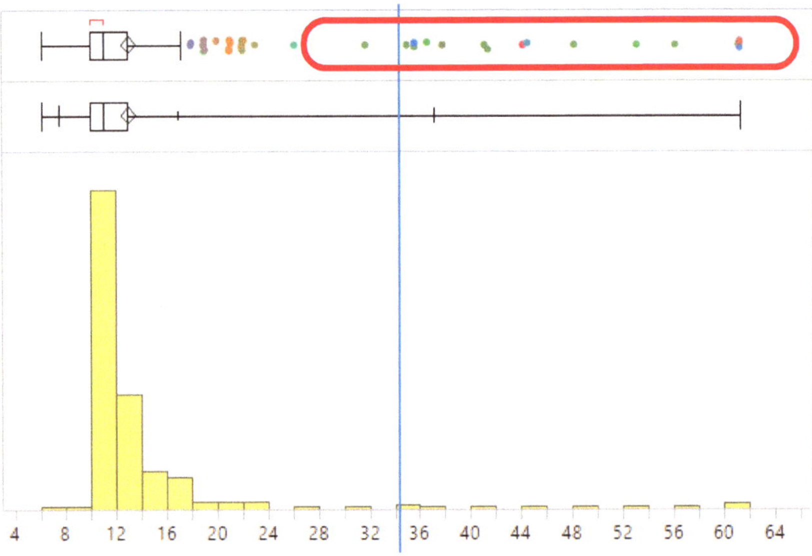

Fig. 7.10 ILI 2 data outliers

Table 7.6 Quantiles details for ILI 2 data

Quantile rank		
100.0%	Maximum	61.26
99.5%		61.2399
97.5%		36.01
90.0%		17.00
75.0%	Quartile	13.006
50.0%	Median	11.05
25.0%	Quartile	10.05
10.0%		10.01
2.5%		10.00
0.5%		7.5857
0.0%	Minimum	6.15

Table 7.7 Summary statistics

Summary statistics	
Mean	13.0443
Std. Dev.	7.0062
Std. Err. mean	0.313328
Upper 95% mean	13.6599
Lower 95% mean	12.4287
N	500

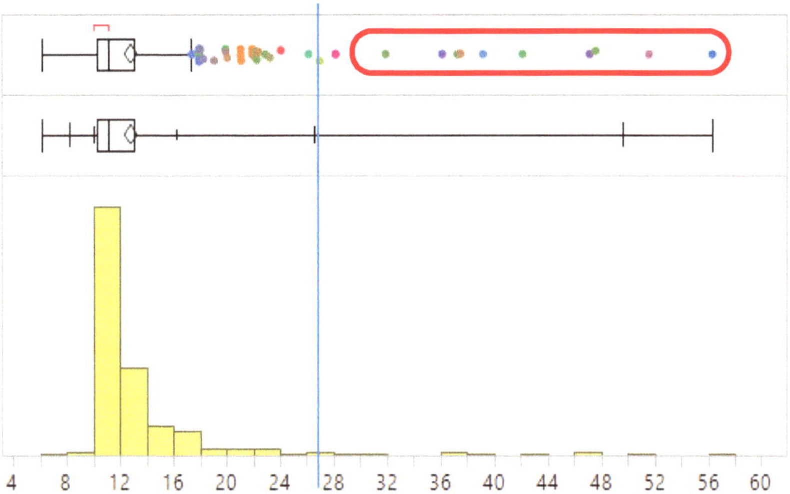

Fig. 7.11 ILI 3 data outliers

Table 7.8 Quantiles details for ILI 3 data

Quantile rank		
100.0%	Maximum	56.31
99.5%		49.575
97.5%		26.54
90.0%		16.259
75.0%	Quartile	13.086
50.0%	Median	11.16
25.0%	Quartile	10.27
10.0%		10.086
2.5%		10.05
0.5%		8.16
0.0%	Minimum	6.16

Table 7.9 Summary statistics

Summary statistics	
Mean	12.8633
Std. Dev.	5.23643
Std. Err. mean	0.234180
Upper 95% mean	13.3234
Lower 95% mean	12.4032
N	500

inspection data. The 97.5th percentile is 29.366. If it is chosen as the threshold, there could be a total of 12 outliers. Table 7.11 shows summary statistics details for ILI 4 raw data.

Figure 7.13 shows analysis results for ILI 5 data with outliers shown inside the red circle. There are 4 outliers identified in ILI 5. Table 7.12 presents quantiles details of ILI 5 inspection data, where the 97.5th percentile is 21.708. There are a total of 12 data values larger than 21.708. Table 7.13 shows summary statistics details for ILI 5 raw data.

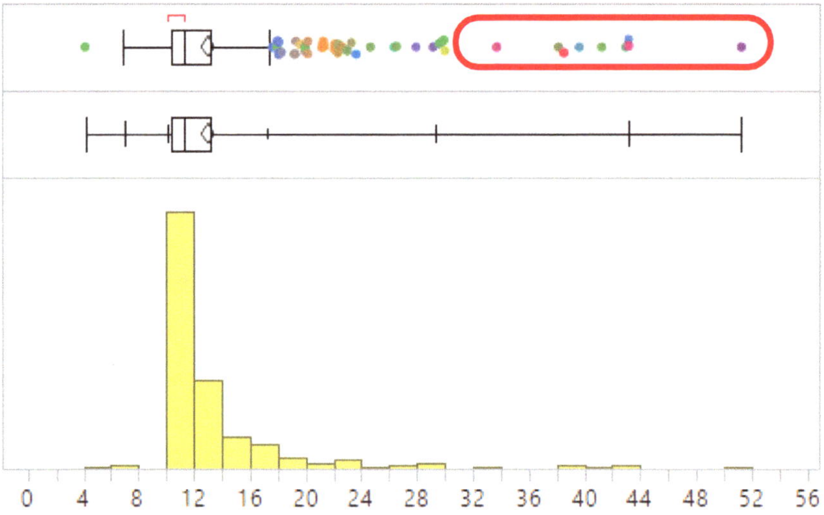

Fig. 7.12 ILI 4 data outliers

Table 7.10 Quantiles details for ILI 4 data

Quantile rank		
100.0%	Maximum	51.26
99.5%		43.15
97.5%		29.366
90.0%		17.16
75.0%	Quartile	13.157
50.0%	Median	11.296
25.0%	Quartile	10.39
10.0%		10.146
2.5%		10.05
0.5%		7.016
0.0%	Minimum	4.12

Table 7.11 Summary statistics

Summary statistics	
Mean	13.0286
Std. Dev.	5.10813
Std. Err. mean	0.228443
Upper 95% mean	13.4774
Lower 95% mean	12.5798
N	500

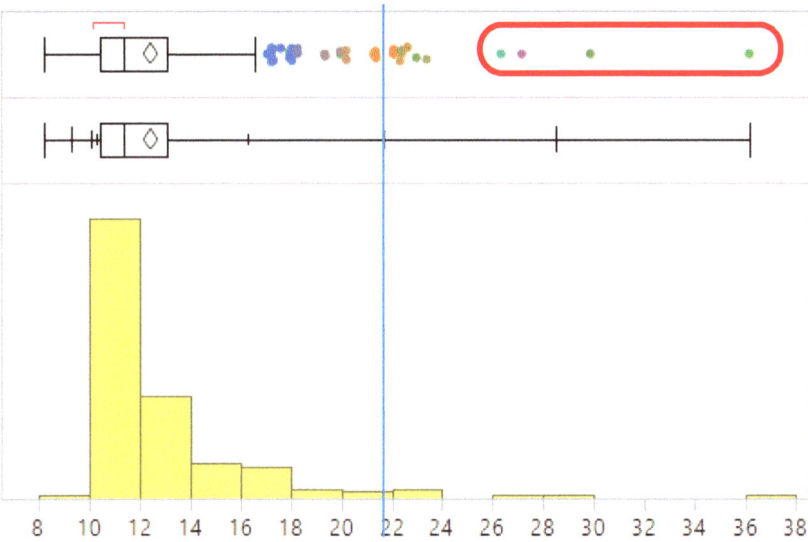

Fig. 7.13 ILI 5 data outliers

Table 7.12 Quantiles details for ILI 5 data

Quantile rank		
100.0%	Maximum	36.15
99.5%		28.491
97.5%		21.708
90.0%		16.3
75.0%	Quartile	13.07
50.0%	Median	11.356
25.0%	Quartile	10.43
10.0%		10.25
2.5%		10.087
0.5%		9.302
0.0%	Minimum	8.16

Table 7.13 Summary statistics

Summary statistics	
Mean	12.4352
Std. Dev.	3.0378
Std. Err. mean	0.135855
Upper 95% mean	12.7021
Lower 95% mean	12.1683
N	500

7.7.1 Z-score Method

Z-score is a commonly used technique to identify outliers in raw data. The Z-score, also known as the standard score or Z-value, measures the number of standard deviations by which a data point deviates from the dataset's mean (Everingham et al. 2016). The Z-score is calculated by

$$z = (x - \mu)/\sigma$$

where, z = Z-score, x is the value being evaluated, μ is the mean of the dataset, and σ is the standard deviation.

Table 7.14 shows a few benefits of using Z-score for identifying outliers in raw ILI data.

The outliers identified by Z-score method are shown in Table 7.15. The threshold value selected is as ± 3.

Table 7.14 Benefits of Z-score to identify outliers

Benefits	Details
Normalization across scales	Allows for consistent comparison of data points regardless of original measurement units
Quantitative assessment of deviation	Provides a numerical measure of how far a data point deviates from the mean, aiding in outlier identification
Identification of extreme values	Flags extreme values based on threshold Z-scores, facilitating targeted outlier detection
Statistical robustness	Based on mean and standard deviation, suitable for various data distributions, ensuring robust outlier detection
Integration with statistical tests	Enhances outlier detection when integrated with statistical models and tests, improving data analysis accuracy

7.7.2 Test of Interquartile Range Method

The interquartile range method is a popular method applied to detection of outliers in a dataset. It is first to determine the first quartile (Q1), the median, and the third quartile (Q3). The second step is to calculate interquartile range (IQR), IQR = Q3 – Q1. The third step is to calculate the upper fence (bound) which is equal to Q3 + (1.5 × IQR) and the lower fence (bound) which equals Q1 – (1.5 × IQR). The fourth step is to identify the outliers, that is, all values falling outside the fences are marked as outliers.

Applying the conventional interquartile range method to outlier detection for ILI 1 to ILI 5 datasets results in the detection results shown in Table 7.16. Clearly, the number of outliers detected is significantly larger than other methods. If the upper fence is calculated using Q3 + 4 × IQR, the number of outliers is still higher than the other methods. The reason why the conventional interquartile range method is not effective in the detection of outliers for the datasets is that the ILI measurement values in each dataset are highly left skewed. This is expressed by the histograms as shown in Figs. 7.9 and 7.13 in this section. Finally, we find a way to determine the upper fence by using the 95th percentile + 3 × IQR. Based on the new upper fences calculated, the detected outliers for each ILI dataset are closely aligned with those identified by using the above methods and machine learning models. Hence, we recommend using the 95th percentile + 3 × IQR to calculate the upper fence when detecting outliers for the ILI data for the pipeline under this study.

Table 7.15 Outliers in raw ILI data, detected by Z-score method

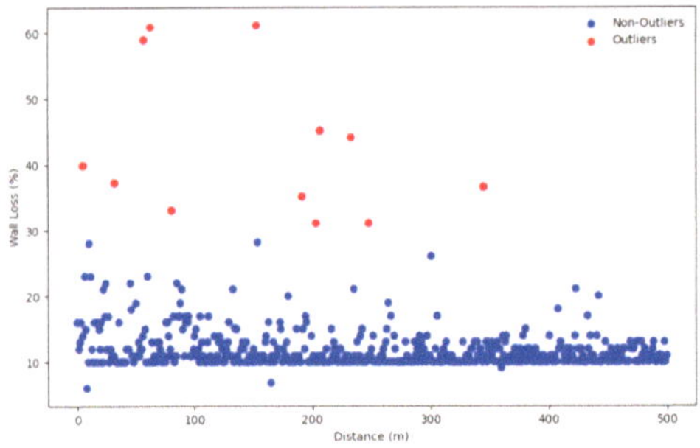	The number of outliers identified by Z-score method for ILI 1 data is 12 which is the same as found by other methods
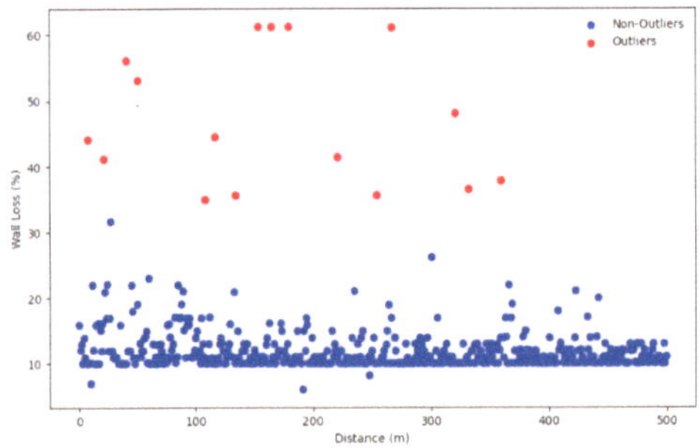	The outliers detected using Z-score method for ILI 2 data are 16, which is well aligned with the findings by other methods
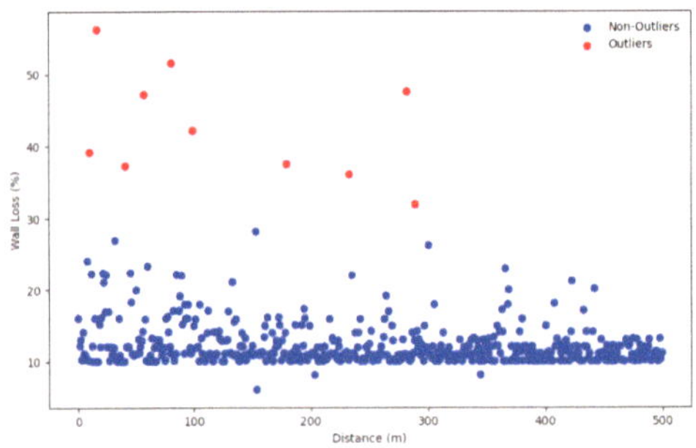	The Z-score method was employed to identify outliers in ILI 3 data. It is noteworthy that 10 outliers were found, demonstrating a perfect correspondence with those detected by other methods

(continued)

Table 7.15 (continued)

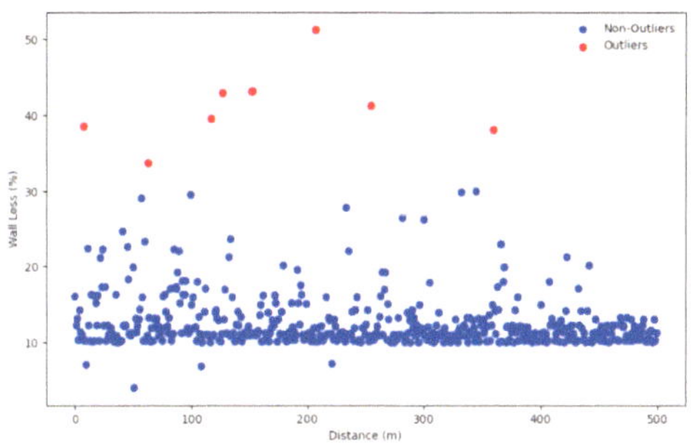	This plot illustrates the identification of outliers in ILI 4 data using Z-score method, a total 8 outliers are identified, which is very close to those identified by other methods
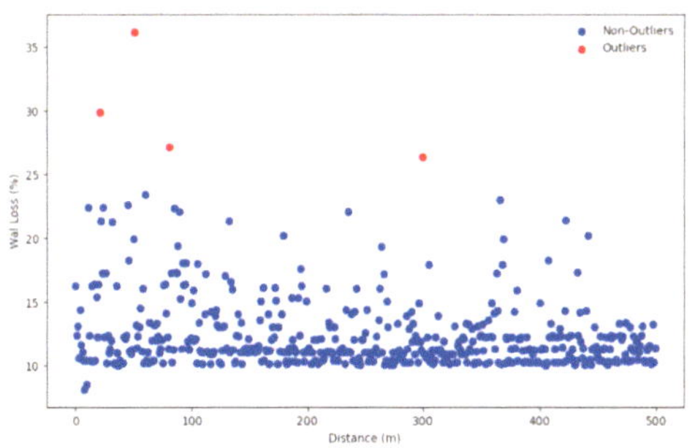	ILI 5 dataset shows 4 outliers by Z-score method, which is the same as the identified by other methods

Table 7.16 Test of interquartile range method for identifying outliers

Determination of threshold	Number of outliers				
	ILI 1	ILI 2	ILI 3	ILI 4	ILI 5
Q3 + 1.5 IQR	46	35	38	48	39
Q3 + 4 IQR	14	17	12	16	5
95th percentile + 3 IQR	12	17	10	9	4

7.8 Machine Learning Models

The application of machine learning (ML) models has become essential in the field of pipeline integrity management for the analysis of ILI raw data. However, ensuring the reliability and accuracy of these models requires rigorous testing and analysis, particularly in identifying outliers. The presence of outliers, anomalies, or unforeseen data points can have a substantial influence on the efficacy and reliability of ML models, potentially resulting in inaccurate findings or compromised safety evaluations. This section discusses the testing of a few models on raw ILI data to identify outliers.

Table 7.17 Approach to choose a threshold value for pipeline ILI raw data

Step	Brief details
Understand data distribution	Begin by analyzing the distribution of your ILI data. Plot histograms or density plots to visualize the distribution. Understanding the data's central tendency and dispersion helps in setting an initial threshold
Consider statistical methods	Statistical methods like Z-score or percentile-based approaches can be used to determine thresholds. Z-score identifies outliers based on how many standard deviations they are from the mean, while percentiles set thresholds based on data distribution
Evaluate impact	Consider the impact of setting different thresholds. A lower threshold may detect more outliers but could also increase false positives. Balancing sensitivity and specificity are essential
Validation and testing	Validate the chosen threshold using historical data or simulated scenarios if available. Testing the threshold helps assess its effectiveness in identifying true anomalies while minimizing false alarms

7.8.1 Random Forest Regression

One of the machine learning algorithms commonly employed to detect outliers in ILI data is Random Forest Regression. This approach is designed to predict a continuous variable by utilizing input data (Zhou et al. 2024). The ensemble approach is a technique that integrates many decision trees to provide predictions.

Random Forest Regression offers valuable insights into the significance of features, aiding in the identification of the most influential features in the prediction process (Suri et al. 2013). Selecting a threshold value is an important step in outlier detection for pipeline ILI raw data.

Refer to Table 7.17 for the guideline in the selection of the threshold. The threshold value for the Random Forest Regression model is set at $\pm$ 3 standard deviations.

An overview of the utilization of Random Forest Regression for the identification of outliers in ILI data is presented in Table 7.18.

The analysis results to identify outliers in raw ILI data using the Random Forest Regression model are shown in Table 7.19 covering the outliers for ILI 1 to ILI 5 data.

7.8.2 Linear Regression Analysis

Linear regression analysis assumes a linear relationship between the variables, making it a valuable tool for prediction and inference. For linear regression analysis, along with the dependent variable (wall loss), we considered the age of the pipeline as an independent variable, that is the number of years of the pipeline in service.

Table 7.18 Overview of random forest regression for outliers' identification

Aspect	Description
Training data	Historical ILI data includes pressure, temperature, flow rate, corrosion levels (anomaly values), etc.
Model training	Ensemble decision trees trained on random subsets of data and features to reduce overfitting and improve generalization
Prediction	Predicts normal behavior of ILI data based on features; data points with significant deviations from predictions are flagged as potential outliers
Anomaly detection	Analyze residuals (differences between actual and predicted values) to identify data points with unusually large residuals, indicating potential anomalies or outliers
Feature importance analysis	It provides insights into which features are most influential in predicting normal ILI data behavior, aiding in understanding potential causes of outliers
Thresholding	Sets thresholds based on residuals or other anomaly detection metrics to refine outlier detection and improve precision in identifying outliers

The threshold value for linear regression analysis is selected as $\pm$ 3 standard deviations. In linear regression, the threshold typically refers to a value used to categorize or filter the data based on the prediction using independent variables. However, in the context of wall loss determination, the threshold is determined as follows:

Threshold in Wall Loss Assessment: This could involve setting a critical wall loss percentage, beyond which the pipeline is considered at risk of failure.
Statistical Threshold: In the analysis, the threshold could also refer to a decision rule based on the regression output. For instance, if the regression model predicts a certain level of wall loss based on independent variables, a threshold might be set to trigger maintenance actions when the predicted wall loss exceeds a specific percentage.

The results of linear regression analysis for outlier detection of ILI 1 to ILI 5 datasets are presented in Table 7.20.

7.8.3 K-Nearest Neighbors (KNN) Model

The KNN model is widely utilized in machine learning for both classification and regression tasks. The capabilities of this model also encompass outlier detection, rendering it a versatile instrument for data analysis, particularly within the realm of pipeline ILI raw data.

The K-nearest neighbors (KNN) algorithm functions by leveraging the concept of similarity, wherein data points are classed or assessed according to their proximity to adjacent points within the feature space. The utilization of KNN in outlier detection

Table 7.19 Outliers in raw ILI data, detected by random forest regression

<table>
<tr><td>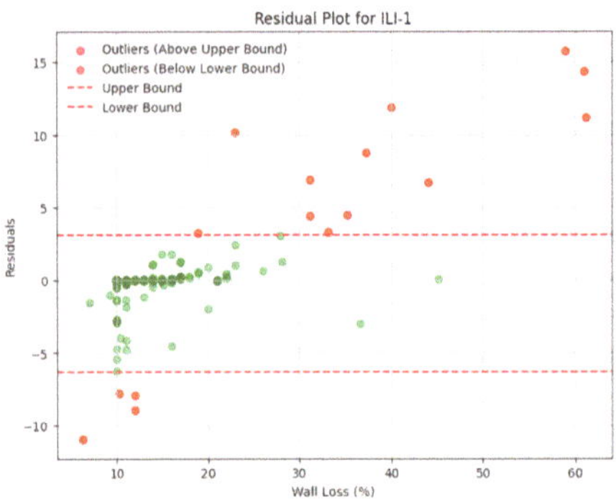
</td><td>A total of 500 data samples were selected for this model analysis. This plot shows the total number of outliers for ILI 1 data, identified by Random Forest Regression. There are 12 outliers with values larger than the Upper bound. This is the same as identified by the distribution and Z-score methods</td></tr>
<tr><td>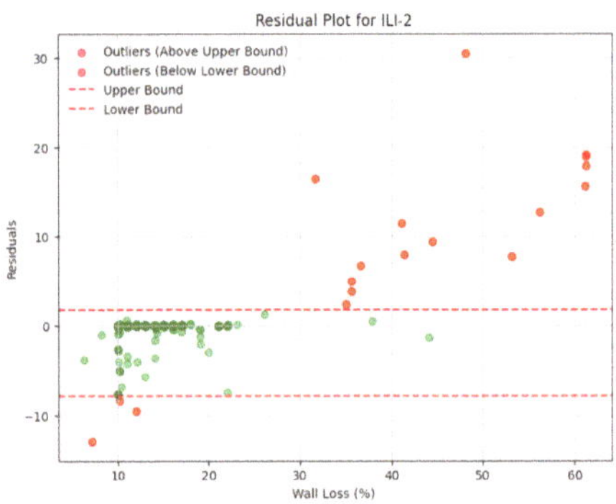
</td><td>For ILI 2 data, a total of 15 outliers above the Upper bound were identified, which is approaching the number of outliers observed using Box plot where it is 17. Using Z-score method, 16 outliers were identified</td></tr>
<tr><td>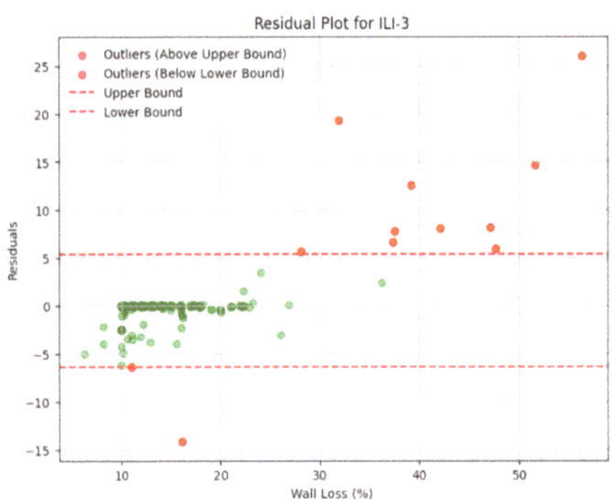
</td><td>This plot shows outliers identified in ILI 3 dataset. The analysis shows a total of 10 outliers above the Upper bound were pinpointed, which are of the same as identified by the distribution and Z-score methods</td></tr>
</table>

(continued)

Table 7.19 (continued)

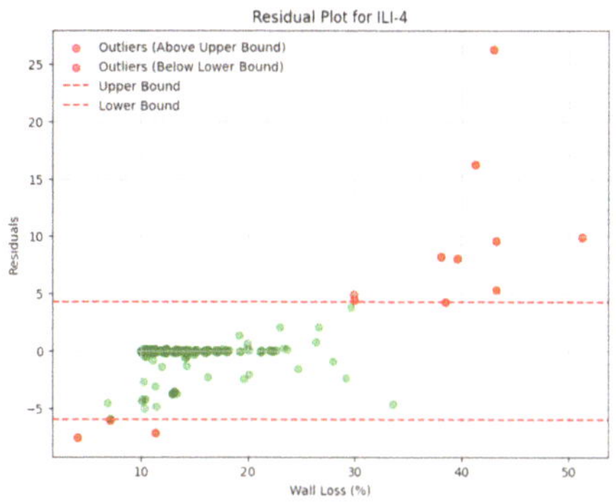	This plot presents the outliers identified within the ILI 4 raw dataset. A total of 10 outliers above the Upper bound are found, which is 9 found by the distribution-based method and 8 found by Z-score method
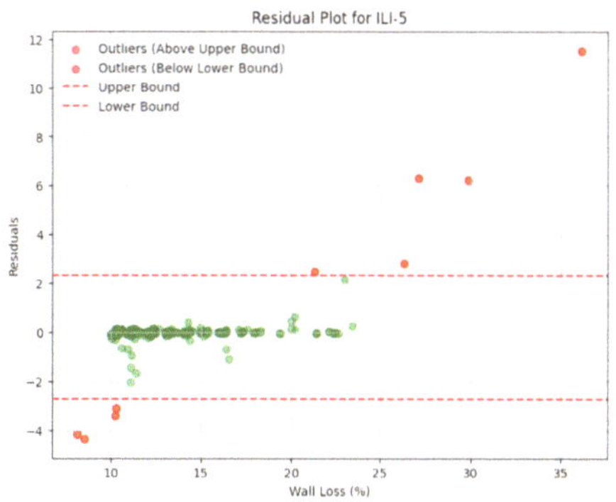	The outliers above the Upper bound detected within ILI 5 raw dataset are 5. This number closely reflects the total outliers given by the distribution and Z-score methods

enables the identification of data points that exhibit substantial deviation from their neighboring data points, hence suggesting the presence of potential anomalies or outliers.

KNN's versatility in managing data with a large number of dimensions and its capacity to adjust to different data distributions make it a powerful tool for identifying outliers in intricate datasets like ILI raw data. Using the KNN method, the outlier detection results for ILI 1 to ILI 5 data are shown in Table 7.21.

7.9 Conclusion

Data quality issues, such as inconsistencies, missing values, missed defect matching, long distance mismatching, girth weld mismatching and outliers, can lead to inaccurate assessments and misinterpretation of inspection results. These issues can have severe consequences, potentially leading to missed anomalies and misinterpretation of defects resulting in safety hazards and operational interruptions. Analyzing and identifying outliers in raw ILI data provides great value in addition to oil and gas

Table 7.20 Outliers in raw ILI data, detected by linear regression analysis

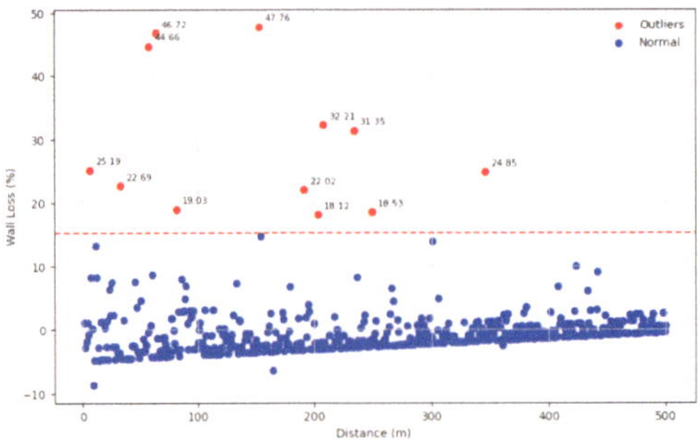	The analysis reveals a total of 12 outliers, consistent with the findings derived from both the distribution and Z-score methods
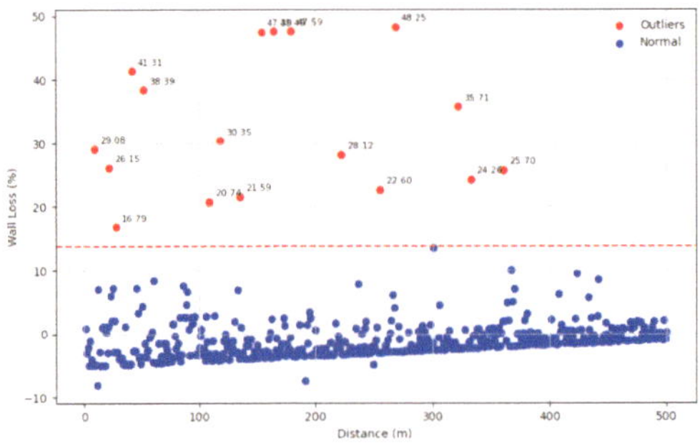	The total number of outliers detected for ILI 2 data is 17. This result is exactly aligned with the outcome from the distribution method, and approaching the findings by Z-score, and Random Forest Regression approaches
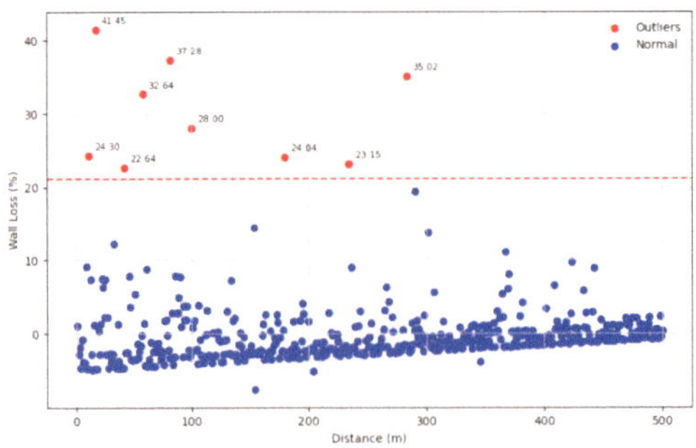	The result for ILI 3 data analysis reveals 9 outliers in raw ILI data, which closely matches the findings from the distribution, Z-score, and Random Forest Model methods

(continued)

Table 7.20 (continued)

<table>
<tr>
<td>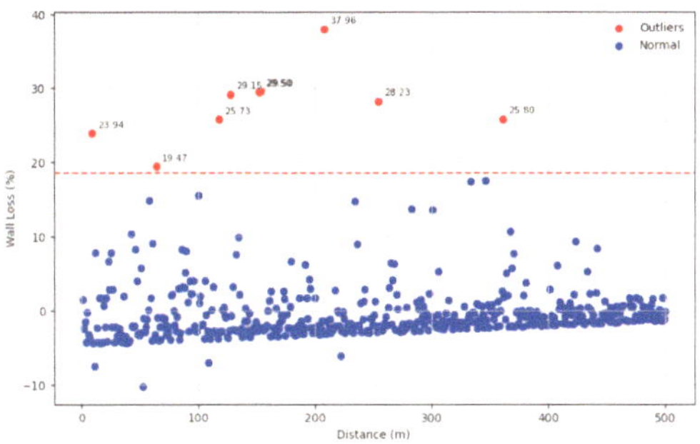
</td>
<td>ILI 4 data showcase 9 outliers, aligning closely with the outcomes observed through the distribution and Z-score methods</td>
</tr>
<tr>
<td>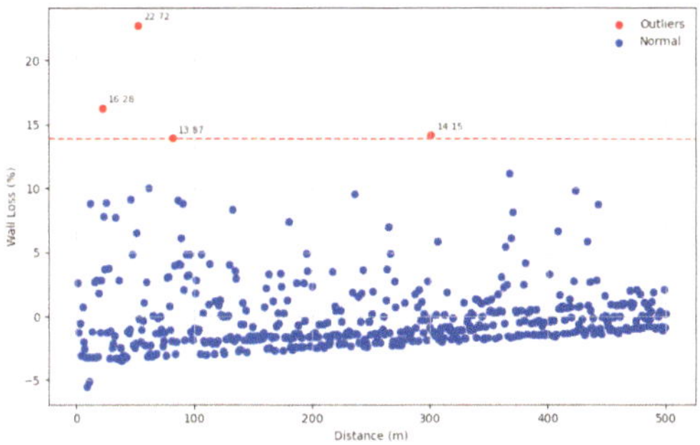
</td>
<td>ILI 5 data show a total of 4 outliers. This result is in close agreement with the findings using the distribution, Z-score, and Random Forest Model methods</td>
</tr>
</table>

pipeline integrity management. This research has examined machine learning models and compared these models with statistical analysis techniques in detecting outliers. Our findings reveal a significant overlap in the results obtained from both statistical analysis and machine learning approaches in detecting outliers of ILI data. Machine learning models, despite their complexity and the need for substantial computational resources, demonstrated a high level of accuracy in identifying outliers, matching the performance of traditional statistical methods. This convergence of results suggests that machine learning models can serve as a robust alternative or complement to statistical techniques, especially in scenarios where the data is vast and complex. Our study also confirms that it is critical to determine a proper threshold in outlier detection for a given dataset for a method to be effective in the detection. This depends on the nature and characteristics of the data in analysis. For example, the conventional interquartile range method is found not effective to detect the outliers for the ILI datasets under study, a new modification of the method to calculate the upper fence by the 95th percentile + 3 IQR is tested to have the same performance as other methods utilized.

Statistical and machine learning methods, including the Z-score method, IQR, Random Forest Regression, Linear Regression, and K-Nearest Neighbours (KNN), are applied to the dataset to identify anomalous observations. The implementation of multiple outlier detection approaches serves two primary objectives:

Table 7.21 Outliers in raw ILI data, detected by KNN technique

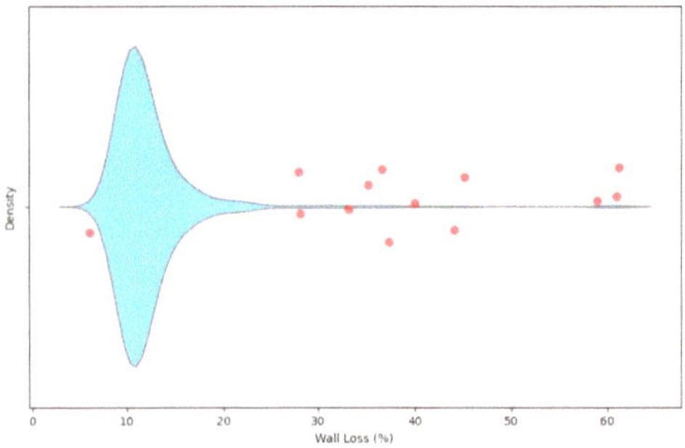	The implementation of the K-nearest neighbors (KNN) algorithm for outlier detection in ILI 1 data is illustrated. The analysis shows the presence of a total of 12 larger outliers, exactly the same as the number of outliers identified by other techniques
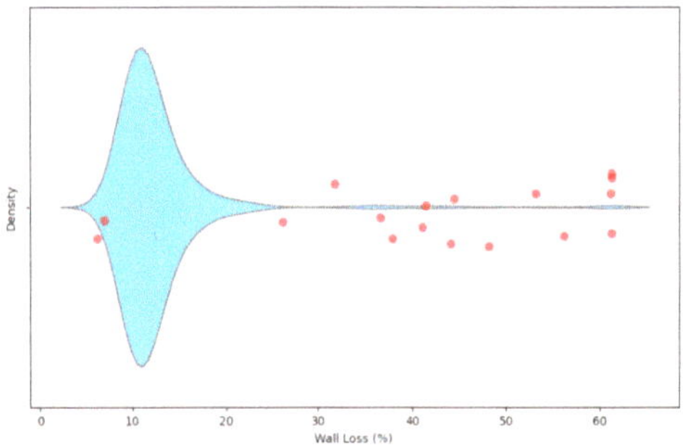	The KNN model detects 15 larger outliers in ILI 2 data. This result is consistent with the findings obtained from other techniques discussed above
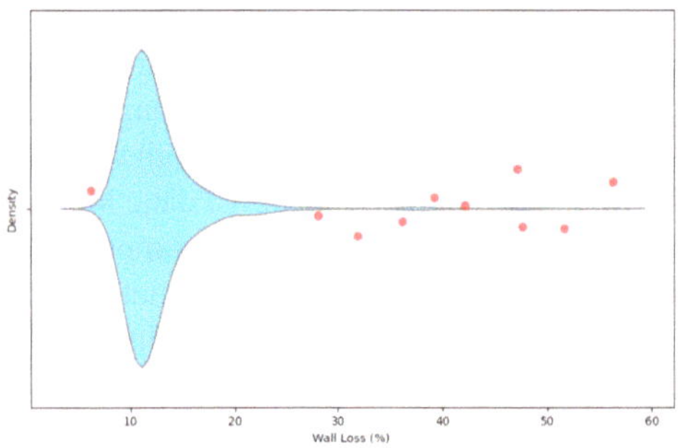	This plot shows the analysis results by utilizing the KNN algorithm to the ILI 3 raw data, showing 9 larger outliers, which aligns with the identifications made by other models and techniques

(continued)

Table 7.21 (continued)

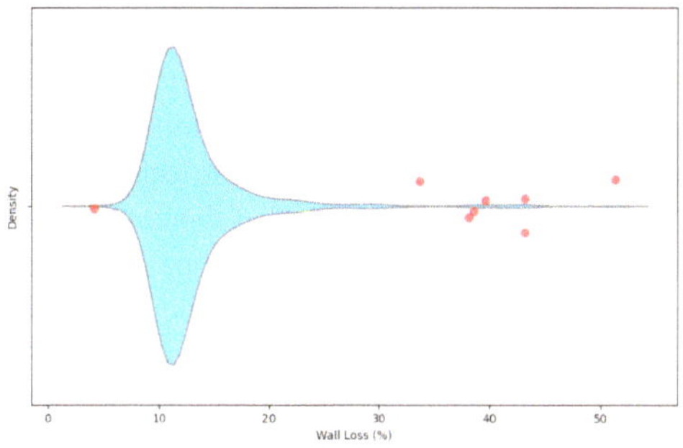	The analysis outcome using the KNN algorithm on the ILI 4 raw dataset reveals 7 larger outliers, which are very close to the total number of outliers detected by other techniques
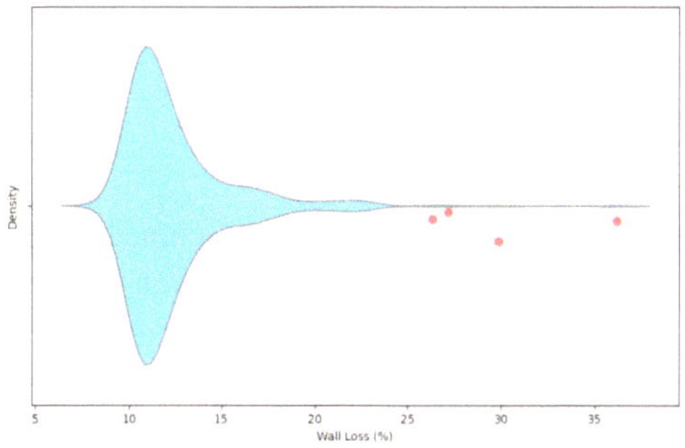	This graph depicts the analysis results of employing the KNN algorithm on ILI 5 raw dataset, revealing 4 larger outliers, which is the same as the number of outliers detected by other methods

- Each outlier detection technique operates based on distinct assumptions and computational principles, leading to variations in their sensitivity to noise and extreme values. By applying multiple approaches to the same dataset, it is possible to evaluate their robustness in distinguishing true anomalies from random fluctuations or measurement errors. Robust methods are characterized by their ability to consistently detect genuine outliers while minimizing false positives and false negatives in the presence of noise. Comparing the performance of these methods allows for the selection of the most reliable technique for ILI data, particularly in real-world scenarios where data imperfections are common.
- Secondly, given that individual methods may yield different sets of outliers due to their inherent assumptions and thresholding criteria, combining multiple approaches can enhance the confidence level of outlier detection. A consensus-based approach, where anomalies are identified by multiple methods, increases the likelihood that detected outliers represent true deviations rather than artifacts of a single method's bias. Additionally, leveraging ensemble techniques such as Random Forest Regression alongside statistical measures (e.g., Z-score, IQR) ensures a more comprehensive assessment of potential outliers. This multi-method strategy strengthens the reliability of outlier detection, which is particularly crucial in public health surveillance, where false positives or missed anomalies can have significant implications.

To test multiple approaches in this study is to propose a method for outlier detection for achieving robust results. On one side, one should be conservative to take all the detected outliers for checking and verification to make sure whether they are the true outliers or false measurements. On the other hand, however, we need to make sure that these required checking and verifications are not to cause much additional work to sort out those which are not the true outliers. The choice of these approaches should be guided by the specific requirements of the analysis, the nature of the dataset, and the available resources. For further study, one may explore hybrid approaches that leverage the strengths of both machine learning and statistical methods, potentially leading to even more robust and reliable outlier detection frameworks. This integration could further enhance data quality, contributing to safer and more efficient infrastructure management.

7.10 Future Perspectives

Regarding the future study, one of the directions is the utilization of machine learning models tailored to the unique characteristics of pipeline ILI raw data. Given the high-dimensional and often noisy nature of raw ILI data, there is potential to develop more sophisticated models that can accurately distinguish between true anomalies and noise. Further research could focus on the integration of hybrid models that combine machine learning techniques with traditional statistical methods. This approach could enhance the robustness and interpretability of outlier detection systems. For instance, machine learning models could be employed to preprocess data, reducing noise and isolating potential outliers, which can then be further validated using statistical analysis.

Comparative studies between machine learning models and traditional statistical methods should be expanded to assess their performance across various pipeline conditions and data types. These comparisons could help identify the strengths and limitations of each approach, guiding the selection of appropriate techniques for specific pipeline integrity challenges.

Other research works may be related to the following aspects:

(a) ***Automated Outlier Detection Algorithms Development***

- Design and implement automatic outlier detection algorithms customized for raw ILI data analysis.
- Utilize machine learning algorithms and statistical approaches to detect anomalies in real-time or near-real-time.

(b) ***Integration with Pipeline Monitoring Platforms***

- Incorporate outlier detection capabilities into existing pipeline monitoring platforms.
- Enable ongoing surveillance of ILI data streams, identifying anomalies as they arise.

(c) ***Enhanced Data Visualization and Reporting***

- Improve data visualization techniques and reporting mechanisms to highlight outliers effectively.
- Develop interactive dashboards, anomaly heatmaps, and trend analysis tools.

(d) ***Predictive Analytics for Outlier Prediction***

- Construct predictive analytics models that utilize historical trends in ILI data, environmental conditions, and operational parameters.
- Forecast potential outliers to enable proactive mitigation methods.

(e) ***Risk-Based Outlier Prioritization***

- Implement risk-based outlier prioritization frameworks.
- Prioritize the removal of outliers based on the severity of potential integrity issues.
- Enhance maintenance efforts and mitigate operational hazards by focusing on high-risk outliers.

(f) ***Continuous Model Refinement and Validation***

- Continuously improve and validate outlier detection models.
- Incorporate feedback from field inspections, maintenance activities, and historical outcomes related to outlier removal.
- Use an iterative technique to ensure the ongoing effectiveness and adaptability of outlier identification algorithms in response to changing data patterns and abnormalities.

Interdisciplinary collaboration between data scientists, pipeline engineers, and industry experts will be crucial in advancing the practical implementation of machine learning-based outlier detection systems. Such collaborations could lead to the development of standardized protocols and best practices, facilitating the widespread adoption of these technologies in the pipeline industry.

To summarize, the progress in outlier detection technologies, collaborative industry initiatives, and risk-based outlier management approaches will be essential in helping pipeline owners examine raw ILI data for outliers and highlight the significance of removing outliers for the overall integrity and operational excellence of the pipeline.

References

Alobaidi MH, Meguid MA, Zayed T (2022) Semi-supervised learning framework for oil and gas pipeline failure detection. Sci Rep 12(1):13758

Asha LN et al (2023) A Quantitative Approach of Measuring Sustainability Risk in Pipeline Infrastructure Systems. Sustainability 15(19):14229

Baesens B, Höppner S, Verdonck T (2021) Data engineering for fraud detection. Decis Support Syst 150:113492

Biswas P, Samanta T (2021) Anomaly detection using ensemble random forest in wireless sensor network. Int J Inf Technol 13(5):2043–2052
Bjørnøy OH, Sigurdsson G, Marley MJ (2004) Updated DNV-RP-F101 for corroded pipelines, 2004. In: International pipeline conference
Brigato L, Iocchi L (2021) A close look at deep learning with small data. In: 2020 25th International conference on pattern recognition (ICPR). IEEE
Brockhaus S et al (2014) In-line inspection (ILI) methods for detecting corrosion in underground pipelines. Underground pipeline corrosion. Elsevier, pp 255–285
Coramik M, Ege Y (2017) Discontinuity inspection in pipelines: a comparison review. Measurement 111:359–373
Correia JB, Abel M, Becker K (2023) Data management in digital twins: a systematic literature review. Knowl Inf Syst:1–32
Dann MR, Dann C (2017) Automated matching of pipeline corrosion features from in-line inspection data. Reliab Eng Syst Saf 162:40–50
Desjardins G (2001) Corrosion rate and severity results from in-line inspection data. In: NACE corrosion. NACE
Everingham Y et al (2016) Accurate prediction of sugarcane yield using a random forest algorithm. Agron Sustain Dev 36:1–9
Ewuzie U, Bolade OP, Egbedina AO (2022) Application of deep learning and machine learning methods in water quality modeling and prediction: a review. In: Current trends and advances in computer-aided intelligent environmental data engineering, p 185–218
Faroukhi AZ et al (2020) An adaptable big data value chain framework for end-to-end big data monetization. Big Data Cogn Comput 4(4):34
Figueroa Barraza J et al (2022) Deep learning health state prognostics of physical assets in the oil and gas industry. Proc Inst Mech Eng, Part o: J Risk Reliab 236(4):598–616
García-Martín J, Gómez-Gil J, Vázquez-Sánchez E (2011) Non-destructive techniques based on eddy current testing. Sensors 11(3):2525–2565
Giglioni V et al (2021) The use of receiver operating characteristic curves and precision-versus-recall curves as performance metrics in unsupervised structural damage classification under changing environment. Eng Struct 246:113029
Gudivada V, Apon A, Ding J (2017) Data quality considerations for big data and machine learning: going beyond data cleaning and transformations. Int J Adv Software 10(1):1–20
Handfield R et al (2020) 4th annual data quality & governance study 2020. https://static1.squarespace.com/static/5ea48ae194c41a2df14fa825/t/602c4b5df20e6f748ef287e3/1613515622616/4th+Annual+Data+Quality+%26+Governance+Study+2020.pdf [cited 9 August 2024]
Hariharan N et al (2017) Multivariate regression model to predict failure of pipelines. Int J Eng Technol 9(5):398–404
Hassan S et al (2022) An assessment of causes and failure likelihood of cross-country pipelines under uncertainty using bayesian networks. Reliab Eng Syst Saf 218:108171
Ho M et al (2020) Inspection and monitoring systems subsea pipelines: a review paper. Struct Health Monit 19(2):606–645
Huyse L, van Roodselaar A (2010) Effects of inline inspection sizing uncertainties on the accuracy of the largest features and corrosion rate statistics. In: Proceedings of the 8th International Pipeline Conference, vol 4, IPC2010-31037, pp 403–413. September 27–October 1, 2010. Calgary, Alberta, Canada
Karkouch A et al (2016) Data quality in internet of things: a state-of-the-art survey. J Netw Comput Appl 73:57–81
Koroteev D, Tekic Z (2021) Artificial intelligence in oil and gas upstream: trends, challenges, and scenarios for the future. Energy and AI 3:100041
Li Y et al (2017) Validation of corrosion growth rate models. In: NACE corrosion. NACE
Ling EH, Rahim RHA (2017) A review on ultrasonic guided wave technology. Aust J Mech Eng
Liu H et al (2019) Matching pipeline in-line inspection data for corrosion characterization. NDT and E Int 101:44–52

Ma Q et al (2021) Pipeline in-line inspection method, instrumentation and data management. Sensors 21(11):3862

McDonald A (2021) Data quality considerations for petrophysical machine-learning models. Petrophysics 62(06):585–613

Meeker WQ, Hong Y (2014) Reliability meets big data: opportunities and challenges. Qual Eng 26(1):102–116

Moreno PJ et al (2012) Development and application of local corrosion growth rates for pipeline integrity assessments. In: CORROSION 2012. NACE International, Salt Lake City, Utah

Olisa SC, Khan MA, Starr A (2021) Review of current guided wave ultrasonic testing (GWUT) limitations and future directions. Sensors 21(3):811

Osisanwo F et al (2017) Supervised machine learning algorithms: classification and comparison. Int J Comp Trends Technol (IJCTT) 48(3):128–138

Ouyang M (2014) Review on modeling and simulation of interdependent critical infrastructure systems. Reliab Eng Syst Saf 121:43–60

Rachman A, Zhang T, Ratnayake RC (2021) Applications of machine learning in pipeline integrity management: a state-of-the-art review. Int J Press Vessels Pip 193:104471

Ragab AMS, Yakoot MSE, Mahmoud O (2021) Application of machine learning algorithms for managing well integrity in gas lift wells. In: SPE Asia Pacific oil and gas conference and exhibition. SPE

Ru G et al (2023) Electromagnetic coupling sensing of pipe in-line inspection system. IEEE Trans Instrum Meas 72:6009815

Salem AM, Yakoot MS, Mahmoud O (2022) Addressing diverse petroleum industry problems using machine learning techniques: literary methodology—spotlight on predicting well integrity failures. ACS Omega 7(3):2504–2519

Soomro AA et al (2022) A review on Bayesian modeling approach to quantify failure risk assessment of oil and gas pipelines due to corrosion. Int J Press Vessels Pip 200:104841

Spencer K et al (2010) A practical application to calculating corrosion growth rates by comparing successive ILI runs from different ILI vendors. In: Proceedings of the 8th International Pipeline Conference, IPC2010-31306, pp 467-473. September 27–October 1, 2010. Calgary, Alberta, Canada

Staberg P (2020) Using anomaly detection to identify rejected media plans in advertisement planning: a feasibility study of using one-class classification and binary classification with artificially generated outliers to identify media plans most likely to be rejected. Thesis for Master in Computer Science, School of Electrical Engineering and Computer Science, KTH Royal Institute of Technology, Sweden

Suri V et al (2013) Spatial outlier detection using improved Z-score test. Int J Eng Sci Technol 5(12):1962

Tabrizchi H, Kuchaki Rafsanjani M (2020) A survey on security challenges in cloud computing: issues, threats, and solutions. J Supercomputing 76(12):9493–9532

Usama M et al (2019) Unsupervised machine learning for networking: techniques, applications and research challenges. IEEE Access 7:65579–65615

Vanaei H, Eslami A, Egbewande A (2017) A review on pipeline corrosion, in-line inspection (ILI), and corrosion growth rate models. Int J Press Vessels Pip 149:43–54

Yin L et al (2018) Detection of corrosion pits based on an analytically optimised eddy current sensor. Insight-Non-Destructive Test Condition Monit 60(10):561–567

Zhang H et al (2020) A modified method for the safety factor parameter: the use of big data to improve petroleum pipeline reliability assessment. Reliab Eng Syst Saf 198:106892

Zhong S, Ghosh J (2003) A unified framework for model-based clustering. J Mach Learn Res 4:1001–1037

Zhou J et al (2009) Reliability-based design and assessment standards for onshore natural gas transmission pipelines. J Pressure Vessel Technol 131(3):031702

Zhou L et al (2017) Machine learning on big data: opportunities and challenges. Neurocomputing 237:350–361

Zhou W, Yan Z, Zhang L (2024) A comparative study of 11 non-linear regression models highlighting autoencoder, DBN, and SVR, enhanced by SHAP importance analysis in soybean branching prediction. Sci Rep 14(1):5905

Chapter 8
Energy Pipeline Defect Growth Prediction Using Degradation Modeling

8.1 Introduction

The continuous operation of pipelines is critical to the oil and gas industry, but the infrastructure is constantly exposed to external and internal stressors that can degrade pipeline integrity over time. Defects such as corrosion, cracks, and mechanical damage can gradually grow, eventually leading to catastrophic failures if left unaddressed. Predicting the growth of these defects is essential for ensuring the safety, reliability, and efficiency of energy pipelines. Energy pipelines contribute significantly to global energy infrastructure, supporting economic development and meeting the growing demand for energy in diverse sectors. However, steel pipes used in these pipelines are susceptible to various defects, including corrosion, cracking, and mechanical damage. If these defects are not properly monitored and repaired, they can lead to public safety concerns and economic losses. Corrosion, as seen in Fig. 8.1, is a common cause of oil and gas line breakdowns (Soomro et al. 2024).

In recent years, pipeline operators have increasingly adopted reliability-based corrosion management programs (Kariyawasam and Peterson 2008; Nessim et al. 2009; Teixeira et al. 2008). These programs usually include three main parts: regular and detailed inspections to identify and measure corrosion defects in a specific pipeline, assessing the likelihood of pipeline failure based on the inspection findings, and repairing any defects if needed to maintain the failure probability of the pipeline below an allowable level for a reference segment of the pipeline (e.g., 1 km) and a specific time frame (e.g., 1 year) (Zhang et al. 2014). To address these concerns, pipeline integrity management programs have been developed to ensure the safe operation of pipelines. These programs coordinate procedures, instruments, and tasks for evaluating pipeline conditions, scheduling inspections, and planning maintenance work to reduce the risk of failure (Xie and Tian 2018).

M. Hussain and T. Zhang, *Big Data Analytics in Energy Pipeline Integrity Management*, Lecture Notes in Energy 104, https://doi.org/10.1007/978-981-96-8019-1_8

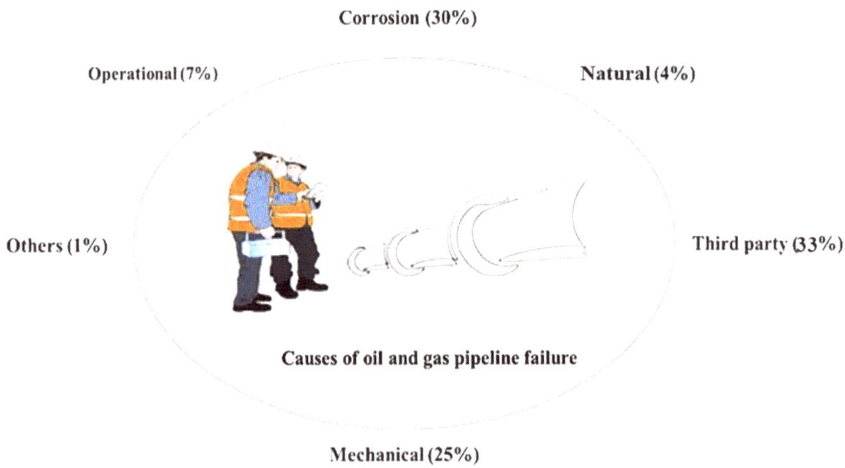

Fig. 8.1 Percentage breakdown of pipeline failure types (Soomro et al. 2022)

Pipeline integrity management generally encompasses three main components: defect detection and identification, defect growth prediction, and risk-based management (Yang et al. 2024). Non-destructive evaluation methods, such as in-line inspection (ILI), are widely employed for locating and identifying anomalies in pipelines. In the oil and gas industry, in-line inspection (ILI) plays a crucial role in the monitoring of pipeline integrity. Periodic in-line inspection (ILI) should be conducted to detect corrosion, which can lead to pipe leakage and failure. The proactive management technique depends on the capacity to forecast the rate at which corrosion will grow over time. In order to assess the rate of corrosion growth, it is necessary to compare the corroded sections of the pipeline with the extracted characteristics from the data obtained from the ILI conducted at various intervals (Liu et al. 2019). Common ILI techniques include Magnetic Flux Leakage (MFL) and ultrasonic tools for corrosion inspection of steel pipes. However, different ILI tools exhibit varying capabilities in identifying corrosion features. Some tools excel in detecting unique geometries like corrosion pits, axial grooving, and general corrosion. On average, ILI tools have an accuracy within $\pm$ 10% of the pipe wall thickness (Vanaei et al. 2017). To predict defect growth and time to failure, ILI needs to be performed periodically, and defects from at least two inspections should be matched to their positions in the pipeline. However, inconsistent coordinate systems used by different ILI tools can lead to unmatched data and inaccuracies (Cui and Wang 2023). Corrosion poses a significant threat to pipeline integrity, and extracting useful information from ILI data on corrosion features is crucial. Corrosion defects can be categorized as single defects or interacting multiple defects. Analyzing interactions between multiple corrosion defects is more complex but essential, as studies have shown that these interactions significantly influence the failure pressure of pipelines (Chiodo and Ruggieri 2009).

Therefore, assessing interacting corrosion defects is imperative for accurate pipeline integrity management.

In addition to considering interacting effects, establishing an appropriate growth model is vital for predicting defect evolution and scheduling future inspection and maintenance activities. Two categories of defect growth models exist: model-based approaches and data-driven approaches. Model-based approaches rely on physical models like finite element models, while data-driven approaches use ILI or sample data to investigate defect propagation. Some researchers have used the Bayesian networks, Markov chains, Weibull distribution, Gumbel distribution to estimate the time dependence of corrosion growth rate of materials (Cui and Wang 2023; Arzaghi et al. 2018).

Data from the multiple years of GIS and ILI inspections are important for understanding the condition of pipelines and achieving pipeline integrity. It is the objective of this study to put forward an entire workflow of analyzing both raw and filtered ILI data. The approach includes graphical depiction for quick assessment of number of corrosions and corrosion depth distributions and for studying the combined effects of adjacent corrosions to deduce the relationship between the defect location and severity. It also includes developing models to predict the growth of corrosion defect depth and density of corrosion defects.

The purpose of this chapter is to present a comprehensive procedure for defect growth prediction using ILI data both in the raw and filtered form. The method includes evaluating the viability of the mathematical models in predicting the growth of corrosion defects in various oil and gas pipes. The primary motivation of this work is to validate the effectiveness of machine learning/artificial intelligence models in prediction of the growth of individual defects using ILI data. It is important to stress that the performance of the model largely depends on the use and understanding of ILI data; timely and informative information is crucial to increasing the effectiveness of these models.

8.1.1 Background

Mitigating corrosion poses a significant challenge for oil and gas companies due to the intricate nature of corrosion defect initiation, stabilization, and progression. To deal with this complex issue, most of the oil and gas companies are working towards sophisticated architectures with intelligent capabilities to estimate the growth of corrosion defects (Ossai 2019). The progression of corrosion in natural gas pipeline systems is characterized as a Markov process by Liu et al. (2018) to support a maintenance optimization technique proposed for pipeline systems affected by external corrosion. The proposed multi-level optimization strategy can further reduce the total cost of natural gas pipeline systems that are susceptible to external corrosion, in comparison to previous methods (Liu et al. 2018). Similarly, Amaya-Gómez et al. (2016) utilize data acquired from ILI to enhance risk-based decision making. They suggested a method that utilizes corrosion growth models to estimate the remaining

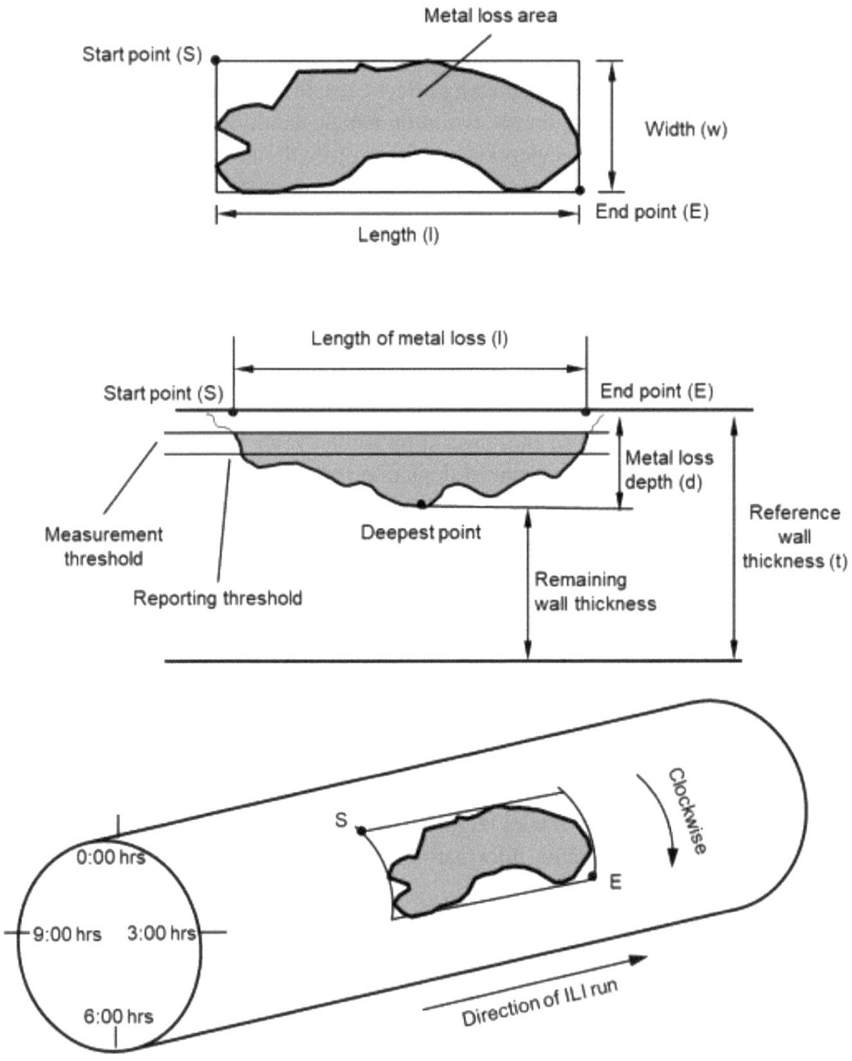

Fig. 8.2 Details of parameters describing location and dimension of metal loss anomaly

lifespan of the pipeline, relying on mechanistic models. The characteristics of anomalies are specified by parameters such as length (l), width (w), and depth (d). Figure 8.2 provides a visual representation of these parameters, defining the starting point (S) and the dimensions of the anomaly along the in-line inspection (ILI) run direction (POF 2021). In this context, the anomaly is encapsulated within a rectangle, with the starting and ending points diagonally located within this boundary. The depth parameter signifies the deepest point identified within the confines of this rectangular representation.

The researchers assessed the reliability of pipelines at various time intervals following exposure to corrosion, calculating the corrosion growth rate over time. Some of the authors (Arzaghi et al. 2018; Cai et al. 1999; Abbas et al. 2018) employed the Markov decision process as a continuous-time, non-homogeneous linear growth pure birth model to estimate the time-dependent corrosion growth. This approach facilitated the evaluation of corrosion defect depth through the utilization of multi-regression and power law models, and methodologies previously employed by other experts in corrosion studies (Caleyo et al. 2009; Ossai et al. 2016).

8.2 In-Line Inspection Technologies

In-Line Inspection (ILI) technologies play a pivotal role in maintaining the integrity of pipelines by enabling the detection and assessment of defects without disrupting regular operations. These technologies encompass various methods, each with its unique approach to identifying flaws such as corrosion, cracks, and deformities within pipeline structures. Magnetic Flux Leakage (MFL), Ultrasonic Testing, Eddy Current Testing, Radiographic Testing, and Acoustic Emission Testing are prominent ILI technologies. The effectiveness of these techniques is contingent upon the geometry of metal loss anomalies. To facilitate anomaly reporting, distinct classes of metal loss anomalies have been delineated, as illustrated in Fig. 8.3 (Ismail et al. 2023).

These defined anomaly classes contribute to a comprehensive understanding of measurement capabilities and enhance the precision of anomaly characterization using the MFL technique. These tools offer diverse capabilities, ranging from high sensitivity to surface defects and real-time monitoring the ability to provide internal views of pipelines. Some of the details along with their merits and demerits are given in Tables 8.1 and 8.2.

8.2.1 Types of ILI Technologies

There are several types of ILI tools that are commonly used in the oil and gas industry, each designed to detect specific types of defects. The most widely used ILI technologies include:

- **Magnetic Flux Leakage (MFL)**: MFL tools use magnetic fields to detect metal loss in pipelines. When an MFL tool passes through a pipeline, it magnetizes the metal. If there is a defect (such as corrosion or pitting), the magnetic field will "leak" around the defect, creating a measurable signal. MFL tools are highly effective at detecting corrosion and wall thinning.
- **Ultrasonic Testing (UT)**: Ultrasonic ILI tools send high-frequency sound waves into the pipeline wall and measure the time it takes to bounce back. Variations in the return signal indicate the presence of defects such as cracks, weld defects, or

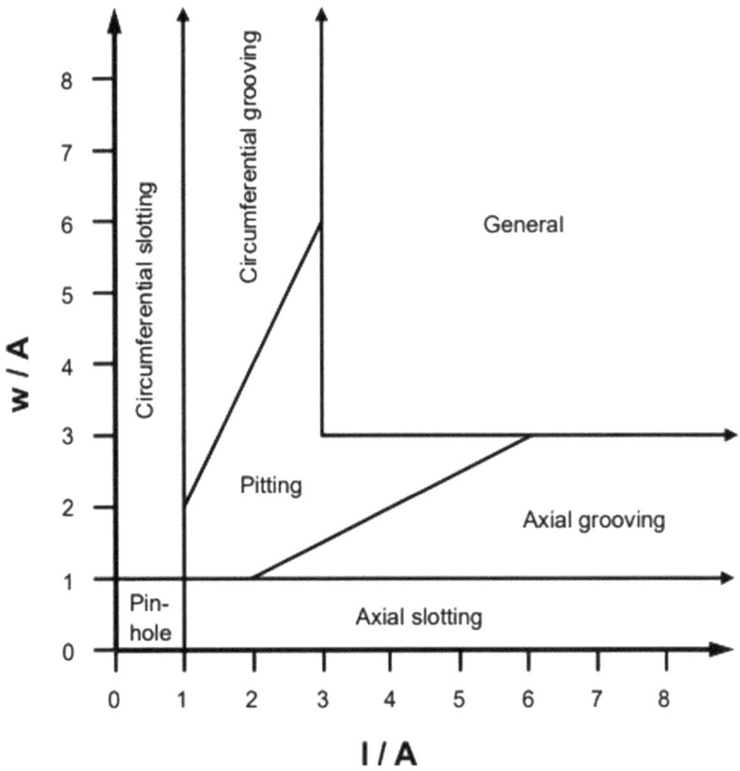

Fig. 8.3 Graphical presentation of surface dimensions of metal loss anomalies

metal loss. UT tools are particularly useful for detecting and measuring the depth of cracks and corrosion.

- **Electromagnetic Acoustic Transducer (EMAT)**: EMAT tools use electromagnetic waves to detect cracks and other stress-related defects. They are often used in conjunction with other ILI technologies to provide a more comprehensive assessment of pipeline conditions.

8.2.2 Data Collected by ILI Tools

ILI tools collect a wide range of data about the condition of the pipeline, including:

- **Wall Thickness**: Measurements of the pipeline's wall thickness provide critical insights into the extent of corrosion or mechanical wear.
- **Corrosion Depth and Length**: ILI tools can measure the depth and length of corrosion defects, allowing operators to assess the severity of corrosion in different pipeline sections.

Table 8.1 In-line inspection tools and details

Technology name	Details	Merits	Demerits
Magnetic flux leakage (MFL)	Utilizes magnetic fields to detect corrosion and defects	• High sensitivity to surface and near-surface defects • Suitable for ferromagnetic materials	• Limited penetration depth • Challenging for non-ferromagnetic materials
Ultrasonic testing (UT)	Sends ultrasonic waves to detect defects and measure thickness (Zang et al. 2023)	• Provides detailed information on defect geometry • Suitable for various materials and thicknesses	• Requires direct contact with the surface • Limited to accessible areas
Eddy Current Testing (ET)	Uses induced eddy currents to detect surface and subsurface flaws	• Sensitive to small cracks and defects • Suitable for non-conductive materials	• Limited penetration depth • Complexity in data interpretation
Electromagnetic acoustic transducer (EMAT)	Generates ultrasonic waves electromagnetically without direct contact	• No need for direct contact with the material • Suitable for various materials and coatings	• Limited sensitivity compared to traditional ultrasonic methods • Initial setup and calibration complexity

Table 8.2 Comparison of ILI technologies

ILI tool	Cracks	Metal losses	Metallurgical changes	Geometry changes
UT	Y	Y	N	N
EMAT	Y	N	N	N
MFL	N	Y	Y	S
ET	Y	N	S	N

S: Some types of the tool can detect this type of flaw

- **Crack Dimensions**: Tools such as EMAT and UT can detect cracks in the pipeline and provide detailed information about their location, length, and depth. This information is essential for understanding the risk of crack propagation.
- **Anomalies in Welds and Joints**: ILI tools can detect defects in pipeline welds and joints, which are common points of failure in pipeline systems.

8.2.3 Importance of ILI Data for Degradation Modeling

The data collected by ILI tools is essential for developing accurate **degradation models**. These models use historical data on defect growth, combined with real-time monitoring data, to predict how quickly defects will grow and when they are likely to reach critical levels. By incorporating ILI data into degradation models, operators can prioritize maintenance activities for the segments of the pipeline that are most at risk.

The primary benefits of using ILI data in degradation modeling include:

- **Early Detection**: ILI tools allow operators to detect defects before they reach dangerous levels, providing time to plan and execute repairs.
- **Accurate Defect Sizing**: The detailed measurements provided by ILI tools allow for more accurate predictions of defect growth, improving the reliability of the degradation models.
- **Improved Maintenance Planning**: By predicting when defects are likely to grow to a critical size, operators can optimize maintenance schedules and reduce the likelihood of unplanned shutdowns.

8.3 Data Analytics in Anomaly Detection and Growth Predictions

Data analytics helps pipeline operators in ensuring the integrity of pipelines by facilitating anomaly detection and growth prediction (Cui and Wang 2023). Anomaly detection is a key aspect of pipeline integrity management, as it helps identify deviations from normal operating conditions that may indicate defects or potential issues. Figure 8.4 highlights the key components of the corrosion prediction process (Xu et al. 2023). Through the analysis of various data sources such as pressure, temperature, flow rates, and corrosion rates, data analytics tools can pinpoint anomalies that may signify structural weaknesses, leaks, or other integrity threats. By leveraging advanced algorithms and machine learning techniques, these tools can continuously monitor vast amounts of data, enabling the early identification of anomalies and potential defects before they escalate into more significant problems (Bhambri and Rani 2024).

In addition to anomaly detection, data analytics contributes significantly to the prediction of growth patterns in pipeline defects. Understanding how defects evolve over time is crucial for proactive maintenance and risk mitigation (Habbal et al. 2024). Data analytics algorithms can analyze historical data on defect growth, considering factors such as material properties, environmental conditions, and process parameters. This analysis enables the development of predictive models that forecast the potential growth of defects, allowing pipeline operators to prioritize and plan maintenance activities effectively (Friederich and Lazarova-Molnar 2024). By adopting data analytics, pipeline integrity management reduces unwanted breakdowns and can

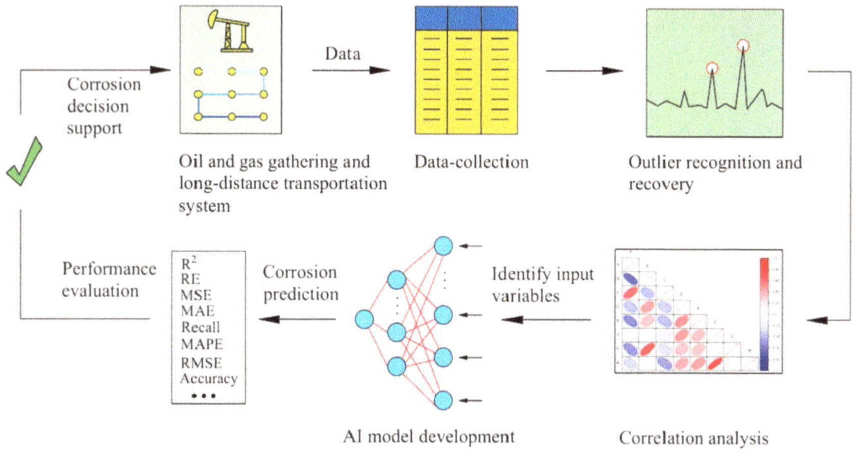

Fig. 8.4 ML-based value-creating analytical framework developed for corrosion prediction (Xu et al. 2023)

move from a reactive maintenance to a proactive and predictive maintenance, which will assist to reduce the likelihood of unplanned downtimes, unwanted failures and optimizing maintenance resources.

Advanced algorithms can swiftly identify irregularities, such as corrosion spots or structural anomalies, in the vast datasets collected during in-line inspections. Machine learning algorithms, statistical modeling, and data visualization tools empower pipeline operators to uncover hidden correlations, predict defect growth rates, and assess the likelihood of pipeline failures. These insights enable proactive maintenance strategies, optimizing resource allocation and preventing costly incidents (Vishnuvardhan et al. 2023).

The application of data analytics in pipeline integrity extends beyond defect detection and prediction (Wang et al. 2023). It also encompasses risk assessment, anomaly identification, and optimization of inspection schedules. By leveraging data-driven insights, pipeline operators can prioritize inspection efforts on high-risk segments, reducing the likelihood of undetected defects and ensuring the integrity of their pipelines.

The primary elements contributing to corrosion in oil and gas pipeline systems are the selected material, the nature of the flowing media, and the ambient conditions. The corrosion data gathered exhibits properties such as multidimensionality, noise interference, non-linearity, and so on. The considerations contribute to the intricacy of predicting corrosion in oil and gas pipelines. Corrosion prediction in numerous industries has been facilitated through thorough inquiry and analysis, employing diverse methodologies. Commonly employed corrosion prediction techniques encompass physical, statistical, and machine learning methodologies.

Figure 8.5 illustrates the initial implementation period and the concise chronological progression of the three techniques, namely, physical, statistical, and machine

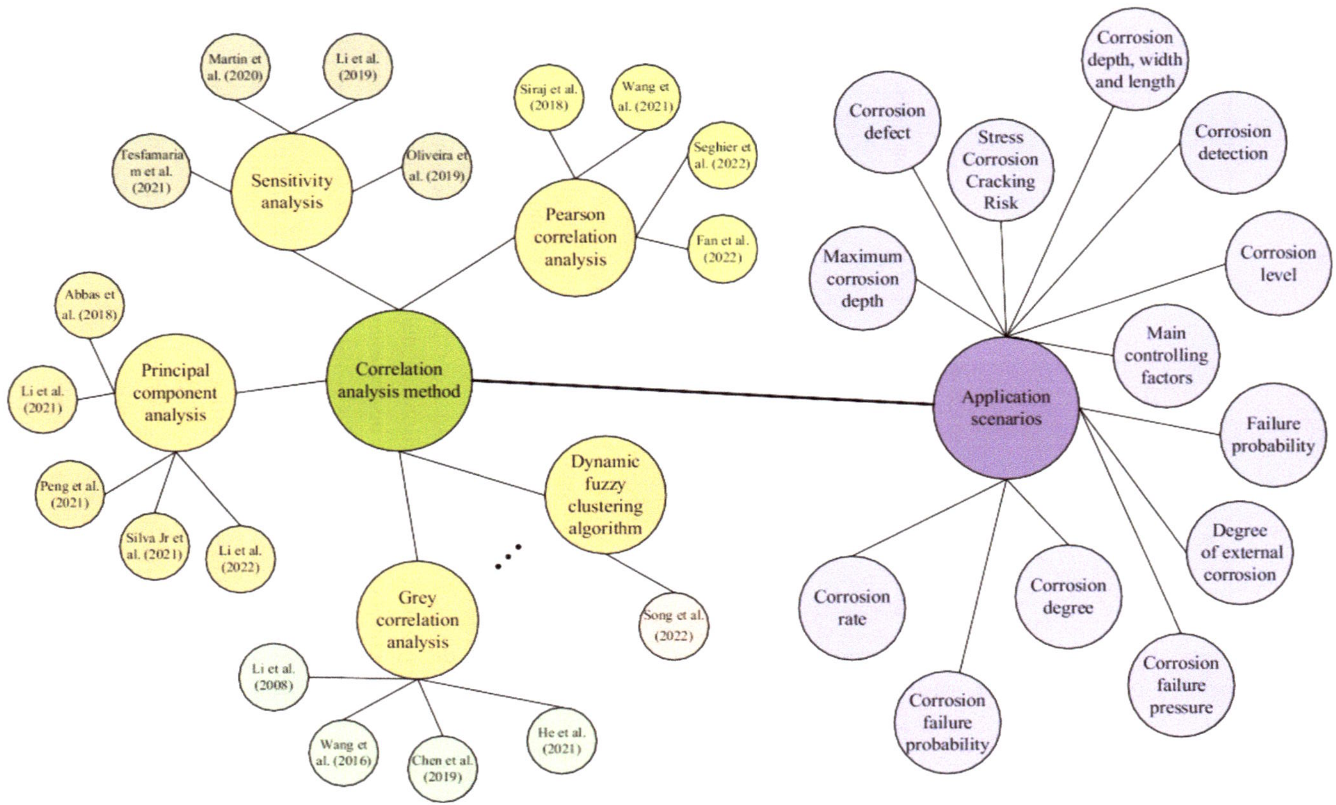

Fig. 8.5 Statistics of different correlation analysis methods (corrosion application)

learning (Xu et al. 2023). This figure illustrates different correlation analysis methods and their application scenarios in corrosion-related studies. It is divided into two main parts, Correlation Analysis Methods (Center—Green Node) and Application Scenarios (Purple Node).

Correlation Analysis Methods, this is the main category and includes different statistical techniques used for analyzing correlations in corrosion studies. A brief detail about the Subcategories of Correlation Analysis Methods is given below.

- **Sensitivity Analysis (Yellow Node)**
  - Involves identifying how different input variables impact an output.
- **Pearson Correlation Analysis (Light Yellow Node)**
  - Measures the strength of the linear relationship between two variables.
- **Principal Component Analysis (PCA) (Yellow-Orange Node)**
  - Reduces the dimensionality of data while preserving variance.
- **Grey Correlation Analysis (Yellow-Orange Node)**
  - Assesses the similarity of different sequences to analyze trends and correlations.
- **Dynamic Fuzzy Clustering Algorithm (Light Orange Node)**
  - Uses fuzzy logic and clustering to analyze patterns.

The second part, **Application Scenarios (Purple Node)**, presents where these correlation analysis methods are applied, mainly in corrosion risk assessment. The Subcategories of Application Scenarios are,

- Stress Corrosion Cracking Risk
- Corrosion Detection
- Corrosion Level
- Failure Probability
- Degree of External Corrosion
- Corrosion Failure Pressure
- Corrosion Rate
- Corrosion Degree
- Maximum Corrosion Depth
- Main Controlling Factors
- Corrosion Defect
- Corrosion Depth, Width, and Length.

Some of the decision-making models that are commonly employed for corrosion prediction are discussed below.

8.4 Degradation Modeling in Pipeline Integrity Management

Degradation modeling involves the use of mathematical models to estimate the progression of defects in pipelines over time (Gorjian et al. 2009). These models allow pipeline operators to predict the future state of defects, such as corrosion, cracks, or mechanical damage, and estimate the time until a defect reaches a critical size that could result in pipeline failure.

The primary goal of degradation modeling is to enable **predictive maintenance**—the ability to schedule repairs before defects grow large enough to pose a risk to pipeline integrity. By understanding how defects grow, operators can avoid costly emergency repairs and unplanned downtime.

8.4.1 Key Parameters in Degradation Models

Degradation models use several key parameters to predict the growth of defects, including:

- **Initial Defect Size**: The size of the defect at the time it is first detected. For example, the depth and length of a corrosion pit or the length of a crack.
- **Growth Rate**: The rate at which the defect grows over time, often influenced by environmental factors (e.g., soil composition, humidity) and operating conditions (e.g., pressure, temperature).
- **Pipeline Material Properties**: A pipeline material's strength and durability determine how resistant the pipeline is to defect propagation.
- **Operating Conditions**: Factors such as **pressure**, **temperature**, and **flow rate** that influence the stress experienced by the pipeline affect the defect growth.

8.4.2 Types of Degradation Models

There are several types of **degradation models** used to predict defect growth in pipelines. Each model has its strengths and weaknesses, and the choice of model depends on the type of defect, the available data, and the operating conditions of the pipeline.

8.4.2.1 Deterministic Linear Growth Model

The **Deterministic Linear Growth Model** assumes that defect growth occurs at a constant rate over time. This model is relatively simple and is commonly used for defects such as **uniform corrosion**, where the metal loss occurs at a steady rate.

Key Assumptions:

- The growth rate of the defect is constant and does not change over time.
- The growth rate can be estimated based on historical data or expert knowledge of the pipeline's operating conditions.

The equation for a deterministic linear growth model is:

$$d(t) = d_0 + r \cdot t$$

where:

- $d(t)$ is the size of the defect at time t,
- d_0 is the initial size of the defect (e.g., depth of corrosion at the time of inspection),
- r is the constant growth rate of the defect.

The deterministic linear model is often used for predicting the progression of **uniform corrosion** in pipelines. For example, if an ILI tool detects that the wall thickness of a pipeline segment has decreased by 1 mm due to corrosion, and the historical growth rate of corrosion in that area is 0.1 mm per year, the model can predict that the pipeline wall thickness will be reduced by another 0.5 mm in five years.

Advantages:

- **Simplicity**: The linear growth model is straightforward to implement and requires minimal data.
- **Predictability**: The model provides a clear, easy-to-understand prediction of defect growth, which is useful for planning maintenance activities.

Challenges:

- **Assumption of Constant Growth**: The main limitation of the linear growth model is that it assumes a constant growth rate, which may not be realistic for all types of defects. In reality, factors such as pressure fluctuations, temperature changes, or environmental conditions can cause the growth rate to vary over time.

8.4.2.2 Stochastic Modeling for Defect Growth

Unlike deterministic models, **stochastic models** account for the uncertainty and variability in defect growth. These models use probability distributions to model the random nature of defect progression, making them particularly useful for predicting the growth of defects like **pitting corrosion** or **cracks**, where growth rates can vary significantly.

Key Assumptions

- Defect growth is influenced by random factors, such as fluctuations in operating conditions or environmental variability.

- The growth rate is not constant but follows a probability distribution.

Types of Stochastic Models

- **Gamma Process Models**: The **gamma process** is commonly used to model cumulative damage processes, such as corrosion, where small increments of degradation occur over time. The Gamma process assumes that the amount of degradation follows a Gamma distribution, with each increment of degradation being positive and non-decreasing.
- **Markov Process Models**: In **Markov models**, the future state of the defect (e.g., its size) depends only on its current state and not on its past history. Markov models are useful for modeling the transition of defects between different states, such as from a "minor defect" to a "severe defect."

Application in Pipeline Integrity Management

Stochastic models are often used to predict the growth of defects where there is significant uncertainty, such as **pitting corrosion** or **stress corrosion cracking (SCC)**. For example, a Gamma process model might be used to predict the depth of a corrosion pit over time, accounting for the fact that the growth rate may vary depending on external factors like soil moisture or the presence of corrosive chemicals.

8.4.2.3 Bayesian Decision Networks (BDNs)

Bayesian Decision Networks (BDNs) can combine degradation data with expert knowledge to assess the risk of a failure. BDNs can incorporate uncertainty in the data and allow operators to update their predictions as new data becomes available. This makes BDNs particularly useful for real-time decision-making in pipeline integrity management.

8.4.2.4 Dynamic Linear Models (DLMs)

Dynamic Linear Models (DLMs) are a type of probabilistic model used to predict defect growth in pipelines (Zhang and Zhou 2014). DLMs account for changes in operating conditions over time, allowing the model to adapt as new data is collected. For example, a DLM might be used to predict how the depth of a corrosion pit will change based on fluctuations in pressure and temperature.

8.4.2.5 Gamma Process for Cumulative Damage

The **Gamma process** is a stochastic model commonly used to model **cumulative damage** in pipelines, such as corrosion or material fatigue (Han et al. 2023). It assumes that degradation occurs in small increments over time and follows a Gamma distribution. This model is particularly useful for predicting the progression of defects that grow slowly and steadily, such as **general corrosion**.

Key Features of the Gamma Process

- **Non-Negative Increments**: The Gamma process assumes that each increment of degradation is non-negative, meaning that defects cannot shrink or heal themselves over time.
- **Cumulative Damage**: The model accumulates damage over time, allowing operators to estimate when a defect will reach a critical size.

8.5 Machine Learning in Pipeline Integrity Management

As pipelines generate vast amounts of data from **in-line inspection (ILI) tools**, **SCADA systems**, and **sensor networks**, **machine learning (ML)** offers a powerful way to analyze this data and predict defect growth (Han et al. 2023). Unlike traditional degradation models, which are often based on predefined mathematical relationships, machine learning models can learn complex patterns in the data, making them particularly effective at forecasting defect progression under variable conditions.

Decision-making in corrosion prediction often involves machine learning models such as Support Vector Machines (SVM), Random Forests, and Neural Networks (Soomro et al. 2024). These models leverage historical corrosion data to identify patterns and make predictions about future corrosion behavior (Hakimian et al. 2023). Machine learning models are adept at handling large and complex datasets, contributing to more accurate and data-driven decision-making. Current research in the prediction of pipeline defects suggests that machine learning regression, particularly employing artificial neural networks (ANN), has yielded the most promising outcomes (Aldosari et al. 2020). Additionally, the application of a support vector machine (SVM) for the three-dimensional (3D) reconstruction of defects has shown promising results (Piao et al. 2019).

8.6 Integration of Machine Learning with Degradation Models

Integrating machine learning models with traditional **degradation models** provides a hybrid approach that leverages the strengths of both methods. Machine learning models can enhance degradation models by providing more accurate predictions based on real-time data, while traditional models offer the benefit of well-established mathematical frameworks for predicting defect growth.

For example, a pipeline operator might use a **Gamma process model** to model the cumulative growth of corrosion and combine it with a **Random Forest** model trained on historical data to predict future corrosion rates. This hybrid approach provides a more comprehensive view of defect growth, allowing operators to make more informed decisions about maintenance and repair schedules.

The earlier discussion in this chapter focuses on degradation models for pipeline condition assessment, which involve several models review with the aim to evaluate and manage the deterioration of pipelines. The authors of this work discussed these models and its association with pipeline condition assessment. These models utilize probabilistic modeling, machine learning, and risk analysis techniques to help prioritize maintenance and repair activities, optimize resource allocation, and ultimately improve the reliability and lifespan of pipeline networks. Table 8.3 shows summary of available models on pipeline condition assessment.

8.7 Application Framework of Machine Learning Methods

As oil and gas pipeline systems advance into intelligent networks, the integration of corrosion-related intelligent electronic devices is becoming increasingly important (Xu et al. 2023). This integration aims to provide stable system operation, economic dependability, and optimal management of the pipeline network. The pipeline network has evolved into an extensive and intricate system, encompassing a diverse array of equipment and a wide range of technology (Aalsalem et al. 2018). Machine learning relies less on specific mathematical models and excels at handling complex problems involving high dimensions, time variations, and non-linear relationships (Dhoni 2023). It is particularly effective at analyzing corrosion data and revealing valuable insights that may otherwise remain hidden. Therefore, machine learning offers a promising solution to overcome technical challenges. Integrating machine learning with oil and gas pipeline systems will gradually combine intelligent sensing and physical state, data-driven and simulation models, and auxiliary decision-making and operational control. This integration will effectively enhance the ability to manage complex systems and improve the safety and efficiency of oil and gas pipeline operations. The primary applications of corrosion prediction in oil and gas pipeline systems are illustrated in Fig. 8.6 (Xu et al. 2023).

These applications primarily involve determining the corrosion rate of the pipeline, identifying leaks, assessing defects, detecting damage, diagnosing faults, studying pipeline erosion, researching corrosion-resistant materials, identifying corrosion images, and constructing corrosion maps. The study of corrosion image recognition and corrosion map creation has mostly been focused on in recent years, with only a limited number of application instances. However, their future development prospects are highly promising.

Figure 8.7 shows an architectural framework for implementing machine learning in the domain of corrosion detection and prediction in oil and gas pipeline systems Xu et al. (2023). This framework encompasses a high-performance computing infrastructure, data aggregation, an algorithm engine, and several business scenarios.

Table 8.3 Summary of available models

Model type	Application to individual defect	Application to population-based defects	Condition to apply	Advantage	Disadvantage
Common Statistical model	No	Yes	At least one ILI measurement data set	Population-based defect estimation or prediction	Not accurate for individual defect growth
Bayesian model	No	Yes	At least two ILI measurement data sets	Population-based defect estimation or prediction	Not accurate for individual defect growth
Gamma process model	Yes	Yes	At least one ILI measurement data set	Population-based defect estimation or prediction	Computationally and data-intensive
Deterministic linear growth model	Yes	No	At least two ILI measurement data set	Simple and easy to interpret	May not capture complex defect growth patterns and uncertainties in the model
Markov decision process	Yes	Yes	At least one ILI measurement data set	Population-based defect estimation or prediction	Inaccurate predictions due to complexity
Dynamic linear model	No	Yes	Multiple ILI datasets are required	Population-based defect estimation or prediction	Not accurate for individual defect growth
Neural network (NN) modeling	Yes	Yes	Multiple ILI measurement data	Population-based defect estimation or prediction	Required multiple datasets to predict anomaly and defect growth
Support vector machines (SVM)	Yes	Yes	Multiple ILI measurement data	Population-based defect estimation or prediction	Required multiple datasets to predict anomaly and defect growth
Random forest	Yes	Yes	Multiple ILI measurement data	Population-based defect estimation or prediction	Required multiple datasets to predict anomaly and defect growth
Gradient boosting	Yes	Yes	Multiple ILI measurement data	Population-based defect estimation or prediction	Required multiple datasets to predict anomaly and defect growth

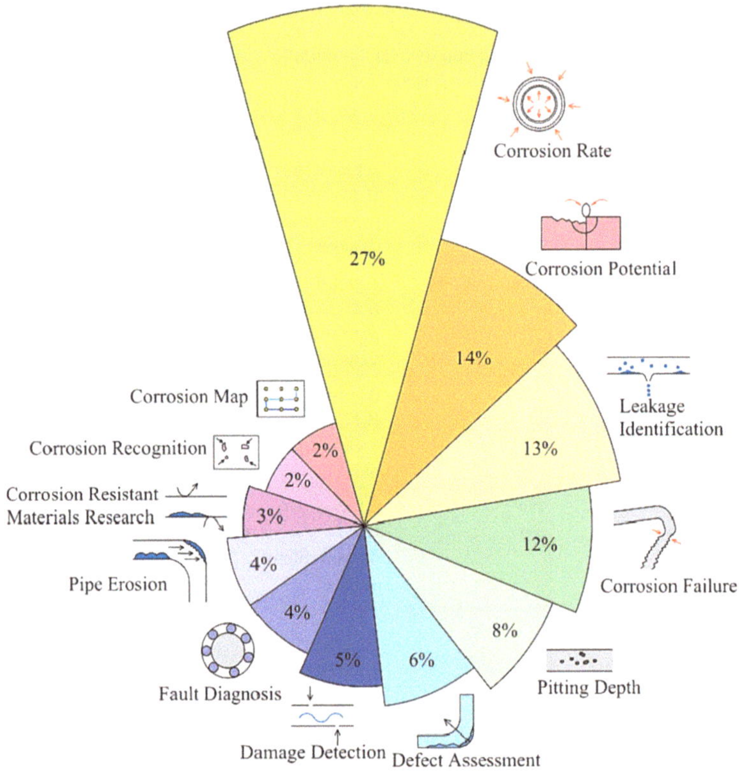

Fig. 8.6 Main application scenarios of machine learning in the field (Xu et al. 2023)

8.8 Methodology and Analysis

This section aims to investigate the growth of corrosion defects in depth in pipelines based on historical operating parametric values and measured historical defect data. It is to demonstrate the instantaneous and time-dependent corrosion defect depth growth. This approach facilitates the estimation of failure probabilities, a critical factor for corrosion risk assessment, as well as informing inspection and maintenance planning strategies.

8.8.1 ILI Data Description and Preparation

The in-line inspection (ILI) dataset utilized in this study was derived from Magnetic Flux Leakage tools employed on two different crude oil pipelines. One pipeline ILI data was collected in the years 2000, 2005, 2010, 2015 and 2020 and the second pipeline in 2003, 2008, 2012, 2016 and 2022. The material of construction for the

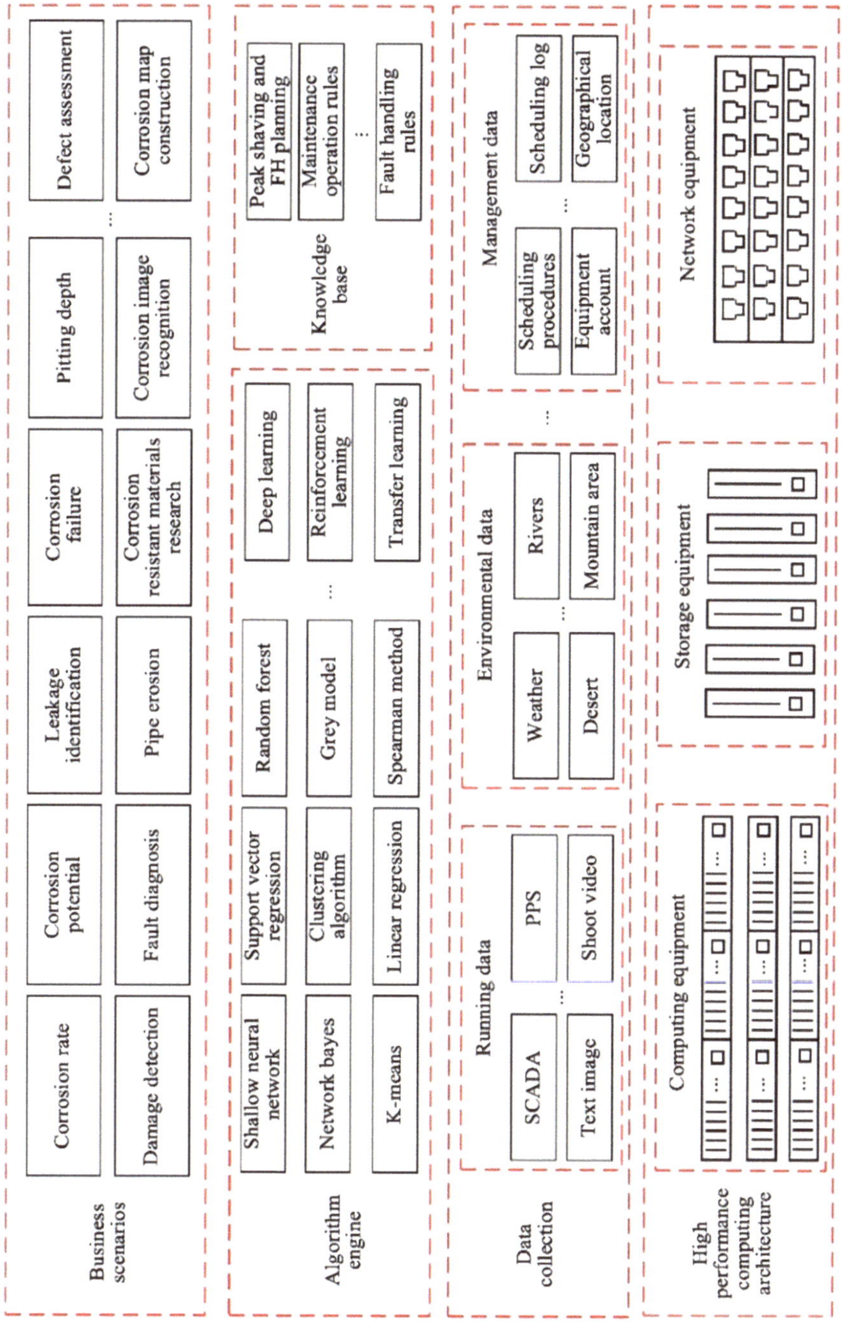

Fig. 8.7 Design framework for the use of machine learning in energy pipeline corrosion (Xu et al. 2023)

pipelines is a mild steel. The two pipelines were constructed in 1979 and 1986 respectively.

Segmentation of the first pipeline into distinct segments (1–10 segments, for example) was conducted, guided by the historical record of replacements and relocations, as delineated in Table 8.4. Segmentation of the second crude oil pipeline into distinct segments (21–25 segments, for example) was conducted, guided by the historical record of replacements and relocations, as delineated in Table 8.5.

Authors also tested another pipeline ILI data, which was analyzed by Zhang et al. (2013). This third pipeline was built in 1972, presently operational in Albert, Canada. The pipeline was examined using high-resolution magnetic flux leakage (MFL) equipment in the years 2000, 2004, 2007, and 2009 and found total 62 external corrosion defects.

This research work focused on all defects in the ILI dataset, collected from three different pipelines. The distribution of corrosion defects across each segment on both the pipelines, exhibited multiple pitting defects, refer to Tables 8.6 and 8.7 for details of number of defects. In Table 8.7, it is noticed that the total number of defects is reduced in the year 2022, compared to previous years. The reason is that a few sections of this pipeline had severe wall losses, and more than 70% of them were replaced during 2021. After repair, the pipeline owner arranged another

Table 8.4 Pipeline 1 (10 segments only for illustration)

Pipeline segment no	Length (meter)	Wall thickness (mm)	Pipe grade	Year commissioned
1	4	9.5	X52	1979
2	11	9.5	X52	1979
3	5	9.5	X52	1979
4	9	9.5	X52	1979
5	13	9.5	X52	1979
6	37	9.5	X52	1979
7	21	9.5	X52	1979
8	8	9.5	X52	1979
9	43	9.5	X52	1979
10	9	9.5	X52	1979

Table 8.5 Pipeline 2 (5 segments only for illustration)

Pipeline segment no	Length (meter)	Wall thickness (mm)	Pipe grade	Year commissioned
21	3	9.5	X52	1986
22	22	9.5	X52	1986
23	12	9.5	X52	1986
24	19	9.5	X52	1986
25	11	9.5	X52	1986

Table 8.6 No of defects on pipeline 1

Inspection year	No of defects
2000	791
2005	1528
2010	1791
2015	2339
2020	3441

Table 8.7 No of defects on pipeline 2

Inspection year	No of defects
2003	921
2008	1791
2012	2691
2016	3341
2022	1545

ILI to develop a baseline inspection for this pipeline and it is noticed that after repair, the total number of defects reduced, making the pipeline more reliable for safe operations.

8.8.2 Data Preprocessing

Data preprocessing techniques used in corrosion prediction primarily encompass tasks such as aligning internal inspection data, normalizing data, identifying, and restoring anomalous data, and so on. Internal inspection data alignment is a crucial need for assessing the development of pipeline problems. To provide an accurate assessment of the defect, it is important to establish a correlation between the data (Zhang et al. 2020). Data normalization aims to mitigate the impact of varying index dimensions, enhance data comparability, prevent inconsistencies in gradient descent, and substantially enhance the training efficiency of the model (Xu et al. 2023). Within the In-Line Inspection (ILI) dataset, various inspection locations encompassed normal points and manufactural bends, in addition to pitting corrosions. To ensure precision in subsequent analyses, the initial step involved filtering and isolating data points related to pitting corrosions.

During the corrosion data collection process, the absence or abnormality of ILI data from the actual field site may occur due to equipment failure in monitoring and detection, errors in data storage, data entry, data storage, and other issues. This can result in the omission of crucial corrosion information. Therefore, it is imperative to conduct research on identifying and restoring anomalous ILI data to ensure that the acquired corrosion data accurately represents the true characteristics of corrosion.

Despite the initial filtering, the filtered data needed systematic arrangement to facilitate the establishment of growth models. In the case of the corrosion growth density model, data points that deviated from the mean value by more than 3 times of the standard deviation were systematically determined as outliers. These outliers were then studied separately to verify if an outlier was a true outlier.

8.9 Model Selection

The authors of this research work tested different models and have chosen below model types for the analysis. In this work, Exponential, Logarithmic, Polynomial (Order 2), Liner and Power Law models with two or three parameters are selected for testing. Other models that use machine learning techniques require more parameters, which make them complex and difficult for pipeline operators to apply in real-world settings due to a lack of training data required.

8.9.1 Exponential Model

An exponential model is a mathematical model that describes a relationship between a dependent variable (y) and an independent variable (x) where the rate of change of the dependent variable is proportional to its current value (Villadsen and Wulff 2021). This characteristic leads to unique behaviors like rapid growth or decay depending on the specific parameters. The general form of an exponential model is:

$$y = ae^{bx}$$

where:

a is the initial value of the dependent variable (y) at $x = 0$
b is the growth or decay factor, where $b > 1$ for growth and $0 < b < 1$ for decay
x is the independent variable.

8.9.2 Linear Model

A linear model is a fundamental concept in statistics and machine learning that represents the relationship between a dependent variable and one or more independent variables in a linear fashion (Dangeti 2017; James et al. 2023). The general form of a linear model for a single independent variable is expressed as:

$$y = mx + b$$

where

y is the dependent variable,
x is the independent variable,
m is the slope of the line, and
b is the y-intercept.

For multiple independent variables, the linear model can be extended as:

$$y = b_0 + b_1x_1 + b_2x_2 + \cdots + b_nx_n$$

where

y is the dependent variable,
$x_1, x_2, \ldots, x_n$ are the independent variables,
b_0 is the intercept term,
$b_1, b_2, \ldots, b_n$ are the coefficients.

Linear models provide a simple and interpretable approach for modeling relationships between variables. While they have limitations, they remain a valuable tool in various applications, especially when interpretability is crucial or when the relationships are predominantly linear.

8.9.3 *Logarithmic Model*

A logarithmic model is a mathematical model that represents the relationship between variables using logarithmic functions (Campo-Meneses et al. 2023). The general form of a logarithmic model is given by:

$$y = a \cdot \log_b(x) + c$$

where

y is the dependent variable,
x is the independent variable,
a is a constant that affects the vertical stretch or compression of the graph,
b is the base of the logarithm, and
c is a vertical shift.

Logarithmic models provide a flexible framework for capturing non-linear relationships between variables (Campo-Meneses et al. 2021). They are particularly useful in scenarios where the rate of change is not constant and may exhibit exponential or logarithmic growth or decay patterns.

8.9.4 Power Law Model

A Power Law model gives a mathematical relationship between two variables, where one variable's value is proportional to a power of the other (Clauset et al. 2009; Machado et al. 2015). Power Law model was applied in a few research papers to estimate of defect growth of pipelines (Al-Amin et al. 2014; Velázquez et al. 2022; Miran et al. 2016). The general form of a Power Law model is expressed as:

$$y = a \cdot x^b$$

where

y is the dependent variable,
x is the independent variable,
a is a scaling factor or normalization constant,
b is the exponent or scaling parameter.

Power Law models are powerful tools for describing the distributional characteristics of various natural and social phenomena. Their ability to capture scale invariance and self-similarity makes them valuable in understanding complex systems and networks.

8.9.5 Polynomial (Order 2) Model

A Polynomial (Order 2) model, often referred to as a quadratic model, is a mathematical representation of a relationship between variables using a second-degree polynomial equation (Makarenkov and Legendre 2002). The general form of a quadratic model is given by:

$$y = ax^2 + bx + c$$

where

y is the dependent variable,
x is the independent variable,
a, b, and c are coefficients with $a \neq 0$,
ax^2 represents the quadratic term,
bx represents the linear term, and
c is the constant term or the y-intercept.

Polynomial (Order 2) models provide a versatile and widely applicable tool for capturing nonlinear relationships between variables. Understanding the curvature, symmetry, and turning points of quadratic functions is crucial for interpreting and applying these models effectively.

8.10 Data Analysis and Interpretation

We collected ILI data from two (02) different pipeline operators and one research paper for analysis. The first pipeline, named "Pipeline 1", has been operational since 1979, while the second pipeline, denoted "Pipeline 2", has been in operation since 1986. The third pipeline referred to as 'Pipeline 3' has been in operation since 1972. The ILI data samples obtained from Pipeline 1 total approximately 13,000, while Pipeline 2 has around 11,500 samples. Following the removal of outliers from the ILI data, Pipeline 1 retained a total of 9890 ILI dataset samples, and Pipeline 2 retained 10,289 ILI dataset samples.

A total of 20 defects were selected for the analysis from each pipeline. Out of 20, 10 defects were selected randomly and 10 from a specific segment. Also, another 3 defects from Zhang et al. (2013) were analyzed. Table 8.8 shows an ILI dataset for one defect, which was randomly chosen and Table 8.9 shows the ILI data for a defect taken from a specific location (Segment 9) on 'Pipeline 1'. Table 8.10 shows data collected from Pipeline 3, 'Pipeline 3'.

Figures 8.8, 8.9, 8.10, 8.11, 8.12, 8.13, 8.14, 8.15, 8.16 and 8.17 depict wall loss (%) over the time (years) using different models for the randomly selected defect from Pipeline 1.

Table 8.11 shows R-square values for different models and model prediction accuracy for each model. A model prediction accuracy is evaluated using the relative error which is calculated by (the model predicted value at the next ILI – the measurement value at the next ILI)/the measurement value at the next ILI. For example, if a model

Table 8.8 ILI data for a randomly selected defect from Pipeline 1

Pipeline segment	Segment distance (meter)	Commissioning date	Time to operations (year)	Wall loss (%)
Random	2134.00	1979	22	10.00
Random	2134.00	1979	27	10.43
Random	2134.00	1979	32	11.50
Random	2134.00	1979	37	11.80
Random	2134.00	1979	42	11.89

Table 8.9 ILI data for a defect selected from Segment 9 of Pipeline 1

Pipeline segment	Segment distance (meter)	Commissioning date	Time to operations (year)	Wall loss (%)
Segment 9	115,942.72	1979	22	13.79
Segment 9	115,942.72	1979	27	13.95
Segment 9	115,942.72	1979	32	14.36
Segment 9	115,942.72	1979	37	14.49
Segment 9	115,942.72	1979	42	14.66

Table 8.10 ILI data for three defects selected from Pipeline 3

Pipeline segment	Segment distance (meter)	Commissioning date	Time to operations (year)	Wall loss (%)
Defect 13	X	1972	28	33
			32	38
			35	37
			37	41
Defect 15	X	1972	28	38
			32	41
			35	51
			37	52
Defect 19	X	1972	28	39
			32	55
			35	59
			37	57

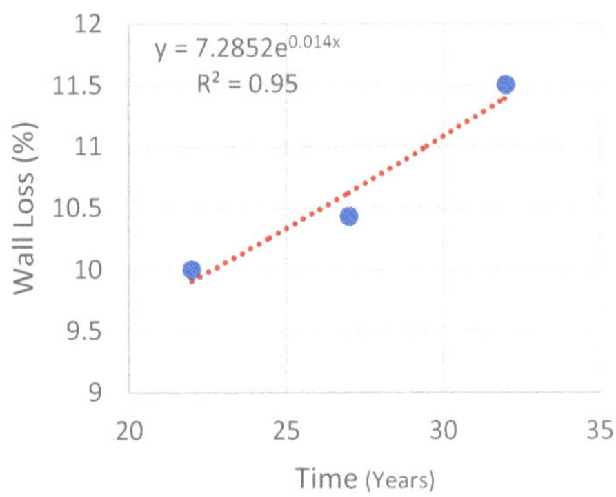

Fig. 8.8 Exponential model developed using the data collected from 3 ILI runs for the randomly selected defect from Pipeline 1

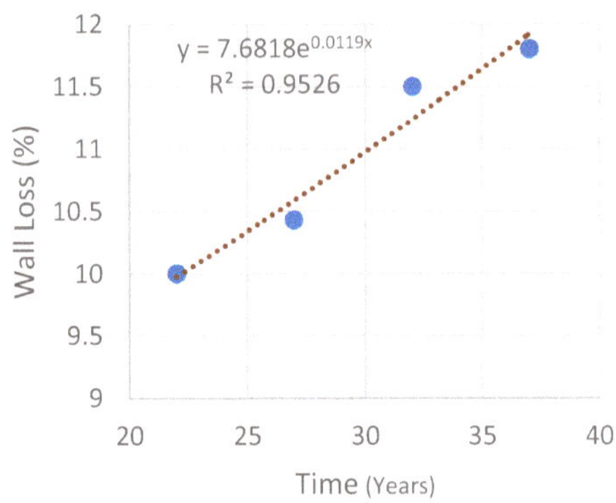

Fig. 8.9 Exponential model developed using the data collected from 4 ILI runs for the randomly selected defect from Pipeline 1

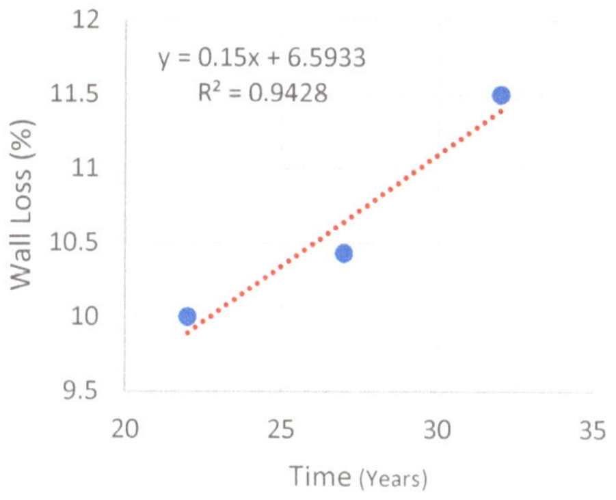

Fig. 8.10 Linear model developed using the data collected from 3 ILI runs for the randomly selected defect from Pipeline 1

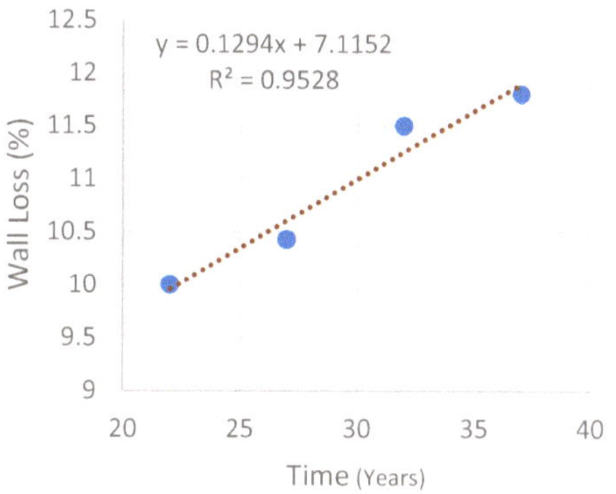

Fig. 8.11 Linear model developed using the data collected from 4 ILI runs for the randomly selected defect from Pipeline 1

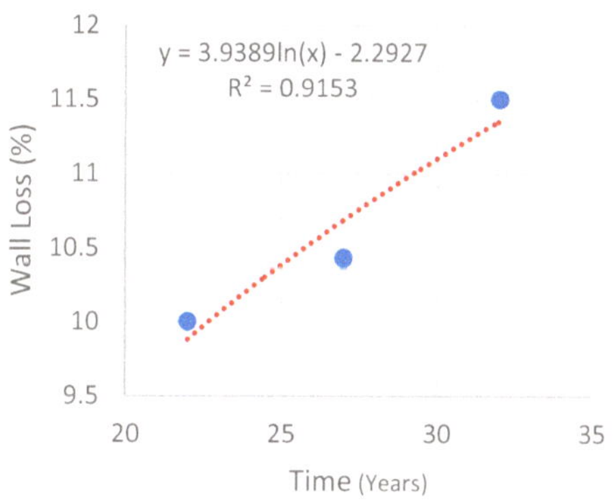

Fig. 8.12 Logarithmic model developed using the data collected from 3 ILI runs for the randomly selected defect from Pipeline 1

is built using the first 3 ILI data (i.e., the data measured in 2000, 2005, and 2010), it is used to predict the defect growth (depth and size) at the time of the next ILI, i.e., 2015 for Pipeline 1.

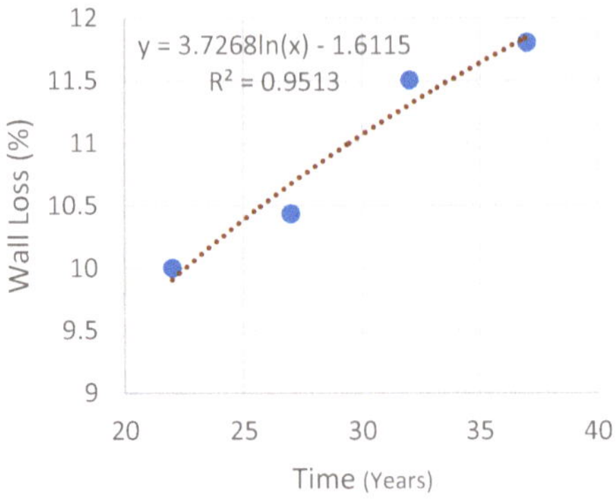

Fig. 8.13 Logarithmic model developed using the data collected from 4 ILI runs for the randomly selected defect from Pipeline 1

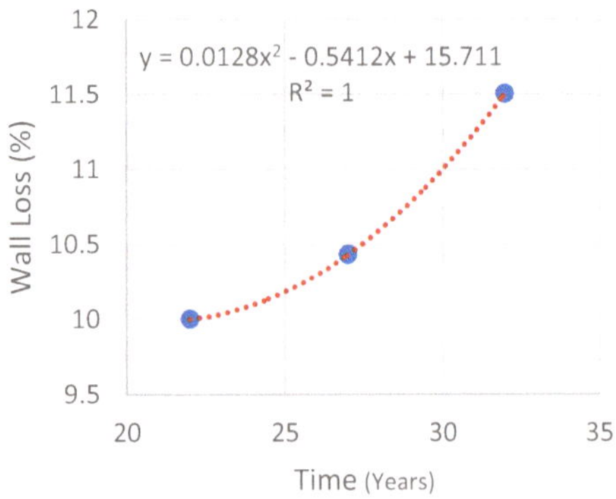

Fig. 8.14 Polynomial (order 2) model developed using the data collected from 3 ILI runs for the randomly selected defect from Pipeline 1

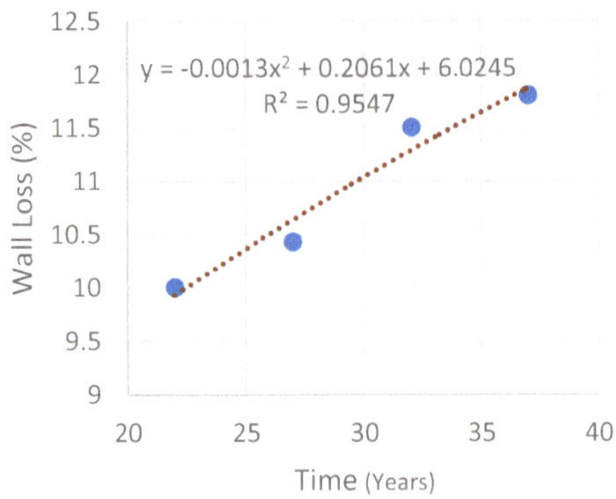

Fig. 8.15 Polynomial (order 2) model developed using the data collected from 4 ILI runs for the randomly selected defect from Pipeline 1

Similarly, Figs. 8.18, 8.19, 8.20, 8.21, 8.22, 8.23, 8.24, 8.25, 8.26 and 8.27 each show wall loss (%) over the time (years) as predicted by the model for the selected defect in Segment 9 from Pipeline 1.

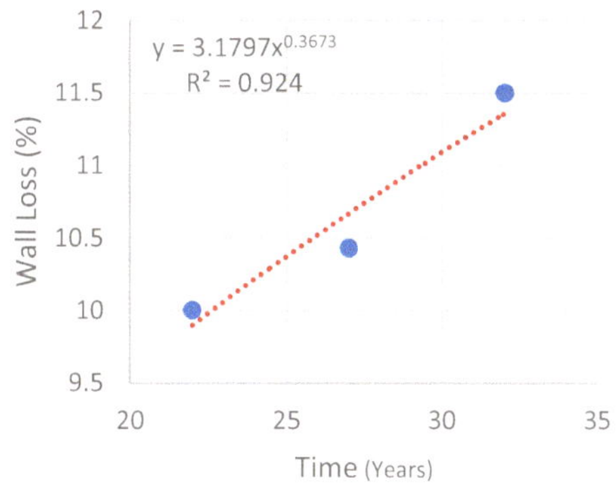

Fig. 8.16 Power law model developed using the data collected from 3 ILI runs for the randomly selected defect from Pipeline 1

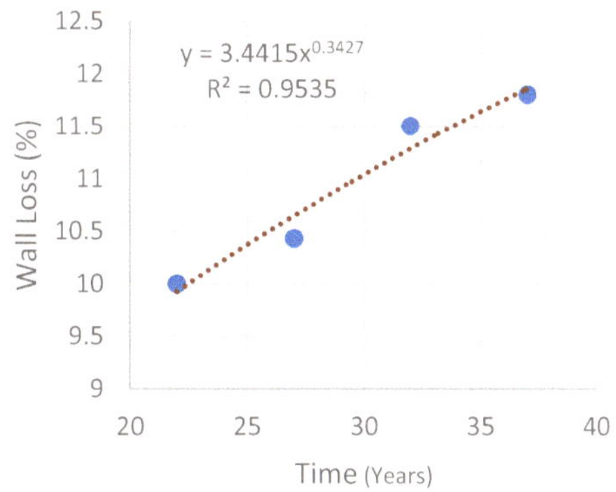

Fig. 8.17 Power law model developed using the data collected from 4 ILI runs for the randomly selected defect from Pipeline 1

Table 8.11 Model prediction accuracy analysis based on the datasets collected from 3 and 4 ILI runs for one randomly selected defect from Pipeline 1

Model	Dataset	Model formula	Relative error (%)	R^2 value
Exponential	3 ILI runs	$y = 7.852e^{0.014x}$	3.639	0.9508
	4 ILI runs	$y = 7.6818e^{0.0199x}$	6.498	0.9526
Logarithmic	3 ILI runs	$y = 3.9389\ln(x) - 2.2927$	1.105	0.9153
	4 ILI runs	$y = 3.7268\ln(x) - 1.6115$	3.600	0.9513
Polynomial (order 2)	3 ILI runs	$y = 0.0128x^2 - 0.5412x + 15.711$	11.95	1.0
	4 ILI runs	$y = -0.0013x^2 + 0.2061x + 6.0245$	4.184	0.9547
Linear	3 ILI runs	$y = 0.150x + 6.5933$	2.909	0.9428
	4 ILI runs	$y = 0.1294x + 7.1152$	5.551	0.9528
Power law	3 ILI runs	$y = 3.1797x^{0.3673}$	1.509	0.9240
	4 ILI runs	$y = 3.4415x^{0.3427}$	4.197	0.9535

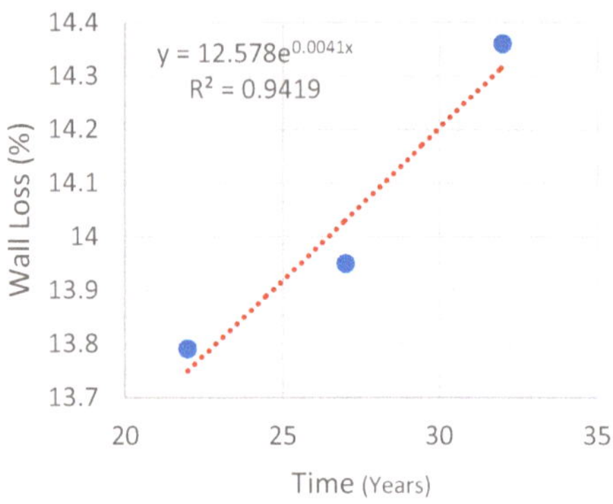

Fig. 8.18 Exponential model developed using the data collected from 3 ILI runs for the selected defect from Pipeline 1

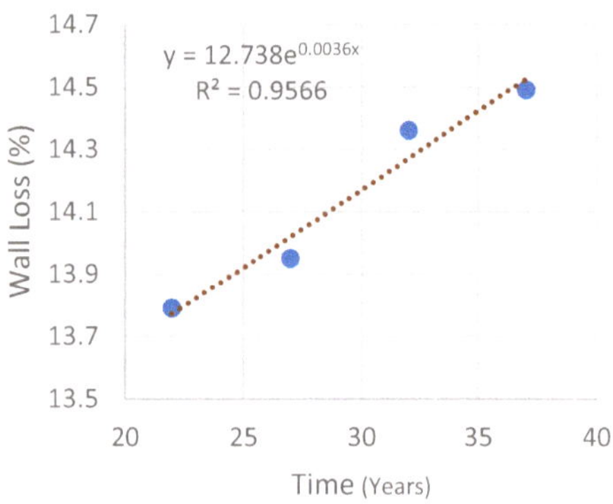

Fig. 8.19 Exponential model developed using the data collected from 4 ILI runs for the selected defect from Pipeline 1

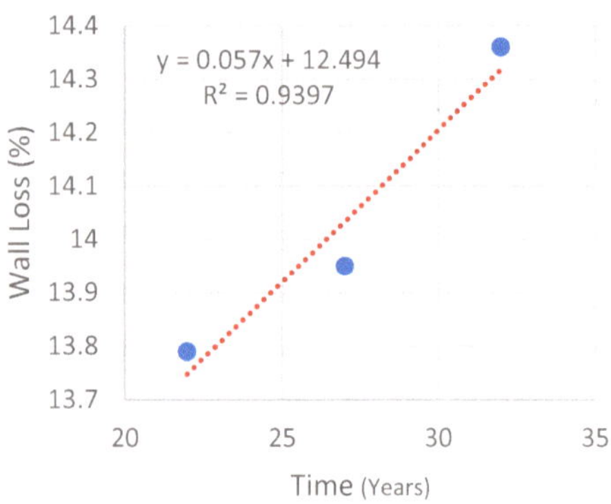

Fig. 8.20 Linear model developed using the data collected from 3 ILI runs for the selected defect from Pipeline 1

Table 8.12 shows R-square values of the models presented in Figures 8.18, 8.19, 8.20, 8.21, 8.22, 8.23, 8.24, 8.25, 8.26, to 8.27, and each model's prediction accuracy in relative error.

To achieve and compare the results, we analyzed another oil pipeline 'Pipeline 2'. Exactly the same method was followed for the analysis, two defect samples were

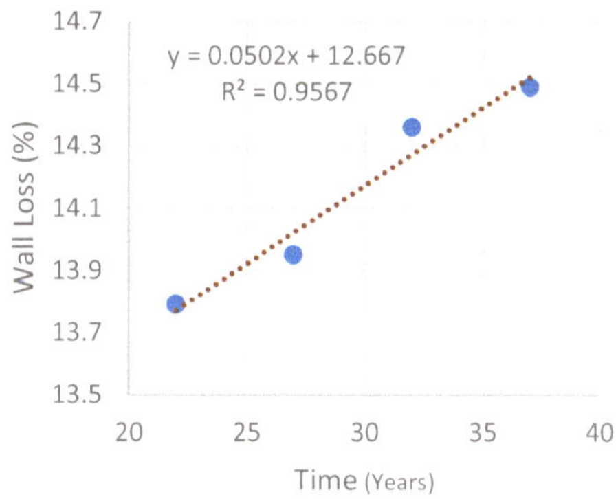

Fig. 8.21 Linear model developed using the data collected from 4 ILI runs for the selected defect from Pipeline 1

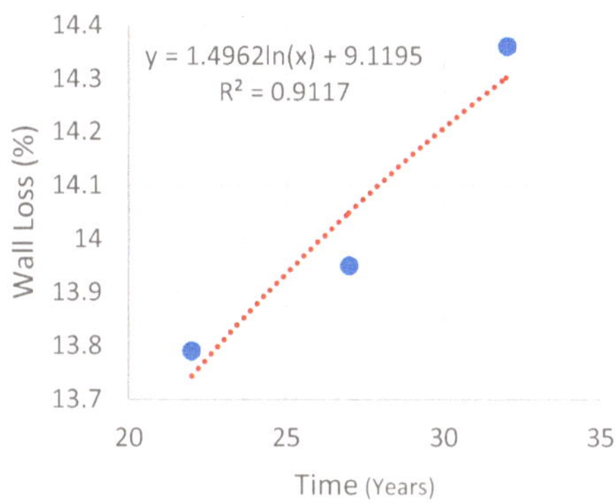

Fig. 8.22 Logarithmic model developed using the data collected from 3 ILI runs for the selected defect from Pipeline 1

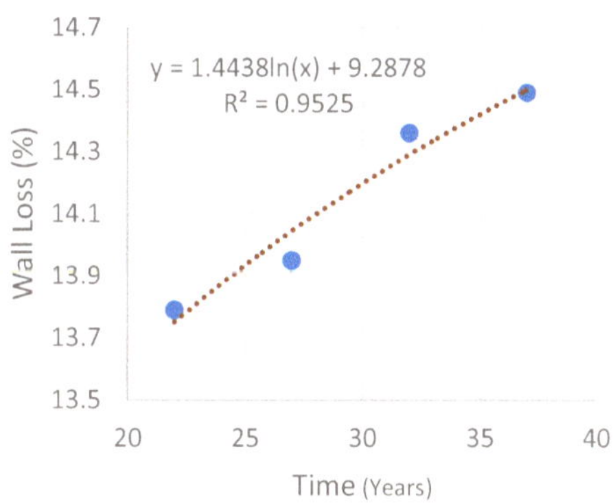

Fig. 8.23 Logarithmic model developed using the data collected from 3 ILI runs for the selected defect from Pipeline 1

selected. One defect sample was chosen randomly and the other from a specific location to understand the corrosion growth. Table 8.13 shows five types of models tested and R-square values of these models for the randomly selected defect from Pipeline 2.

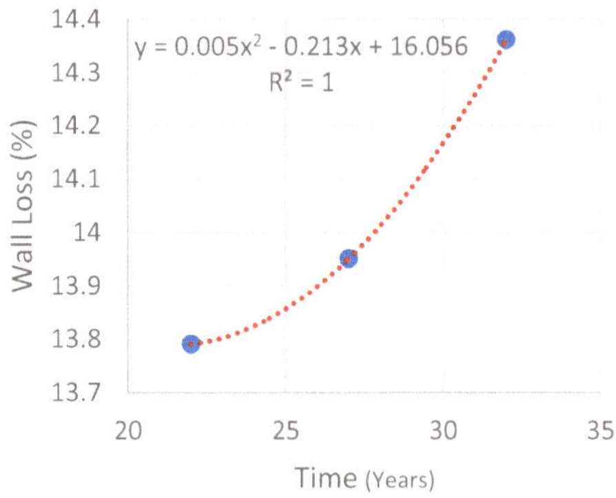

Fig. 8.24 Polynomial (order 2) model developed using the data collected from 3 ILI runs for the selected defect from Pipeline 1

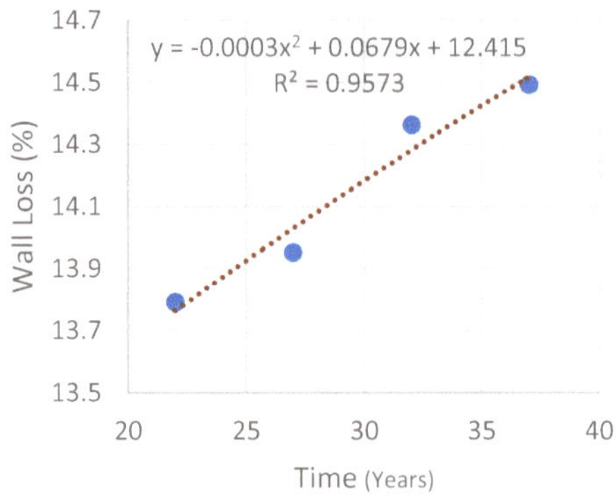

Fig. 8.25 Polynomial (order 2) model developed using the data collected from 4 ILI runs for the selected defect from Pipeline 1

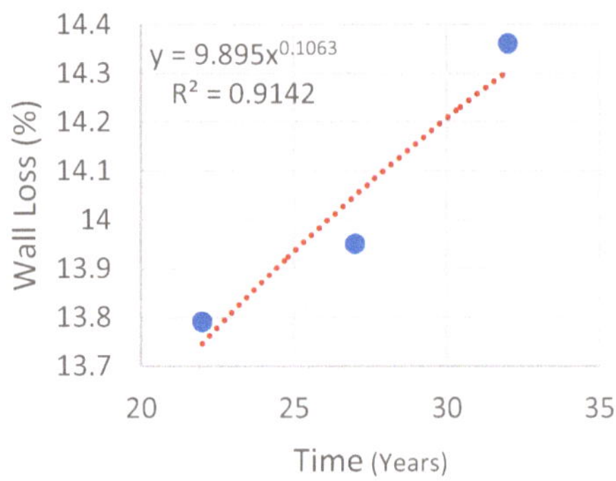

Fig. 8.26 Power law model developed using the data collected from 3 ILI runs for the selected defect from Pipeline 1

Similarly, we tested the third pipeline, using 'Pipeline 3' ILI data. Table 8.15 shows R-square values of the models tested for Defect 13.

Table 8.16 presents R-square values of the models tested for Defect 15.

Similarly for Defect 19, R-square values of the five models are shown in Table 8.17. The R-square values of the five models for Defect 15 are higher than

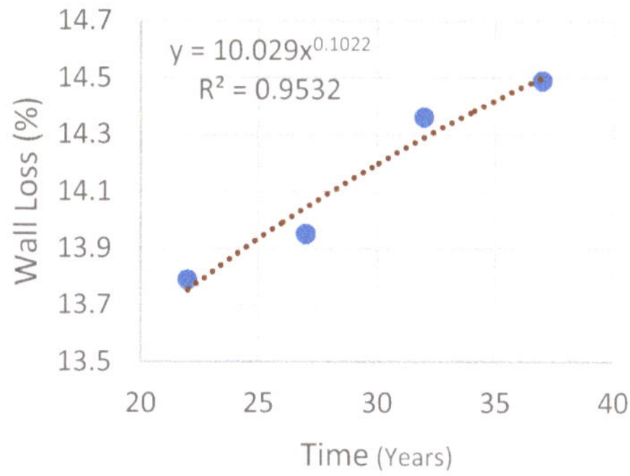

Fig. 8.27 Power law model developed using the data collected from 4 ILI runs for the selected defect from Pipeline 1

Table 8.12 Model prediction accuracy analysis based on the datasets collected from 3 and 4 ILI runs for one defect chosen from a specific location of Pipeline 1

Model	Dataset	Model formula	Relative error (%)	R^2 value
Exponential	3 ILI runs	$y = 12.578e^{0.0041x}$	1.024	0.9419
	4 ILI runs	$y = 12.738e^{0.0036x}$	1.072	0.9566
Logarithmic	3 ILI runs	$y = 1.4962\ln(x) + 9.1195$	0.222	0.9117
	4 ILI runs	$y = 1.4438\ln(x) + 9.2878$	0.165	0.9525
Polynomial (order 2)	3 ILI runs	$y = 0.005x^2 - 0.213x + 16.056$	3.658	1.0
	4 ILI runs	$y = -0.0003x^2 + 0.0679x + 12.415$	0.529	0.9573
Linear	3 ILI runs	$y = 0.057x + 12.494$	0.780	0.9397
	4 ILI runs	$y = 0.0502x + 12.667$	0.787	0.9567
Power law	3 ILI runs	$y = 9.895x^{0.1063}$	0.241	0.9142
	4 ILI runs	$y = 10.029x^{0.1022}$	0.235	0.9532

that for Defect 13 and Defect 19. The R-square values of the models for both Defect 13 and Defect 19 are quite lower.

8.11 Model Fitting Analysis

Model fitting analysis is the process of adjusting a mathematical or computational model to capture the underlying patterns and relationships within a set of data. The primary goal of model fitting is to find the optimal parameters for the chosen model that best describe the observed data. This process requires a careful balance between a model's complexity and its ability to generalize to new, unseen data (Kratzer et al.

Table 8.13 Model fitting analysis using ILI data collected from 3 and 4 ILI runs for one randomly selected defect from Pipeline 2

Model	Dataset	Model formula	R^2 value
Exponential	3 ILI runs	$y = 14.906e^{0.0022x}$	0.9717
	4 ILI runs	$y = 14.875e^{0.0024x}$	0.9871
Logarithmic	3 ILI runs	$y = 0.7320\ln(x) + 13.410$	0.9478
	4 ILI runs	$y = 0.8332\ln(x) + 13.109$	0.9639
Polynomial (order 2)	3 ILI runs	$y = 0.0024x^2 - 0.0672x + 15.951$	1.0
	4 ILI runs	$y = 0.0008x^2 - 0.0007x + 15.257$	0.9937
Linear	3 ILI runs	$y = 0.0352x + 14.888$	0.9707
	4 ILI runs	$y = 0.0370x + 14.852$	0.9864
Power law	3 ILI runs	$y = 13.563x^{0.0468}$	0.9490
	4 ILI runs	$y = 13.314x^{0.053}$	0.9655

Table 8.14 Model fitting analysis using ILI data collected from 3 and 4 ILI runs for one defect chosen from a specific location of Pipeline 2

Model	Dataset	Model formula	R^2 value
Exponential	3 ILI runs	$y = 7.664e^{0.0103x}$	0.9896
	4 ILI runs	$y = 7.747e^{0.0098x}$	0.9923
Logarithmic	3 ILI runs	$y = 2.0733\ln(x) + 3.2508$	0.970
	4 ILI runs	$y = 2.1699\ln(x) + 2.9638$	0.9837
Polynomial (order 2)	3 ILI runs	$y = 0.0045x^2 - 0.0935x + 9.449$	1.0
	4 ILI runs		
Linear	3 ILI runs	$y = 0.0993x + 7.446$	0.9868
	4 ILI runs	$y = 0.0956x + 7.5187$	0.9926
Power law	3 ILI runs	$y = 4.9469x^{0.2163}$	0.9742
	4 ILI runs	$y = 4.8508x^{0.2229}$	0.9870

Table 8.15 Model comparison for defect 13 of Pipeline 3

Model type	Dataset	Model formula	R^2 value
Exponential	4 ILIs	$y = 19.817\ e^{0.0191x}$	$R^2 = 0.7994$
Logarithmic	4 ILIs	$y = 22.763\ \ln(x) - 42.092$	$R^2 = 0.8015$
Polynomial (order 2)	4 ILIs	$y = -0.0052x^2 + 1.0455x + 8.6288$	$R^2 = 0.8007$
Linear	4 ILIs	$y = 0.7065x + 14.06$	$R^2 = 0.8004$
Power law	4 ILIs	$y = 4.3131x^{0.6178}$	$R^2 = 0.8010$

Table 8.16 Model comparison for defect 15 of Pipeline 3

Model type	Dataset	Model formula	R^2 value
Exponential	4 ILIs	$y = 11.436\ e^{0.0416x}$	$R^2 = 0.9619$
Logarithmic	4 ILIs	$y = 59.301\ \ln(x) - 161.4$	$R^2 = 0.9533$
Polynomial (order 2)	4 ILIs	$y = 0.0394x^2 - 0.7045x + 25.545$	$R^2 = 0.9624$
Linear	4 ILIs	$y = 1.8478x - 15.353$	$R^2 = 0.9595$
Power law	4 ILIs	$y = 0.4232x^{1.3378}$	$R^2 = 0.9506$

Table 8.17 Model comparison for defect 19 of Pipeline 3

Model type	Dataset	Model formula	R^2 value
Exponential	4 ILIs	$y = 11.155\ e^{0.0463x}$	$R^2 = 0.7452$
Logarithmic	4 ILIs	$y = 72.052\ \ln(x) - 199.29$	$R^2 = 0.8264$
Polynomial (order 2)	4 ILIs	*Not adequate*	
Linear	4 ILIs	$y = 2.1957x - 20.207$	$R^2 = 0.7956$
Power law	4 ILIs	$y = 0.2538x^{1.5218}$	$R^2 = 0.7786$

2023). In terms of pipeline integrity management, this type of analysis serves as a critical tool for assessing the condition of the pipeline, predicting potential issues, and making informed decisions for maintenance and repair (Razvarz et al. 2021). Model fitting analysis enables the development of predictive models that can capture and quantify the effects of factors such as corrosion, stress, and material degradation on the pipeline (Xiao et al. 2024). By fitting models to the ILI data at different intervals, trends and patterns in the pipeline's condition can be identified, allowing for proactive management strategies.

Following the same procedure as described above, we conducted model fitting analysis for 10 randomly selected defect samples from Pipeline 1 and another 10 defect samples selected from a specific location/segment from Pipeline 1. Each model's prediction accuracy (i.e., the relative error) is shown in Tables 8.18, 8.19, 8.20 and 8.21. Similarly, we tested the models for 10 randomly selected defect samples and another 10 defect samples selected from a specific location/segment from Pipeline 2, and the analysis results are shown in Tables 8.22, 8.23, 8.24 and 8.25.

In Table 8.24, it is found that the model prediction errors for Defect ID 001 and 002 are all large and all negative. That indicates that the model predicted value is smaller than the ILI measurement. For this case, a pipeline operation engineer should carry out a careful check on the data processing to make sure there is no error created in the data preprocessing. If there is no error involved in the data preprocessing, the pipeline operation engineer should closely monitor these defects, as they are exhibiting accelerated progression. The field excavation activities may be scheduled if they grow extremely fast. On the other hand, the model prediction results indicate a potential threat to the pipeline, drawing the operator's attention.

Table 8.18 Model prediction errors for the randomly selected defects from Pipeline 1 using the data from 3 ILI runs, followed by the mean and standard deviation (Std dev) of the prediction errors

Defect ID	Model prediction relative errors for the models built using the data from 3 ILIs				
	Logarithmic (%)	Quadratic (%)	Exponential (%)	Linear (%)	Power law (%)
001	0.366	3.723	2.435	1.787	0.705
002	0.237	3.909	0.757	0.833	0.287
003	3.100	**5.698**	**6.078**	**5.003**	3.823
004	0.301	4.958	0.980	1.059	0.366
005	1.100	**11.947**(*)	3.639	2.909	1.509
006	3.410	**6.203**	**6.577**	**5.456**	4.271
007	– 2.104	1.837	– 1.449	– 1.410	– 2.032
008	– 3.460	**6.924**	– 1.829	– 2.129	– 3.293
009	– 0.324	– 4.327	0.238	0.276	– 0.197
010	0.270	4.391	1.046	0.936	0.304
Mean	0.290	4.526	1.847	1.472	0.574
Std dev	2.067	4.108	2.855	2.451	2.301

Table 8.19 Model prediction errors for the randomly selected defects from Pipeline 1 using the data from 4 ILI runs, followed by the mean and standard deviation (Std dev) of the prediction errors

Defect ID	Model prediction relative errors for the models built using the data from 4 ILIs				
	Logarithmic (%)	Quadratic (%)	Exponential (%)	Linear (%)	Power law (%)
001	1.880	2.833	4.224	3.552	2.252
002	0.177	0.565	0.926	0.840	0.221
003	2.260	**73.199**(*)	**5.274**	4.116	3.052
004	0.223	0.714	1.394	1.058	0.313
005	3.600	4.184	**6.498**	**5.551**	4.197
006	3.410	– 1.078	**6.839**	**5.409**	4.309
007	– 0.822	2.799	0.329	0.279	– 0.711
008	0.463	7.281	3.043	2.453	0.850
009	0.887	0.251	1.926	1.836	0.897
010	0.198	0.633	1.119	0.942	– 3.748
Mean	1.228	9.138	3.157	2.604	1.163
Std dev	1.489	22.637	2.241	1.944	2.433

For Pipeline 3, the model prediction errors of the five types of models in the test for Defect 13, Defect 15, and Defect 19 are shown in Table 8.26. For Pipeline 3, there were four in-line inspections at years to operation of 28, 32, 35, and 37 years. Therefore, we built the models for each of Defect 13, Defect 15, and Defect 19, using the ILI data collected from 4 ILI runs. At Year 38, that is, one year after the 4th ILI, there was a field excavation so that those defects were able to be measured.

Table 8.20 Model prediction errors for the defects chosen from a segment on Pipeline 1 using the data from 3 ILI runs, followed by the mean and standard deviation (Std dev) of the prediction errors

Defect ID	Model prediction relative errors for the models built using the data from 3 ILIs				
	Logarithmic (%)	Polynomial (order 2) (%)	Exponential (%)	Linear (%)	Power law (%)
001	− 3.871	**− 7.210**	− 2.855	− 3.091	− 3.686
002	3.335	**6.124**	**6.491**	**5.377**	4.193
003	− 0.425	**14.768**(*)	1.366	1.158	− 0.222
004	0.216	3.555	0.686	0.758	0.264
005	− 0.036	**49.813**(*)	0.885	0.620	0.077
006	0.303	0.523	0.468	0.523	0.293
007	0.230	3.791	0.410	0.808	0.253
008	1.064	**176.252**(*)	2.651	2.098	1.348
009	0.222	3.658	1.024	0.780	0.241
010	− 2.285	0.902	1.988	0.180	− 0.867
Mean	− 0.125	**25.218**(*)	1.311	0.921	0.189
Std dev	1.907	**55.313**(*)	2.324	2.062	1.935

Table 8.21 Model prediction errors for the defects chosen from a segment on Pipeline 1 using the data from 4 ILI runs, followed by the mean and standard deviation (Std dev) of the prediction errors

Defect ID	Model prediction relative errors for the models built using the data from 4 ILIs				
	Logarithmic (%)	Polynomial (order 2) (%)	Exponential (%)	Linear (%)	Power law (%)
001	0.378	4.746	2.197	1.942	0.637
002	**5.012**	0.556	**8.301**	**9.705**	**5.918**
003	1.689	**5.427**	3.909	3.504	2.067
004	0.161	0.515	0.634	0.765	0.210
005	0.012	0.799	2.041	1.945	1.235
006	0.161	0.263	0.409	0.393	0.169
007	0.030	0.406	0.828	0.673	0.097
008	1.379	1.351	3.084	2.572	1.697
009	0.165	0.529	1.072	0.787	0.235
010	4.206	**7.535**	**10.172**(*)	**7.535**	**6.145**
Mean	1.319	2.213	3.265	2.982	1.841
Std dev	1.837	2.653	3.366	3.167	2.312

Table 8.22 Model prediction errors for the randomly selected defects from Pipeline 2 using the data from 3 ILI runs, followed by the mean and standard deviation (Std dev) of the prediction errors

Defect ID	Model prediction relative errors for the models built using the data from 3 ILIs				
	Logarithmic (%)	Quadratic (%)	Exponential (%)	Linear (%)	Power law (%)
001	1.264	**7.074**	2.236	2.233	1.346
002	– 0.144	**16.502**(*)	0.257	0.263	– 0.158
003	0.00941	1.0648	0.556	0.457	2.572
004	0.510	1.309	1.148	1.067	0.554
005	– 0.369	1.389	**53.778**(*)	– 0.161	– 0.366
006	0.6651	**– 18.37**(*)	**– 5.423**	**– 5.059**	**– 5.803**
007	– 0.023	**18.395**(*)	0.387	0.236	– 0.196
008	– 0.948	1.730	– 0.996	– 0.666	– 0.986
009	0.201	1.773	0.597	0.599	0.244
010	0.101	0.903	0.356	0.296	0.113
Mean	0.127	3.176	5.292	– 0.073	– 0.270
Std dev	0.600	**10.050**(*)	**17.159**(*)	1.911	2.183

Table 8.23 Model prediction errors for the randomly selected defects from Pipeline 2 using the data from 4 ILI runs, followed by the mean and standard deviation (Std dev) of the prediction errors

Defect ID	Model prediction relative errors for the models built using the data from 4 ILIs				
	Logarithmic (%)	Quadratic (%)	Exponential (%)	Linear (%)	Power law (%)
001	0.486	0.983	2.185	1.805	0.658
002	– 0.255	– 0.172	0.558	0.505	– 0.190
003	0.178	0.425	0.884	0.950	0.247
004	1.120	– 0.280	1.957	1.967	1.198
005	– 0.210	0.755	0.154	0.150	– 0.238
006	0.020	0.051	0.533	0.363	0.043
007	1.267	4.661	2.594	2.503	1.368
008	– 0.099	1.193	0.661	0.162	– 0.045
009	0.022	0.301	0.697	0.624	0.044
010	0.011	0.900	0.410	0.320	0.035
Mean	0.254	0.882	1.063	0.935	0.312
Std dev	0.538	1.418	0.851	0.849	0.570

The field-measured values of a defect are regarded as the true values of the defect including depth, length, width, and volume. We utilized a model prediction value of the wall loss (%) of a defect to compare with the field-measured value of the wall loss (%) to evaluate the model prediction accuracy such as those relative errors shown in Table 8.26.

Table 8.24 Model prediction errors for the defects chosen from a segment on Pipeline 2 using the data from 3 ILI runs, followed by the mean and standard deviation (Std dev) of the prediction errors

Defect ID	Model prediction relative errors for the models built using the data from 3 ILIs				
	Logarithmic (%)	Polynomial (order 2) (%)	Exponential (%)	Linear (%)	Power law (%)
001	**– 24.525**(*)	**– 12.189**(*)	**– 24.071**(*)	**– 24.215**(*)	**– 24.502**(*)
002	**– 7.843**	**– 8.014**	**– 7.418**	**– 7.535**	**– 7.834**
003	– 0.051	4.437	0.363	0.359	– 0.053
004	1.113	– 0.0123	1.532	1.597	1.162
005	2.515	2.853	0.984	0.945	0.373
006	– 2.133	0.871	– 1.472	– 1.533	– 2.087
007	– 0.585	– 0.241	– 0.437	– 0.330	– 0.566
008	0.824	2.395	1.379	1.319	0.854
009	0.411	3.562	1.216	1.178	0.487
010	– 2.133	0.871	– 1.472	– 1.533	– 2.087
Mean	– 3.241	– 0.547	– 2.940	– 2.975	– 3.425
Std dev	**7.999**	**5.350**	**7.884**	**7.931**	**7.859**

Table 8.25 Model prediction errors for the defects chosen from a segment on Pipeline 2 using the data from 4 ILI runs, followed by the mean and standard deviation (Std dev) of the prediction errors

Defect ID	Model prediction relative errors for the models built using the data from 4 ILIs				
	Logarithmic (%)	Polynomial (order 2) (%)	Exponential (%)	Linear (%)	Power law (%)
001	– 3.226	**45.635**(*)	3.106	2.162	– 3.278
002	– 1.177	**– 126.640**(*)	1.054	0.905	– 1.009
003	– 0.657	1.267	– 0.135	– 0.015	– 0.632
004	0.967	– 2.176	1.644	1.654	1.045
005	0.031	0.118	1.012	0.982	0.103
006	– 2.963	2.436	– 1.359	– 1.579	– 1.403
007	– 0.770	– 0.385	– 0.299	– 0.275	– 0.756
008	1.664	**40.233**(*)	2.526	2.361	1.744
009	0.039	0.147	1.576	1.219	0.177
010	– 2.963	2.436	– 1.359	– 1.579	– 1.403
Mean	– 0.906	– 3.693	0.777	0.584	– 0.541
Std dev	1.696	**46.692**(*)	1.532	1.411	1.411

Table 8.26 Summary of relative errors of model predictions for defects 13, 15 and 19 of Pipeline 3

Defect ID	Model prediction relative errors for the models built using the data of the first 4 ILIs				
	Logarithmic (%)	Polynomial (order 2) (%)	Exponential (%)	Linear (%)	Power law (%)
13	3.064	3.415	3.67	3.562	3.32
15	**11.525**(*)	**14.307** (*)	**14.098**(*)	**12.656**(*)	**12.837**(*)
19	– 3.376	n.a	– 0.309	– 2.724	– 0.987

In the above tables, a value shown in Bold indicates the absolute value of the relative error is equal to or larger than 5%, while a value marked with an asterisk (*) indicates a very large prediction error, i.e., the relative error is larger than 10.00%. A model prediction with a relative error of less than 5% is considered good, while predictions with an relative error below 3% are regarded as highly accurate. From the analysis results shown in the above tables, it is found that the ILI measurement data for 'Pipeline 1' and 'Pipeline 2' possesses good consistence while the ILI measurement data for 'Pipeline 3' indicates a slightly poor quality.

In this study, it is also verified that a model with the highest R-square value does not always give a better prediction accuracy than other models. This is evidenced by the results shown in Tables 8.11, 8.12, and other tables (see Tables 8.18, 8.19, 8.20, 8.21, 8.22, 8.23, 8.24, to 8.25). Therefore, relying on R-square value to select the best model for defect growth prediction is not reliable. Hence, it is not recommended in model selection for this task.

Through the testing and validation analysis, we recognized that Logarithmic model yields the best prediction performance overall, and then Power Law, Linear, and Exponential model. The Polynomial (Order 2) model's prediction is not reliable, sometimes, it can give extremely large prediction errors, so that this model is not suggested to be applied.

By making a comparison of model prediction accuracy with the results shown in Zhang et al. (2013), it finds the following: the prediction relative error given in Zhang et al. (2013) is below 1.0% for Defect 13, however, it is in 3.0–4.0% using the simple models in this study; the prediction relative error given in Zhang et al. (2013) is 4.72% for Defect 15, however, it is in 11.5–14.5% using the simple models in this study; the prediction relative error given in Zhang et al. (2013) is around –11% for Defect 19, however, it is in the range of –0.3 ~ –3.4% using the simple models in this study. The model presented in Zhang et al. (2013) is complicated and it is difficult for a pipeline integrity management engineer to use it. Therefore, the simple models discussed in this work are competitive to the complex model as presented in Zhang et al. (2013) and Zhang and Zhou (2014).

For Defect 13 of Pipeline 3, the model prediction relative errors of these five models are approaching but the Logarithmic model gives the best prediction accuracy. For Defect 15 of Pipeline 3, Logarithmic model gives the smallest prediction error, but the relative errors of the model prediction of these five models are all above 10%.

For Defect 19 of Pipeline 3, Exponential model gives the best prediction followed by Power Law, Linear, and Logarithmic model. It is interesting to note that the R-square values of the five models built for Defect 15 are higher than that for Defect 13 and Defect 19. However, the model prediction accuracy for Defect 15 is quite lower than those for the other two defects. Therefore, it is evidenced that we cannot reply on the R-square value in model selection for the purpose of achieving a better prediction of the model.

The research findings above show the importance of tailoring model selection using very basic details of individual pipelines. Using these simple models, it will assist pipeline operators to understand their pipeline's current condition and predict any corrosion or anomaly growth using ILI data. This research explores and recommends a comprehensive model fitting approach for more accurate predictions of pipeline anomalies/defects using simple models.

The research also contributes to the industry's understanding of model generalizability by validating results across different pipeline datasets. Through model fitting and analysis, it is found that Logarithmic model provides better overall prediction performance for defect progression, followed by the Power Law model and other models tested in this work. Practitioners are encouraged to adopt a cross-validation approach, testing predictive models on diverse datasets to ensure the reliability of anomaly predictions across various pipeline configurations and operational conditions.

8.12 Implications of the Findings

The implications of these findings hold significant promise for the practical application of pipeline defect prediction in the asset management world. The identification of robust predictive models, such as Logarithmic and Power Law models, for different temporal scales and specific pipeline segments offers a valuable guide for industry practitioners using ILI data. Incorporating these models into routine maintenance planning can help pipeline asset managers to proactively address potential defects, thereby minimizing the risk of unplanned disruptions and enhancing overall pipeline integrity. Furthermore, this research highlights the value of employing these simple models to predict the progression of individual defects in energy pipelines, providing practitioners with a more comprehensive and reliable approach to model testing and validation. As a result, pipeline companies are encouraged to integrate these validated models into their predictive maintenance frameworks, tailoring them to the unique characteristics of their assets, like pipeline and associated infrastructures. Ultimately, the practical application of these research findings has the potential to optimize resource allocation, improve operational efficiency, and elevate the overall safety and reliability of energy pipeline systems.

8.13 Conclusion and Future Outlook

This research work has addressed the critical issue of predicting pipeline defects, specifically focusing on corrosion, through the utilization of In-line inspection (ILI) data. The study systematically tested and validated available and simple mathematical models, including Exponential, Logarithmic, Polynomial (Order 2), Liner, and Power Law models, using three distinct pipelines' ILI data. The results demonstrate that Logarithmic and Power Law models are particularly effective with high accuracy in the context of oil and gas pipelines, showing their strength and acceptance in the oil and gas industry.

The comprehensive testing, analysis and validation of these models were conducted on ILI datasets for 47 selected defects. Notably, the research findings emphasize the better predictive performance of Logarithmic and Power Law models, outperforming other selected models overall in the analysis. They are simple and easy to use. This conclusion is further supported by model fitting analysis.

The study emphasizes the importance of tailoring model selection based on the specific characteristics of individual pipelines. The results suggest that pipeline operators can benefit from employing these simple models to gain awareness into the current condition of their pipelines and forecast corrosion or anomaly growth using ILI data. Furthermore, the research recommends a thorough model fitting analysis, particularly focusing on Logarithmic and Power Law models, to enhance the accuracy of predictions.

In essence, this research contributes valuable insights into the pipeline integrity management system (PIMS), providing a practical and effective approach for predicting corrosion growth.

The future of **pipeline integrity management** will be shaped by the integration of **machine learning**, **degradation modeling**, and **IoT** technologies. As these technologies continue to evolve, they will provide operators with more accurate, real-time predictions of defect growth, enabling **proactive maintenance** and reducing the risk of pipeline failures. By embracing these technologies, the oil and gas industry can improve the **safety, reliability,** and **efficiency** of its pipeline networks, ensuring the continued flow of energy while minimizing the risk of catastrophic failures.

References

Aalsalem MY et al (2018) Wireless sensor networks in oil and gas industry: recent advances, taxonomy, requirements, and open challenges. J Netw Comput Appl 113:87–97

Abbas MH, Norman R, Charles A (2018) Neural network modelling of high pressure CO_2 corrosion in pipeline steels. Process Saf Environ Prot 119:36–45

Al-Amin M et al (2014) Hierarchical Bayesian corrosion growth model based on in-line inspection data. J Pressure Vessel Technol 136(4):041401

Aldosari H, Elfouly R, Ammar R (2020) Evaluation of machine learning-based regression techniques for prediction of oil and gas pipelines defect. In: International conference on computational science and computational intelligence (CSCI). IEEE

Amaya-Gómez R, Sánchez-Silva M, Muñoz F (2016) Pattern recognition techniques implementation on data from In-Line Inspection (ILI). J Loss Prev Process Ind 44:735–747
Arzaghi E et al (2018) Developing a dynamic model for pitting and corrosion-fatigue damage of subsea pipelines. Ocean Eng 150:391–396
Bhambri P, Rani S (2024) Challenges, opportunities, and the future of industrial engineering with IoT and AI. In: Integration of AI-based manufacturing and industrial engineering systems with the Internet of Things. pp 1–18
Cai J, Cottis R, Lyon S (1999) Phenomenological modelling of atmospheric corrosion using an artificial neural network. Corros Sci 41(10):2001–2030
Caleyo F et al (2009) Markov chain modelling of pitting corrosion in underground pipelines. Corros Sci 51(9):2197–2207
Campo-Meneses KG et al (2021) Mathematical connections activated in high school students' practice solving tasks on the exponential and logarithmic functions. EURASIA J Math Sci Tech Ed 17(9):em1998
Campo-Meneses KG, García-García J, Moll VF (2023) Mathematical connections associated with the exponential and logarithmic functions promoted in the mathematics curriculum. Int J Instr 16(4)
Chiodo MS, Ruggieri C (2009) Failure assessments of corroded pipelines with axial defects using stress-based criteria: numerical studies and verification analyses. Int J Press Vessels Pip 86(2–3):164–176
Clauset A, Shalizi CR, Newman ME (2009) Power-law distributions in empirical data. SIAM Rev 51(4):661–703
Cui B, Wang H (2023) Analysis and prediction of pipeline corrosion defects based on data analytics of in-line inspection. J Infrastruct Preserv Resil 4(1):14
Dangeti P (2017) Statistics for machine learning. Packt Publishing Ltd
Dhoni P (2023) Exploring the synergy between generative AI, data and analytics in the modern age. Authorea Preprints, TechRxiv. August 29, 2023
Friederich J, Lazarova-Molnar S (2024) Reliability assessment of manufacturing systems: a comprehensive overview, challenges and opportunities. J Manuf Syst 72:38–58
Gorjian N et al (2010) A review on degradation models in reliability analysis. In: Engineering asset lifecycle management: proceedings of the 4th world congress on engineering asset management (WCEAM 2009), 28–30 September 2009. Springer
Habbal A, Ali MK, Abuzaraida MA (2024) Artificial intelligence trust, risk and security management (AI TRiSM): frameworks, applications, challenges and future research directions. Expert Syst Appl 240:122442
Hakimian S et al (2023) Application of machine learning for the classification of corrosion behavior in different environments for material selection of stainless steels. Comput Mater Sci 228:112352
Han Z, Li X, Chen G (2023) A stochastic model for RUL prediction of subsea pipeline subject to corrosion-fatigue degradation. Process Saf Environ Prot 178:739–747
Ismail MFH et al (2023) Machine-learning-based classification for pipeline corrosion with Monte Carlo probabilistic analysis. Energies 16(8):3589
James G et al (2023) Linear regression. An introduction to statistical learning: with applications in python. Springer, pp 69–134
Kariyawasam S, Peterson W (2008) Revised corrosion management with reliability based excavation criteria. In: Proceedings of the 7th International Pipeline Conference, IPC2008-64536, pp 489–500. September 29–October 3, 2008 Calgary, Alberta, Canada
Kratzer G et al (2023) Additive Bayesian network modeling with the R package abn. J Stat Softw 105:1–41
Liu X et al (2018) Multi-level optimization of maintenance plan for natural gas pipeline systems subject to external corrosion. J Nat Gas Sci Eng 50:64–73
Liu H et al (2019) Matching pipeline in-line inspection data for corrosion characterization. NDT E Int 101:44–52

Machado JT, Pinto CM, Lopes AM (2015) A review on the characterization of signals and systems by power law distributions. Signal Process 107:246–253
Makarenkov V, Legendre P (2002) Nonlinear redundancy analysis and canonical correspondence analysis based on polynomial regression. Ecology 83(4):1146–1161
Miran SA, Huang Q, Castaneda H (2016) Time-dependent reliability analysis of corroded buried pipelines considering external defects. J Infrastruct Syst 22(3):04016019
Nessim M et al (2009) Reliability based design and assessment for location-specific failure threats with application to natural gas pipelines. J Pressure Vessel Technol 131(4):041701
Ossai CI (2019) A data-driven machine learning approach for corrosion risk assessment—a comparative study. Big Data Cognitive Comput 3(2):28
Ossai CI, Boswell B, Davies I (2016) Markov chain modelling for time evolution of internal pitting corrosion distribution of oil and gas pipelines. Eng Fail Anal 60:209–228
Piao G et al (2019) Fast reconstruction of 3-D defect profile from MFL signals using key physics-based parameters and SVM. NDT E Int 103:26–38
POF (2021) Specifications and requirements for in-line inspection of pipelines. https://pipelineoperators.org/cdn/276341a3-f5e6-49cf-9897-de6ab41bdd5a/POF%20100%20Specifications%20and%20requirements%20for%20ILI%20-%20Nov%202021.pdf [cited 9 May 2024]
Razvarz S, Jafari R, Gegov A (2021) Flow modelling and control in pipeline systems. Stud Syst Decis Control 321:25–57
Soomro AA et al (2022) A review on Bayesian modeling approach to quantify failure risk assessment of oil and gas pipelines due to corrosion. Int J Pressure Vessels Pip 104841
Soomro AA et al (2024) Analysis of machine learning models and data sources to forecast burst pressure of petroleum corroded pipelines: a comprehensive review. Eng Fail Anal 155:107747
Teixeira A et al (2008) Reliability of pipelines with corrosion defects. Int J Press Vessels Pip 85(4):228–237
Vanaei H, Eslami A, Egbewande A (2017) A review on pipeline corrosion, in-line inspection (ILI), and corrosion growth rate models. Int J Press Vessels Pip 149:43–54
Velázquez JC et al (2022) Probabilistic and statistical techniques to study the impact of localized corrosion defects in oil and gas pipelines: a review. Metals 12(4):576
Villadsen AR, Wulff JN (2021) Statistical myths about log-transformed dependent variables and how to better estimate exponential models. Br J Manag 32(3):779–796
Vishnuvardhan S, Murthy AR, Choudhary A (2023) A review on pipeline failures, defects in pipelines and their assessment and fatigue life prediction methods. Int J Press Vessels Pip 201:104853
Wang Q et al (2023) Evolution of corrosion prediction models for oil and gas pipelines: from empirical-driven to data-driven. Eng Fail Anal 146:107097
Xiao R et al (2024) Improving failure modeling for gas transmission pipelines: a survival analysis and machine learning integrated approach. Reliab Eng Syst Saf 241:109672
Xie M, Tian Z (2018) A review on pipeline integrity management utilizing in-line inspection data. Eng Fail Anal 92:222–239
Xu L et al (2023) The research progress and prospect of data mining methods on corrosion prediction of oil and gas pipelines. Eng Fail Anal 144:106951
Yang Y et al (2024) Intelligent methods for pipeline operation and integrity. J Pipeline Syst Eng Pract 15(1):04023053
Zang X et al (2023) Ultrasonic guided wave techniques and applications in pipeline defect detection: a review. Int J Pressure Vessels Pip 206:105033
Zhang S, Zhou W (2014) Bayesian dynamic linear model for growth of corrosion defects on energy pipelines. Reliab Eng Syst Saf 128:24–31
Zhang S, Zhou W, Qin H (2013) Inverse Gaussian process-based corrosion growth model for energy pipelines considering the sizing error in inspection data. Corros Sci 73:309–320
Zhang S et al (2014) Time-dependent corrosion growth modeling using multiple in-line inspection data. J Pressure Vessel Technol 136(4):041202

Zhang H et al (2020) A modified method for the safety factor parameter: the use of big data to improve petroleum pipeline reliability assessment. Reliab Eng Syst Saf 198:106892

Chapter 9
Predictive Maintenance and Pipeline Integrity

9.1 Introduction to Predictive Maintenance

The integration of different data types to efficiently assess the condition of energy pipelines is one of the fundamental components of pipeline integrity management. In order to develop a thorough model to assess the effect of urban environments on pipeline safety, the work on spatial data fusion by Shen et al. (2024) highlights the significance of integrating the physical, temporal, and semantic features of subterranean pipes. In addition to improving pipeline monitoring accuracy, this strategy makes it easier for stakeholders to share information effectively, which is essential for timely decision-making during maintenance tasks.

Reliability engineering plays a pivotal role in pipeline design and maintenance strategies. Omoya et al. (2019) highlights the necessity of incorporating reliability concepts into pipeline standards and codes, advocating for an integrated application of reliability engineering tools throughout the pipeline's lifecycle. This integration spans from design for reliability (DFR) to end-of-life considerations, ensuring that pipelines are not only designed to prevent failures but also managed in a way that maximizes their operational lifespan.

Corrosion represents an imminent risk to pipeline integrity, and understanding its effects is vital for the development of appropriate maintenance strategies. Sulaiman and Tan (2018), utilize a Bayesian network to analyze the dynamic reliability of corroded pipelines, providing a probabilistic framework that accounts for uncertainties related to corrosion defect growth. This method allows operators to predict the likelihood of failure due to corrosion, enabling more informed maintenance decisions that can preempt catastrophic failures.

Learning from data has grown into an innovative development in pipeline integrity management. The comprehensive review by Amadhe et al. (2024) discusses various predictive and optimization techniques that leverage machine learning to enhance pipeline integrity management processes. These techniques include predictive analytics for maintenance scheduling, anomaly detection in operational data, and

M. Hussain and T. Zhang, *Big Data Analytics in Energy Pipeline Integrity Management*, Lecture Notes in Energy 104, https://doi.org/10.1007/978-981-96-8019-1_9

optimization of inspection routines, all of which contribute to a proactive maintenance culture that minimizes downtime and operational risks.

In-line inspection (ILI) methods are crucial for the ongoing assessment of pipeline integrity. Ma et al. (2021) detail the instrumentation and data management practices associated with ILI, emphasizing the need for tailored detection equipment that aligns with the specific characteristics of the pipeline being inspected. The effective management of historical data from these inspections allows for the identification of high-risk areas within the pipeline network, facilitating targeted maintenance efforts. The development of robotic technologies for pipeline inspection represents a significant advancement in the field. James and Rajasekaran (2023) describe the design of an external modular robot capable of detecting defects on pipelines, which underscores the importance of regular integrity checks as part of a comprehensive maintenance strategy. Such innovations not only enhance the efficiency of inspections but also reduce the risks associated with manual inspections in hazardous environments.

Mahmoodzadeh et al. (2020) stress the significance of adaptive maintenance techniques and real-time monitoring such as CBM for dry pipelines that are prone to internal corrosion. This method optimizes resource allocation and reduces operational disruptions by enabling maintenance activities to be customized depending on the pipeline's real condition rather than just following preset schedules. The integration of advanced monitoring techniques, such as distributed fiber-optic sensors, has significantly revolutionized pipeline integrity management. In order to detect leaks and structural damage quickly, Naeem et al. (2023) investigate the use of internally deployed fibers for high-sensitivity pipeline infrastructure monitoring. In addition to improving pipeline operations' safety, this technology offers useful data for predictive maintenance models.

Pipeline integrity is utterly threatened by geohazards, especially in regions that are prone to earthquakes and landslides. The difficulties of evaluating the safety of corroded energy pipelines in the event of a landslide are discussed by Zhang et al. (2023), who offer insights into risk assessment techniques that take geological and corrosion aspects into consideration. Developing efficient maintenance plans that reduce the risks associated with geohazards requires an understanding of how environmental factors and pipeline integrity interact.

The value of data analytics in pipeline integrity monitoring is crucial. The work of Qin et al. (2022) on risk assessment methods for gathering pipelines illustrates the importance of utilizing data-driven approaches to inform maintenance decisions. By employing information entropy-based methods, operators can better understand the risks associated with various pipeline segments, allowing for more strategic allocation of maintenance resources.

The development of integrated health monitoring systems that integrate predictive analytics and real-time data is essential for efficient pipeline management. Chalgham et al. (2020) propose a holistic approach to pipeline integrity management with their system-level prognosis and health monitoring platform, which combines real-time data with predictive failure models. From operational data to predictive modeling, this interdisciplinary approach guarantees that every facet of pipeline health is taken into account.

The importance of regular inspections and maintenance cannot be overlooked. In their analysis of current developments in pipeline monitoring and leak detection systems, Adegboye et al. (2019) point out the advantages and disadvantages of different approaches. In order to improve the efficacy of their maintenance plans, operators can use this comparison study as a reference for choosing the right technologies for their unique operational situations.

9.2 Evolution of Maintenance Strategies

Over time, technological developments, data analytics, and a better knowledge of the elements influencing pipeline reliability have all contributed to a substantial transition in the evolution of maintenance techniques in pipeline integrity management. The goal of this development is to improve the safety and effectiveness of pipeline operations by moving away from conventional, reactive maintenance techniques and toward more proactive, predictive approaches.

In the past, pipeline maintenance techniques were primarily reactive, concentrating on fixing problems after they happened. Because operators were forced to respond to events rather than avoid them, this strategy frequently resulted in severe downtime and expensive repairs. The management of pipeline integrity has undergone a substantial evolution owing to the adoption of predictive maintenance methodologies. Predictive maintenance utilizes data analytics, machine learning, and real-time monitoring to foresee probable breakdowns, enabling prompt interventions that can reduce operational costs and alleviate hazards (Sulaiman and Tan 2018; Amadhe et al. 2024).

The use of machine learning algorithms into pipeline integrity management systems signifies a notable progression in predictive maintenance. These algorithms evaluate substantial amounts of operational data to identify trends and anomalies that may indicate impending failures. As an example, Amadhe et al. (2024) stresses the significance of developing a thorough pipeline integrity management (PIM) technique that includes monitoring, testing, inspection, and maintenance. Instead of relying on fixed maintenance intervals that may not reflect the pipeline's actual condition operators can adjust maintenance plans based on its real-time state using machine learning techniques. A strong foundation for comprehending how corrosion flaws affect pipeline integrity has been made available via the use of Bayesian networks in dynamic reliability analysis. In order to inform maintenance methods, Sulaiman and Tan's (2018) research demonstrates how probabilistic models may estimate the chance of failure and account for uncertainty in corrosion progression. Operators may prioritize maintenance tasks according to the unique hazards connected to various pipeline segments according to this data-driven approach, which facilitates better decision-makings.

The development of maintenance approaches has been profoundly impacted by the development of Maintenance Prioritization Models (MPM). The strategic use of integrity monitoring techniques can greatly improve pipeline integrity programs' efficacy and save maintenance costs, according to Burdylo and Aelst (2000). Operators can better allocate resources and carry out targeted maintenance interventions by determining which sections of the pipeline are most vulnerable to failure. Pipeline maintenance procedures have been transformed by the use of modern monitoring technologies in addition to predictive maintenance and prioritizing models. Methods like In-Line Inspection (ILI) are now crucial for evaluating the integrity of pipelines. In order to improve the accuracy of integrity assessments, Ma et al. (2021) explain the significance of creating detecting equipment that is particular to the features of the pipeline being inspected. Proactive maintenance initiatives can be facilitated by identifying high-risk locations through the integration of historical data from these inspections.

The efficiency of risk assessment techniques in locating and assessing high-risk pipelines is demonstrated by the work of Qin et al. (2022). Operators can deploy maintenance resources more effectively by applying information entropy-based methodologies to acquire a better understanding of the risks related to different pipeline sections. The use of Internet of Things (IoT) technology has significantly improved the capacity to monitor and maintain pipeline integrity. IoT-based solutions provide real-time data collection and analysis, hence improving operational efficiency and enabling prompt actions (Priyanka et al. 2021). Continuous evaluation of pipeline conditions is made possible by the deployment of smart sensors and monitoring equipment, providing operators with valuable insights to support maintenance decisions.

The evolution of maintenance techniques is expected to accelerate as the sector continues to embrace digital transformation. The integration of machine learning and artificial intelligence (AI) into pipeline integrity monitoring systems is expected to enhance predictive capabilities, enabling operators to predict faults with greater precision. For example, Hussain et al. (2024) summarized prediction methods for gas pipeline stress corrosion cracking highlighting how AI may enhance pipeline fault identification and prediction.

9.3 What is Predictive Maintenance?

Predictive maintenance (PdM) is a proactive maintenance technique that employs data analytics and real-time monitoring to determine the optimal timing for equipment maintenance, as illustrated in Fig. 9.1. By preventing unplanned malfunctions and streamlining maintenance plans, this strategy seeks to improve operational effectiveness and lower downtime and repair expenses. Technological advancements, particularly in Big Data, artificial intelligence (AI), and the Internet of Things (IoT), have significantly influenced the evolution of predictive maintenance (Krupitzer et al. 2002).

Fundamentally, predictive maintenance entails using a variety of sensors and data collection techniques to continuously monitor the state of the equipment. Organizations can find trends and abnormalities that might point to future problems by

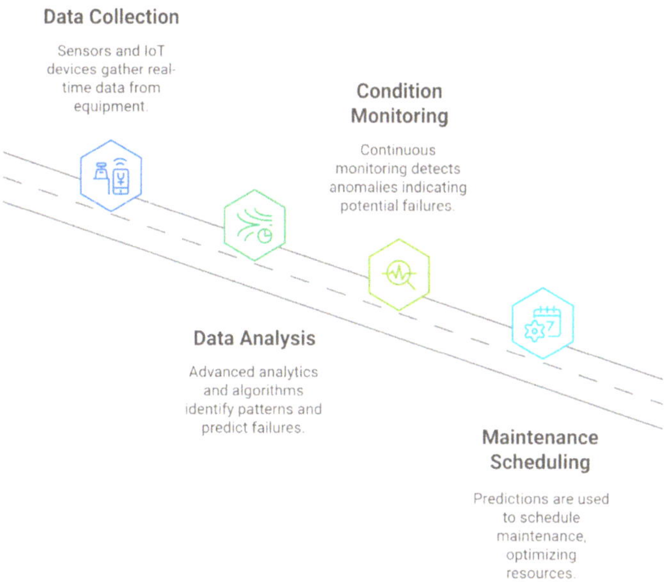

Fig. 9.1 Predictive maintenance process

examining this data. Tagawa et al. (2021), for example, illustrates the relevance of sophisticated sensor technologies in predictive maintenance by showing how acoustic anomaly detection may be used to identify mechanical breakdowns in noisy industrial environments. This feature minimizes operational disruptions by enabling prompt interventions prior to equipment failures.

Predictive maintenance systems are now much more successful because of the incorporation of machine learning algorithms. These algorithms can process substantial data volumes to generate insights into equipment condition and predict future faults. To make well-informed maintenance decisions, Krupitzer et al. (2002) highlights that predictive maintenance should be seen as a component of a broader integrated system that uses data from several sources. In addition to increasing prediction accuracy, this all-encompassing approach makes it easier to create more complex maintenance plans. Predictive maintenance has been applied in multiple industrial sectors, such as manufacturing, transportation, and energy. For instance, predictive maintenance techniques are used in the railway industry to track component status and anticipate breakdowns, guaranteeing dependability and safety (Binder et al. 2023). According to Jain et al. (2022), predictive maintenance is also essential in the automobile sector for identifying the condition of vehicles and averting malfunctions.

Predictive maintenance has advantages beyond merely preventing failures. Organizations can save a lot of money by tailoring maintenance plans to the actual conditions of their equipment, some of the benefits are shown in Fig. 9.2. In enterprises where equipment malfunction may lead to substantial financial detriment, this is

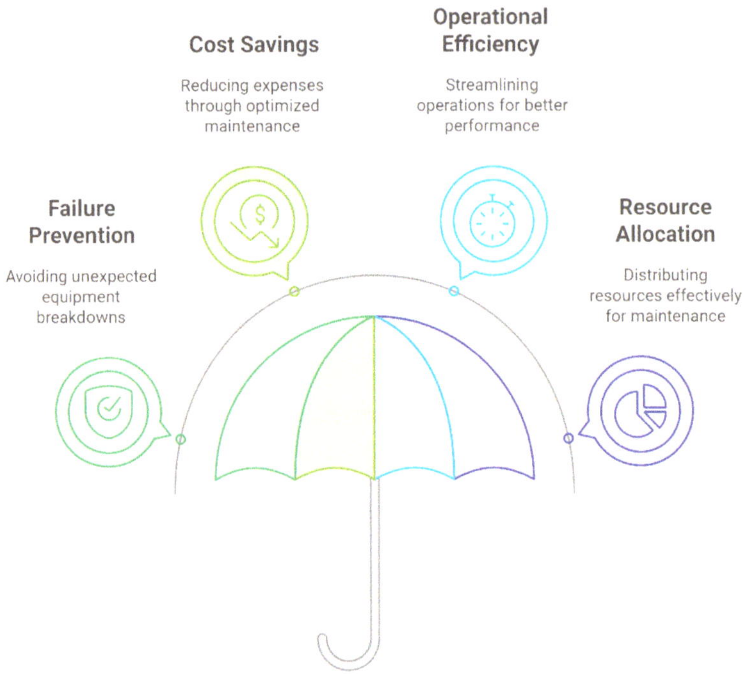

Fig. 9.2 The multifaceted benefits of predictive maintenance

particularly vital. Mohammed et al. (2023) discuss how predictive maintenance can save maintenance expenses and enhance operational efficiency across many applications. Furthermore, by anticipating maintenance requirements, businesses may better manage resources and make sure that maintenance is done when it is most needed.

Predictive maintenance's role is anticipated to expand as more companies embrace Industry 4.0 concepts, as illustrates in Fig. 9.3. More advanced predictive maintenance solutions are being made possible by the confluence of Big Data, AI, and IoT technology. The integration of IoT devices enables real-time equipment monitoring, while machine learning algorithms analyze the data to generate actionable insights (Cardoso and Ferreira 2020; Kisić et al. 2017). This synergy promotes the creation of intelligent manufacturing processes in addition to improving the predictive power of maintenance systems.

Fig. 9.3 Predictive maintenance in Industry 4.0

9.3.1 *Key Differences Between Predictive and Preventive Maintenance*

Predictive maintenance is data-driven and takes place based on actual pipeline conditions, whereas preventive maintenance depends on preset timetables for repairs and inspections. Predictive maintenance brings several key benefits:

- **Reduced Downtime**: Predictive maintenance allows operators to schedule repairs only when necessary, minimizing the need for unnecessary shutdowns.
- **Cost Savings**: Predictive maintenance lowers the expenses of emergency repairs and unscheduled downtime by addressing problems before they become serious failures.
- **Improved Safety**: Early detection of potential failures improves pipeline safety, reducing the risk of leaks, ruptures, or other catastrophic events.
- **Optimized Resource Allocation**: Maintenance teams can focus their efforts on equipment that truly needs attention, improving overall productivity.

9.3.2 *The Role of Data in Predictive Maintenance*

Both historical and real-time data are essential for the efficacy of predictive maintenance. Pipeline operators collect data from various sources, including:

- **IoT Sensors**: Pressure, temperature, and vibration sensors provide real-time data on the operating conditions of the pipeline.
- **Historical Inspection Data**: Data from previous inspections, such as corrosion levels and wall thickness measurements, helps build predictive models based on past trends.

- **Environmental Data**: External factors, such as temperature fluctuations, humidity, and geological activity, also impact pipeline integrity and are included in predictive models.

Operators can create models that anticipate possible faults and optimize maintenance schedules by integrating data from many sources using predictive analytics.

9.4 Benefits of Predictive Maintenance in Pipeline Systems

In pipeline systems, predictive maintenance (PdM) has several advantages that include cost-effectiveness, safety, and operational efficiency. PdM assists enterprises in predicting future breakdowns and optimizing maintenance schedules by the application of modern technologies, including data analytics, machine learning, and real-time monitoring. The reduction in unplanned downtime is one of the primary advantages of predictive maintenance. Conventional maintenance methods frequently depend on planned examinations and repairs, which may result in needless outages. Predictive maintenance, on the other hand, aims to track the real state of pipes and forecast when maintenance will be required. Bęben and Steliga (2023) highlights that companies can use artificial intelligence models to identify high-risk pipelines and prioritize inspection and maintenance tasks by monitoring for corrosion and compiling an extensive database of pipeline conditions. This proactive strategy reduces the probability of unforeseen breakdowns, therefore improving operational continuity.

Another important advantage of predictive maintenance is cost reduction. Organizations can avoid the high expenses of emergency repairs and prolonged downtime by preventing problems before they happen. Although their primary focus is on scheduling operations rather than the direct cost benefits from predictive maintenance, Cafaro et al. (2011) describe how optimizing maintenance schedules based on predictive analytics can result in considerable reductions in operational expenses. As a result, although the reference endorses efficiency in general, predictive maintenance cost benefits are not particularly covered.

Additionally, predictive maintenance makes pipeline operations safer. The oil and gas sector is especially susceptible to disastrous breakdowns that could endanger public safety and cause environmental harm. Organizations can spot possible problems before they become major ones by putting predictive maintenance techniques into practice. In order to prevent needless losses brought on by pipeline breakdowns, Zheng et al. (2024) explain how predictive maintenance can evaluate pipeline dependability and carry out timely repair actions based on current conditions. In addition to preserving the environment, this proactive strategy also protects the health and safety of surrounding communities and employees.

Predictive maintenance solutions that incorporate data-driven technology and machine learning enable ongoing maintenance practice improvement. Organizations can improve their comprehension of pipeline behavior under different conditions and improve their predictive models as they gather more data over time. More precise forecasts and well-informed maintenance choices are the results of this iterative approach. Heidarydashtarjandi et al. (2022) highlight the value of optimizing maintenance plans for corroded pipelines using Markov Decision Processes, illustrating how data-driven methods can greatly enhance maintenance results.

The increased lifespan of assets is a noteworthy advantage of predictive maintenance. Organizations can prolong the operational life of pipeline infrastructure by proactively addressing maintenance needs and preventing significant degradation. In the context of aging pipeline systems, where proactive maintenance can lessen the effects of corrosion and other deterioration mechanisms, this is especially crucial. According to Mahmoodzadeh et al. (2020), condition-based maintenance can successfully control internal corrosion in dry gas pipelines using reinforcement learning support, hence extending the pipelines' service life.

Improved regulatory compliance is made possible via predictive maintenance. Operators of pipelines must adhere to strict safety and environmental protection rules. Organizations can show their dedication to preserving pipeline integrity and lowering risks by putting predictive maintenance techniques into practice. This proactive strategy improves the organization's standing among stakeholders in addition to helping it comply with regulatory standards. During regulatory audits, having data-driven proof of maintenance operations and pipeline conditions can be quite helpful. Along with these advantages, predictive maintenance improves organizational decision-making. Predictive maintenance solutions provide operators and management with real-time data on pipeline conditions, facilitating informed decisions regarding maintenance priorities and resource allocation. Organizations can evaluate the potential impacts of various maintenance methods and choose the optimal course of action by integrating sophisticated analytics into maintenance planning. This data-driven decision-making process is crucial for optimizing maintenance operations and ensuring the long-term sustainability of pipeline systems.

The oil and gas industry's shift to smart infrastructure is aided by predictive maintenance. The incorporation of predictive maintenance into pipeline management is consistent with the larger trend of Industry 4.0, which is characterized by the industry's growing adoption of digital technologies. Organizations can develop networked systems that improve operational responsiveness and efficiency by utilizing IoT devices, sensors, and advanced analytics. Odimarha et al. (2024) focuses on more general uses of machine learning in supply chain optimization rather than particularly addressing predictive maintenance in pipelines.

9.5 Predictive Maintenance Algorithms and Models

Predictive maintenance (PdM) forecasts equipment breakdowns and optimizes maintenance schedules by using data-driven models and algorithms. The diverse predictive maintenance models and algorithms have been created to improve the dependability and effectiveness of maintenance procedures in a variety of sectors, especially pipeline systems.

Machine learning (ML) approaches can examine vast datasets and find patterns that traditional methods would miss; they have become increasingly popular in predictive maintenance. The use of machine learning models for predictive maintenance is covered by Hichri et al. (2022), who highlight how these models may describe system behavior and anticipate future conditions using past data. Predictive maintenance that incorporates machine learning (ML) improves decision-making by enabling more precise forecasts of equipment health and possible faults.

Deep learning models, namely, Long Short-Term Memory (LSTM) networks, have been employed in predictive maintenance to evaluate the remaining useful life (RUL) and predict the need for maintenance. Wang et al. (2020) demonstrate the efficacy of LSTM networks in facilitating proactive and predictive maintenance for high-speed railway power equipment. The LSTM model is especially well-suited for applications where time-series data is common, like pipeline condition monitoring over time, because of its capacity to learn from sequential data. Using probabilistic models to estimate RUL and optimize maintenance choices is another important method of predictive maintenance. The study on dynamic grouping maintenance optimization emphasizes how crucial it is to forecast RUL via data processing and analysis (Cao et al. 2024). By understanding the probabilistic nature of equipment degradation, organizations can implement maintenance strategies that minimize costs while maximizing equipment availability and reliability.

To improve predictive maintenance policies, multi-objective optimization techniques have been created. In order to predict component failures and optimize maintenance schedules, Pisacane et al. (2021) propose a data-driven predictive maintenance policy that makes use of previous data. This method is beneficial for companies looking to enhance their maintenance procedures since it not only concentrates on reducing downtime but also considers the total cost-effectiveness of maintenance activities.

The application of hybrid models that combine different predictive techniques has also been explored. Three primary strategies for predictive maintenance are identified by Khan et al. (2025) through literature review: hybrid, data-driven, and physical modeling. By combining the advantages of data-driven and physical methods, hybrid models provide more accurate maintenance decisions and more reliable forecasts. Apart from these models, reinforcement learning has become a potent instrument for making predictive maintenance decisions. Feng and Li (2022) present a decision approach based on reinforcement learning that combines predictive maintenance and production system modeling. This approach enhances the overall effectiveness of maintenance operations by facilitating the real-time adjustment of maintenance techniques according to data and system performance. A probabilistic framework for simulating the uncertainty related to equipment breakdowns is offered by the application of Bayesian networks in predictive maintenance. Partovi et al.'s (2022) research highlights the potential of Bayesian networks to improve maintenance efficiency by incorporating various factors that influence equipment health and performance. This model enables organizations to make data-driven decisions regarding maintenance actions, ultimately leading to better resource allocation and reduced operational risks.

Predictive maintenance methods have become increasingly prevalent when combined with Internet of Things (IoT) technologies. Organizations can improve the precision of their prediction models by employing IoT devices to gather real-time data on equipment status. Liu et al. (2013) describe the creation of a predictive maintenance model employing hidden semi-Markov processes that integrates real-time data from Internet of Things (IoT) sensors to enhance maintenance scheduling. This link enhances maintenance results by enabling timely interventions and ongoing surveillance.

9.5.1 Regression Models

Regression models are commonly used in predictive maintenance to estimate the relationship between different variables and predict outcomes. In the context of pipeline integrity, regression models can be used to predict when a component, such as a pipeline segment, is likely to fail based on factors like age, material, operating pressure, and environmental conditions (Achouch et al. 2022).

- **Linear Regression**: Linear regression models predict failure based on a linear relationship between variables. For example, a linear regression model might predict pipeline corrosion as a function of time and operating temperature.
- **Multiple Regression**: In order to estimate the probability of failure, multiple regression models take into account a number of variables. These models are especially helpful in complicated pipeline systems where a number of variables interact to affect pipeline health, including flow rate, pressure, and ambient temperature.

9.5.2 Time Series Analysis

Because pipelines produce continuous data throughout time, time series analysis is especially useful for them. To predict future behavior, time series models examine patterns and trends in the data. These models can be used to identify trends that point to upcoming breakdowns in predictive maintenance.

- **ARIMA (Auto-Regressive Integrated Moving Average)**: A well-liked time series model for forecasting future values from historical data is ARIMA (Albeladi et al. 2023). ARIMA models can be used in pipeline integrity management to

forecast anomalies that could result in failures by applying them to temperature or pressure data.
- **Exponential Smoothing**: Another time series method that prioritizes recent data points is exponential smoothing, which makes it perfect for pipeline condition monitoring in real time (Biem et al. 2013).

9.5.3 Anomaly Detection Algorithms

Algorithms for anomaly detection find odd patterns or behaviors in data that could point to possible problems. When it comes to identifying early indicators of pipeline flaws like leaks, corrosion, or structural fatigue, these algorithms are especially helpful.

- **Clustering Algorithms**: Clustering algorithms group data points with similar characteristics and identify outliers that deviate from the norm. These outliers may represent anomalies that require further investigation.
- **Support Vector Machines (SVM)**: SVM is a machine learning algorithm that classifies data points and detects deviations from normal operating conditions. In pipeline integrity management, SVM can be used to detect subtle anomalies in sensor data that may indicate early-stage defects.

9.5.4 Survival Analysis

A statistical method called survival analysis is used to calculate how long it will take for an event, such an equipment breakdown, to occur. Operators can optimize their maintenance schedules by using survival analysis to estimate the remaining useful life (RUL) of pipeline components in predictive maintenance (Rossouw and Doorsamy 2021).

- **Weibull Analysis**: Weibull analysis is a common method of survival analysis used to model failure rates. It is particularly effective in predicting failures in systems with varying levels of reliability, such as pipelines exposed to different environmental conditions.
- **Cox Proportional Hazards Model**: The Cox model estimates the effect of different variables on the likelihood of failure over time, making it useful for pipelines with multiple interacting risk factors.

9.6 Predictive Maintenance Technologies

Predictive maintenance (PdM) technology encompasses an array of tools and procedures that enable firms to monitor equipment status, analyze data, and anticipate maintenance needs. This section addresses three fundamental components of predictive maintenance technologies: data collecting and analysis tools, sensor technologies for monitoring, and the roles of artificial intelligence (including machine learning) in predictive maintenance.

9.6.1 Sensor Technologies for Monitoring

Predictive maintenance is based on sensor technologies, which supply the information required to track the state of infrastructure and machinery. Real-time data on variables including temperature, vibration, pressure, and acoustic emissions are recorded using a variety of sensor types. For example, Pech et al. (2021) emphasizes the use of intelligent sensors based on nanoparticles and microparticles, which allow for direct monitoring within the item being monitored, improving data collecting accuracy and dependability. These sophisticated sensors can identify minute variations in the state of the equipment that can point to possible malfunctions. Furthermore, piezoelectric sensors have become more popular in applications related to structural health monitoring (SHM). The basic features and current state of piezoelectric materials and sensors for SHM are covered by Ju et al. (2023), who highlight their capacity to continually gather and examine data about the structural integrity of diverse systems. In fields where structural safety and dependability are crucial, including civil engineering and aerospace, this skill is very useful.

Another essential piece of technology for predictive maintenance is wireless sensor networks, or WSNs. These networks are made up of dispersed autonomous devices that keep an eye on environmental or physical conditions. In order to identify anomalies like leaks or bursts, Almazyad et al. (2014), for instance, talk about integrating WSNs into the monitoring of vital infrastructure, such as gas and water pipelines. WSN's adoption increases flexibility and lowers installation costs by enabling continuous monitoring without requiring substantial wiring.

Predictive maintenance has been transformed by the incorporation of IoT technologies in addition to conventional sensors. IoT-enabled sensors may transmit data in real time and monitor equipment remotely, enabling businesses to keep tabs on equipment conditions from any location. Maktoubian and Ansari (2019) demonstrate the potential for real-time medical device monitoring by demonstrating how IoT architecture may be used for preventative maintenance in healthcare organizations. By guaranteeing that vital equipment operates at peak efficiency, this technology not only increases maintenance efficiency but also improves patient safety.

9.6.2 Data Acquisition and Analysis Tools

The management of vast quantities of data generated by sensor technology necessitates the utilization of data collecting and analysis tools. Utilizing these instruments, organizations may collect, store, and analyze data to obtain valuable insights regarding the status of their equipment and the necessary maintenance. Advanced data collection systems can amalgamate sensors' data to deliver a comprehensive overview of equipment performance.

When it comes to evaluating sensor data, machine learning techniques are essential. For example, Kubik et al. (2022) highlight how crucial data translation and preprocessing are for improving machine learning models' ability to anticipate wear stages in manufacturing processes. Organizations may increase the precision of their predictive maintenance models and make better decisions by adding pertinent information to datasets. Furthermore, predictive maintenance initiatives can be strengthened using statistical process control (SPC) methodologies. Hsu et al. (2020) show how SPC and machine learning can be used to diagnose faults and perform predictive maintenance on wind turbines. With this method, system performance can be continuously monitored and possible problems can be found before they become serious failures.

Another important development is the incorporation of Big Data Analytics into predictive maintenance. Organizations can create predictive models that anticipate equipment breakdowns and optimize maintenance plans by utilizing both historical and real-time sensor data. For instance, a probabilistic method of calculating the equipment's remaining useful life (RUL) is made possible by the application of Gaussian process modeling in predictive maintenance, which enables more informed maintenance choices (Shao and Kumral 2024).

9.6.3 Machine Learning and AI in Predictive Maintenance

Machine learning (ML) technologies have revolutionized predictive maintenance by allowing firms to analyze intricate datasets and discern trends that signal probable issues. These technologies can improve the precision of predictive models and the overall efficiency of maintenance programs.

Predictive maintenance systems powered by AI are able to automatically evaluate data from a variety of sources, such as sensors, operational data, and maintenance logs from the past. Achouch et al. (2022), for example, give a summary of predictive maintenance models that use ML methods to reduce machine downtime and related expenses. Over time, these systems can adjust and enhance their predicting abilities by consistently gaining knowledge from fresh data.

Another interesting strategy for predictive maintenance is learning reinforcement. By combining predictive maintenance and production system modeling, Feng and Li (2022) present a reinforcement learning-based decision model that enables dynamic modifications based on real-time data. This method improves the capacity to adapt to shifting circumstances and modify maintenance plans as necessary. Predictive maintenance has also been seen an increase in the use of digital twins, which are virtual replicas of physical assets. By simulating how equipment will behave in different scenarios, digital twins help businesses anticipate malfunctions and plan for optimal maintenance. In their discussion of the use of digital twins in IoT-enabled induction motor monitoring, Santos et al. (2022) emphasize the possibility of predictive maintenance and real-time monitoring.

9.7 Integrating Prediction Maintenance with Pipeline Integrity Management

For pipeline systems to operate safely and effectively, predictive maintenance (PdM) and pipeline integrity management must be combined. This integration improves pipeline longevity and dependability by utilizing cutting-edge technology, real-time monitoring systems, and early warning systems. Two essential elements of this integration are covered in this section: early warning systems for failure prevention and real-time monitoring systems.

9.7.1 Real-Time Monitoring Systems

Systems for real-time monitoring are essential for combining pipeline integrity management with predictive maintenance. These systems continuously gather information on pipeline conditions using a variety of sensor technologies, giving operators real-time health monitoring capabilities. For example, Tejedor et al. (2018) outline a fiber-optic surveillance system that continuously checks the integrity of pipelines by identifying possible dangers in actual field settings. This system is a prime example of how cutting-edge monitoring tools may supply vital information for preventative maintenance plans.

Distributed Acoustic Sensing (DAS) technology is particularly effective for real-time monitoring of pipelines. Stajanča et al. (2018) demonstrate the use of DAS to detect leak-induced vibrations, highlighting its capability to provide timely information about the pipeline's condition. By analyzing the acoustic signals captured by DAS, operators can identify anomalies that may indicate leaks or structural issues, allowing for prompt maintenance interventions. Furthermore, real-time monitoring systems that incorporate Internet of Things (IoT) technologies improve the capacity to gather and examine data from various sources. Wireless data transmission from IoT-enabled sensors makes remote monitoring possible and eliminates the need for manual checks. This feature is essential for long-distance pipelines since it enables constant monitoring without requiring physical access to the system. In their discussion of the Internet of Things' use in water pipeline leak detection, Liu et al. (2019) highlight how useful it is for supplying real-time data for predictive maintenance.

Machine learning algorithms can be used to examine the data gathered from real-time monitoring systems to spot trends and anticipate possible malfunctions. For instance, predictive models that can predict equipment deterioration and maintenance requirements can be created by combining machine learning techniques with sensor data. Optimizing maintenance schedules and reducing downtime require the integration of data analytics with real-time monitoring.

9.7.2 Early Warning Systems for Failure Prevention

Early warning systems, which provide notifications about possible faults before they happen, are essential to the predictive maintenance architecture. These systems evaluate the quality of pipeline systems and pinpoint factors that could cause failures using data from real-time monitoring. Organizations can take proactive steps to address problems before they become serious ones by putting early warning systems in place.

The development of early warning systems heavily relies on machine learning methods. Hussain et al. (2024), for example, discussed the use of machine learning approaches to forecast gas pipeline stress corrosion cracking, emphasizing how these models might offer early warnings about possible problems. These algorithms can detect danger factors and produce alarms when conditions depart from typical operating settings by examining historical data and real-time sensor readings. Moreover, early warning systems are more successful when predictive analytics is incorporated into them. Organizations can evaluate the probability of failure by using statistical models and machine learning algorithms, which take into account past trends and present circumstances. By enabling prompt interventions, this predictive capability lowers the possibility of catastrophic failures and the expenses that come with them.

The accuracy of early warning systems can be further improved by utilizing ensemble learning techniques. Organizations can decrease the possibility of false alarms and produce more accurate predictions by merging many predictive models. With this method, operators can prioritize maintenance tasks according to the seriousness of hazards that have been discovered and gain a more nuanced view of pipeline health. Using Bayesian networks in conjunction with machine learning can provide early warning systems on a probabilistic basis. In order to adequately model uncertainties related to pipeline integrity, Woldesellasse and Tesfamariam (2025) investigate the application of Bayesian belief networks to evaluate the risk of gas pipeline failures. By integrating various factors that influence pipeline health, Bayesian networks can generate comprehensive risk assessments and facilitate informed decision-making regarding maintenance strategies.

9.8 Challenges and Limitations

There are obstacles and restrictions associated with the application of predictive maintenance (PdM) in pipeline systems, refer to Fig. 9.4. These challenges may impair the efficacy of predictive maintenance plans and have an influence on pipeline integrity management as a whole. Three main issues are covered in this section: operational and technological obstacles, cost and resource limitations, and data availability and quality.

9.8.1 Data Quality and Availability

The availability and quality of data are important variables that affect how well predictive maintenance programs work. Since the integrity and completeness of the input data are crucial to the efficacy of machine learning algorithms and statistical analyses, high-quality data is necessary for creating precise predictive models. However, gathering, managing, and storing data present difficulties for many firms.

The incompleteness and inaccuracy of data gathered from multiple sources is one major problem. For example, environmental conditions, noise, or calibration problems can alter sensor data, resulting in inaccurate observations. Substandard data quality can result in erroneous projections and ill-informed maintenance choices.

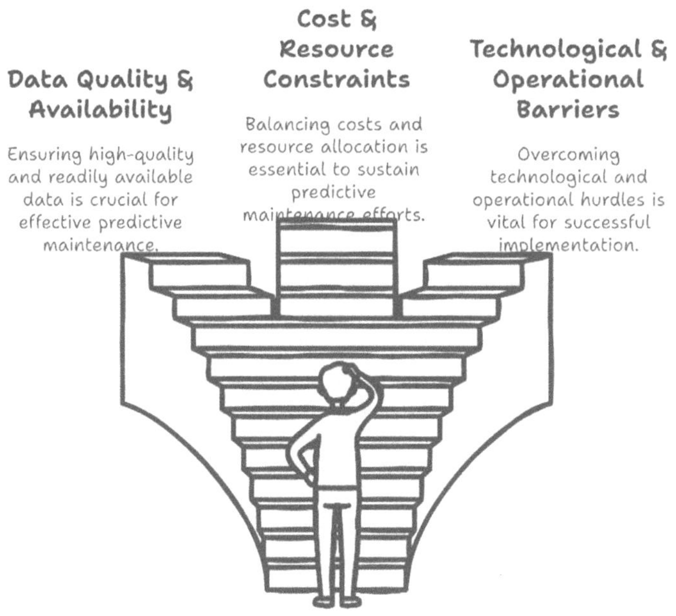

Fig. 9.4 Challenges in predictive maintenance implementation

To ensure the precision and reliability of data utilized for predictive maintenance, organizations must invest in robust data validation and cleansing protocols. Additionally, historical data availability is essential for predictive model training. The creation of efficient predictive maintenance plans may be hampered by the fact that many pipeline operators do not have access to adequate historical data on equipment performance and failure events.

The integration of data from many systems and sources is another difficulty. Pipeline operators frequently use a variety of monitoring methods, each of which produces data with a unique structure and format. The creation of cohesive predictive maintenance models may be hampered by this fragmentation, which can also make data analysis more difficult. Organizations should think about putting in place standardized data formats and integration platforms that facilitate easy data sharing and analysis amongst various systems in order to solve this problem.

9.8.2 Cost and Resource Constraints

Implementing predictive maintenance in pipeline systems is severely hampered by financial and resource limitations. Although lowering unscheduled downtime and maintenance costs through predictive maintenance might result in long-term cost benefits, putting PdM technology into practice can come with a hefty upfront cost. Businesses may find it challenging to defend these up-front expenses, especially in sectors with narrow profit margins.

The expenditures related to implementing modern sensor technologies, data collecting systems, and analytics platforms may be prohibitive for certain enterprises. The efficacy of predictive maintenance programs may be restricted by resource limitations, such as staffing and experience. PdM implementation calls for knowledgeable staff with expertise in maintenance management, machine learning, and data analytics. However, a lot of businesses have trouble locating and keeping skilled experts in these fields. The effective deployment and operation of PdM systems may be hampered by the lack of qualified personnel in the field of predictive maintenance. Companies must spend money on training and development initiatives to increase internal knowledge and make sure that employees are prepared to handle predictive maintenance technologies.

Organizational resources may be strained by the continuous operating expenses related to predictive maintenance system maintenance. Organizations with tight resources may find it difficult to provide the specialized staff and infrastructure needed for ongoing monitoring, data analysis, and model changes. To solve this problem, businesses should carry out in-depth cost–benefit assessments to assess predictive maintenance's long-term worth and pinpoint possible cost-saving measures.

9.8.3 *Technological and Operational Barriers*

The effective incorporation of predictive maintenance into pipeline integrity management may be hindered by operational and technological obstacles. The quick speed at which technology is developing is one major obstacle that can make it hard for businesses to stay up to date with the newest tools and techniques. Operators may become perplexed regarding the suitable procedures and solutions due to the continuous advancement of predictive maintenance technologies.

Operational difficulties may arise due to the intricacy of predictive maintenance systems. It takes careful planning and coordination to integrate multiple technologies, including sensors, data analytics platforms, and machine learning algorithms. It could be challenging for organizations to integrate new predictive maintenance technologies with their current workflows and processes. During the implementation stage, inefficiencies and operational disruptions may result from a lack of defined procedures and protocols. Additionally, the adoption of predictive maintenance techniques can be significantly influenced by company culture. PdM initiatives may not be implemented successfully if management and staff are resistant to change. Employers must encourage staff members to adopt new technology and processes by cultivating an innovative and continuous improvement culture. This cultural change may entail offering assistance and training to staff members so that they can comprehend the advantages of predictive maintenance.

References

Achouch M et al (2022) On predictive maintenance in industry 4.0: overview, models, and challenges. Appl Sci 12(16):8081

Adegboye MA, Fung W-K, Karnik A (2019) Recent advances in pipeline monitoring and oil leakage detection technologies: principles and approaches. Sensors 19(11):2548

Albeladi K, Zafar B, Mueen A (2023) Time series forecasting using LSTM and ARIMA. Int J Adv Comput Sci Appl 14(1):313–320

Almazyad AS et al (2014) A proposed scalable design and simulation of wireless sensor network-based long-distance water pipeline leakage monitoring system. Sensors 14(2):3557–3577

Amadhe FO, Anjorin RO, Uwoghiren FO (2024) Advancements in machine learning for pipeline integrity management: a comprehensive review of predictive and optimization techniques. Cognizance J Multi Stud 4(11):129–138

Bęben D, Steliga T (2023) Monitoring and preventing failures of transmission pipelines at oil and natural gas plants. Energies 16(18):6640

Biem A et al (2013) Real-time analysis and management of big time-series data. IBM J Res Dev 57(3/4):8:1–8:12

Binder M, Mezhuyev V, Tschandl M (2023) Predictive maintenance for railway domain: a systematic literature review. IEEE Eng Manage Rev 51(2):120–140

Burdylo RS, Van Aelst AL (2000) Enhanced pipeline integrity at lower cost: a pipeline integrity prioritization model. In: Proceedings of the 3rd International Pipeline Conference, vol 2, IPC2000-207. October 1–5, 2000. Calgary, Alberta, Canada. American Society of Mechanical Engineers

Cafaro VG et al (2011) Detailed scheduling of operations in single-source refined products pipelines. Ind Eng Chem Res 50(10):6240–6259

Cao X et al (2024) Dynamic grouping maintenance optimization by considering the probabilistic remaining useful life prediction of multiple equipment. Eksploatacja Niezawodność 26(3):187793

Cardoso D, Ferreira L (2020) Application of predictive maintenance concepts using artificial intelligence tools. Appl Sci 11(1):18

Chalgham W, Wu K-Y, Mosleh A (2020) System-level prognosis and health monitoring modeling framework and software implementation for gas pipeline system integrity management. J Nat Gas Sci Eng 84:103671

Dos Santos JF et al (2022) Digital twin-based monitoring system of induction motors using IoT sensors and thermo-magnetic finite element analysis. IEEE Access 11:1682–1693

Feng M, Li Y (2022) Predictive maintenance decision making based on reinforcement learning in multistage production systems. IEEE Access 10:18910–18921

Heidarydashtarjandi R, Prasad-Rao J, Groth KM (2022) Optimal maintenance policy for corroded oil and gas pipelines using markov decision processes. Int J Prognostics Health Manage 13(1)

Hichri B et al (2022) Predictive maintenance based on machine learning model. In: IFIP international conference on artificial intelligence applications and innovations. Springer

Hussain M et al (2024) Review of prediction of stress corrosion cracking in gas pipelines using machine learning. Machines 12(1):42

Jain M et al (2022) Systematic literature review on predictive maintenance of vehicles and diagnosis of vehicle's health using machine learning techniques. Comput Intell 38(6):1990–2008

James S, Rajasekaran A (2023) Design and development of an external modular Pi-XTS robot for detecting presence of defects in pipelines. Preprints. Research Square

Ju M et al (2023) Piezoelectric materials and sensors for structural health monitoring: fundamental aspects, current status, and future perspectives. Sensors 23(1):543

Khan ZU, Nasim B, Rasheed Z (2025) Generative AI-based predictive maintenance in aviation: a systematic literature review. CEAS Aeronaut J 1–19

Kisić E, Đurović Ž, Petrović V (2017) Predictive maintenance based on control charts applied at thermoelectric power plant. In: Recent improvements of power plants management and technology. pp 27–48

Krupitzer C et al (2020) A survey on predictive maintenance for industry 4.0. arXiv preprint arXiv: 2002.08224

Kubik C, Knauer SM, Groche P (2022) Smart sheet metal forming: importance of data acquisition, preprocessing and transformation on the performance of a multiclass support vector machine for predicting wear states during blanking. J Intell Manuf 33(1):259–282

Liu Q, Dong M, Peng Y (2013) A dynamic predictive maintenance model considering spare parts inventory based on hidden semi-Markov model. Proc Inst Mech Eng C J Mech Eng Sci 227(9):2090–2103

Liu Y et al (2019) Water pipeline leakage detection based on machine learning and wireless sensor networks. Sensors 19(23):5086

Ma Q et al (2021) Pipeline in-line inspection method, instrumentation and data management. Sensors 21(11):3862

Mahmoodzadeh Z et al (2020) Condition-based maintenance with reinforcement learning for dry gas pipeline subject to internal corrosion. Sensors 20(19):5708

Maktoubian J, Ansari K (2019) An IoT architecture for preventive maintenance of medical devices in healthcare organizations. Heal Technol 9:233–243

Mohammed NA, Abdulateef OF, Hamad AH (2023) An IoT and machine learning-based predictive maintenance system for electrical motors. J Eur Systèmes Automatisés 56(4):651

Naeem K et al (2023) High-sensitivity distributed pipelines infrastructure monitoring with internal deployed fibers and Rayleigh enhancement. In: Optical waveguide and laser sensors II. SPIE

Odimarha AC, Ayodeji SA, Abaku EA (2024) Machine learning's influence on supply chain and logistics optimization in the oil and gas sector: a comprehensive analysis. Comput Sci IT Res J 5(3):725–740

Omoya OA, Papadopoulou KA, Lou E (2019) Reliability engineering application to pipeline design. Int J Qual Reliab Manage 36(9):1644–1662

Partovi M et al (2022) Predictable maintenance: a Bayesian network-based model. Int J Reliab Risk Safety: Theory Appl 5(2):97–105

Pech M, Vrchota J, Bednář J (2021) Predictive maintenance and intelligent sensors in smart factory. Sensors 21(4):1470

Pisacane O et al (2021) Data-driven predictive maintenance policy based on multi-objective optimization approaches for the component repairing problem. Eng Optim 53(10):1752–1771

Priyanka EB, Maheswari C, Thangavel S (2021) A smart-integrated IoT module for intelligent transportation in oil industry. Int J Numer Model Electron Networks Devices Fields 34(3):e2731

Qin G et al (2022) An information entropy-based risk assessment method for multiple-media gathering pipelines. J Infrastruct Preserv Resil 3(1):19

Rossouw E, Doorsamy W (2021) Predictive maintenance framework for cathodic protection systems using data analytics. Energies 14(18):5805

Shao Z, Kumral M (2024) Implementing Gaussian process modelling in predictive maintenance of mining machineries. Min Technol 133(4):348–368

Shen T et al (2024) Spatial data fusion model design and research for an underground pipeline in urban environment scene modeling. Appl Sci 14(15):6760

Stajanca P et al (2018) Detection of leak-induced pipeline vibrations using fiber—optic distributed acoustic sensing. Sensors 18(9):2841

Sulaiman NSA, Tan H (2018) Dynamic reliability analysis of corroded pipeline using Bayesian Network. Int J Eng Technol (UAE) 7(4):210–215

Tagawa Y, Maskeliūnas R, Damaševičius R (2021) Acoustic anomaly detection of mechanical failures in noisy real-life factory environments. Electronics 10(19):2329

Tejedor J et al (2018) Real field deployment of a smart fiber-optic surveillance system for pipeline integrity threat detection: architectural issues and blind field test results. J Lightwave Technol 36(4):1052–1062

Wang Q, Bu S, He Z (2020) Achieving predictive and proactive maintenance for high-speed railway power equipment with LSTM-RNN. IEEE Trans Industr Inf 16(10):6509–6517

Woldesellasse H, Tesfamariam S (2025) Risk assessment of gas pipeline using an integrated Bayesian belief network and GIS: using Bayesian neural networks for external pitting corrosion modelling. Can J Chem Eng 103(1):98–109

Zhang P et al (2023) Safety analysis and condition assessment of corroded energy pipelines under landslide disasters. Appl Sci 13(23):12880

Zheng Y et al (2024) Pipeline reliability assessment and predictive maintenance considering multi-crack dependent degradation. Proc Inst Mech Eng Part b: J Eng Manuf 238(8):1122–1144

Chapter 10
Machine Learning Applications in Pipeline Integrity Management

10.1 Introduction

Machine learning (ML) has revolutionized pipeline integrity management with its advanced methodologies in risk assessment, anomaly detection, and predictive maintenance. This chapter elucidates the methodologies, strategies, and discoveries from recent research, exploring various applications of machine learning to ensure the integrity and reliability of pipeline systems.

Predictive maintenance is the key reason for incorporating machine learning into pipeline integrity management. Conventional maintenance methods frequently depend on planned examinations and reactive actions, which might result in unanticipated malfunctions and expensive downtime. Machine learning makes it possible to anticipate possible problems before they happen, allowing for prompt interventions, by utilizing data-driven methodologies. For example, research has shown how well a variety of machine learning algorithms, such as support vector machines (SVMs) and artificial neural networks (ANNs), can forecast the burst pressure of corroded pipes using historical data (Lo et al. 2021; Hussain et al. 2024). A more sophisticated knowledge of pipeline behavior under varied circumstances is made possible by these prediction models using large datasets that cover a variety of operational characteristics and previous failure records.

The evaluation of pipeline corrosion hazards is another area in which machine learning is being used. Studies show that machine learning methods, especially principal component analysis (PCA) and regression models, can greatly improve corrosion depth prediction accuracy (Ossai 2019; Wang et al. 2023). These models may forecast future corrosion rates and identify significant risk factors by examining environmental conditions and previous corrosion data. This information can then be used to influence resource allocation and maintenance schedules. Because it directly affects the pipeline's structural integrity, the ability to anticipate the maximum corrosion depth is essential for preserving pipeline safety and operating effectiveness (Wang et al. 2023).

M. Hussain and T. Zhang, *Big Data Analytics in Energy Pipeline Integrity Management*, Lecture Notes in Energy 104, https://doi.org/10.1007/978-981-96-8019-1_10

In addition to corrosion risk assessment and predictive maintenance, machine learning is crucial for identifying anomalies in pipeline systems. The advent of advanced sensor technology enables continuous monitoring of pipeline conditions, generating vast amounts of data that can be analyzed by machine learning algorithms. Real-time leak and anomaly detection has been explored with convolutional neural networks (CNNs) and recurrent neural networks (RNNs) (Rayhana et al. 2024; Şahin and Yüce 2023). These models can swiftly respond to emerging issues as they are trained on historical data to discern patterns indicative of potential failures. The ability to detect and classify pipeline defects is enhanced by integrating distributed fiber sensors with machine learning analytics, providing a comprehensive method for integrity monitoring (Zhang et al. 2023).

The area of pipeline integrity management has also undergone a revolution thanks to the introduction of automated machine learning (AutoML) techniques. There is less need for intensive manual involvement when machine learning pipelines are designed and optimized automatically thanks to autoML frameworks like the Tree-based Pipeline Optimization Tool (TPOT) (Gijsbers et al. 1801; Olson and Moore 2016). These technologies allow practitioners to implement efficient machine learning solutions with little experience by streamlining the feature selection, model training, and hyperparameter tweaking processes. In the context of pipeline integrity monitoring, where prompt and precise forecasts are crucial for averting failures and guaranteeing safety, the capacity to automate these procedures is especially advantageous.

Machine learning models become more predictive when domain information is included into them. Researchers can create more reliable models that take into consideration the intricacies of pipeline systems by fusing data-driven methodologies with expert insights. For example, it has been demonstrated that the accuracy of failure predictions is increased by using multi-parameter models that take into account a variety of parameters, including material qualities and environmental circumstances. In addition to improving prediction accuracy, this comprehensive method helps identify important risk factors that would not be visible through data analysis alone.

The use of machine learning in risk assessment frameworks emphasizes its significance in pipeline integrity management. Machine learning models can assess the probability of pipeline failures by using relative risk score methodologies, which take into account a variety of factors such as corrosion rates, material qualities, and environmental impacts (Vanitha et al. 2023; Cen et al. 2023). Operators may more efficiently manage resources and prioritize maintenance tasks thanks to this probabilistic method, which provides a more thorough understanding of the hazards related to pipeline operations.

10.2 Types of Machine Learning Models Used in Pipeline Integrity Management

The utilization of machine learning (ML) in pipeline integrity management has garnered considerable attention, with diverse models implemented to improve predictive capacities, risk evaluation, and anomaly identification. Figure 10.1 presents the available machine learning models in Pipeline Integrity Management.

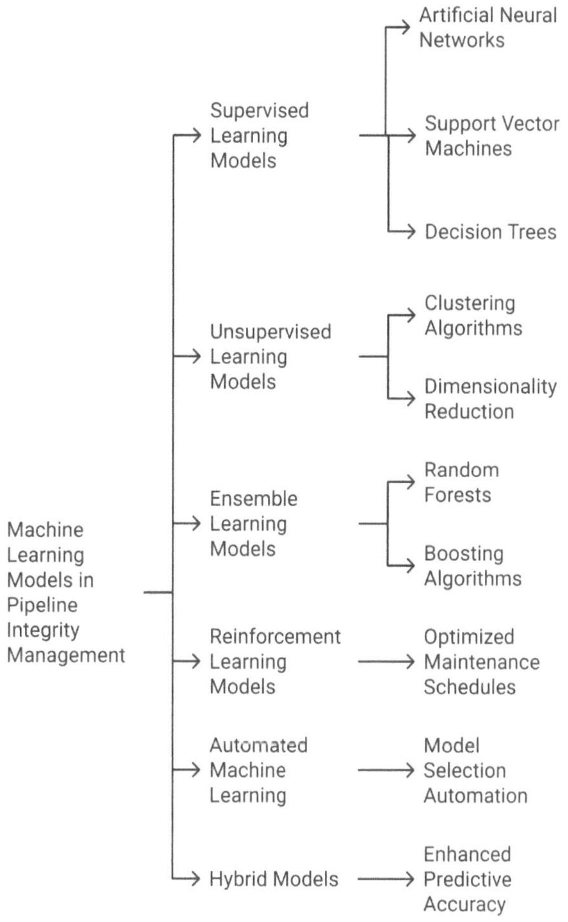

Fig. 10.1 Machine learning models in pipeline integrity management

10.2.1 Supervised Learning Models

Supervised learning is one of the most prevalent models for pipeline integrity control. To train algorithms for result prediction based on input features, labeled datasets are required. Decision trees, support vector machines, and artificial neural networks (ANNs) are some of the most prevalent supervised learning techniques. ANNs have demonstrated impressive success in forecasting pipeline failures and evaluating corrosion risks, especially when it comes to deep learning systems. For instance, research has shown that ANNs are capable of accurately simulating the link between a number of operational characteristics and the probability of corrosion-related pipeline failure (Lo et al. 2021). Because of their adaptability, ANNs can capture intricate nonlinear correlations in data, which makes them appropriate for situations where more conventional models would not be able to perform well.

Another effective supervised learning method for pipeline integrity monitoring categorization jobs is Support Vector Machines (SVMs). By identifying the best hyperplane, SVMs can successfully separate classes in situations when the data is high-dimensional. Based on sensor data, they have been used to categorize pipeline conditions, allowing operators to spot possible leaks or structural flaws (Şahin and Yüce 2023; Vanitha et al. 2023). Because of their resilience and interpretability, decision trees and ensemble techniques like random forests are also frequently used. The developed models facilitate informed decision-making by highlighting the key variables that influence pipeline integrity (Amadhe et al. 2024; Alobaidi et al. 2022).

Supervised learning models are trained on labeled datasets, wherein the input data is linked to known outcomes. These models acquire the ability to associate inputs with outputs and are frequently employed for classification and regression tasks (Nasteski 2017).

- **Classification Models**: Classification models are used to categorize data points into predefined classes (Osisanwo et al. 2017). In pipeline integrity management, classification models can be used to detect anomalies, such as identifying whether a specific sensor reading indicates normal operation or a potential failure.
- **Regression Models**: Regression models are employed to forecast continuous values, such as calculating the remaining usable life (RUL) of a pipeline component or projecting future pressure levels based on existing data (Shaik et al. 2022; Lei et al. 2018).

10.2.2 Unsupervised Learning Models

Models for unsupervised learning are essential for anomaly identification and exploratory data analysis. These models find patterns or groupings within the dataset rather than requiring labeled data. Operational data indicates that clustering techniques like K-means and hierarchical clustering are commonly employed to categorize pipeline conditions. Clustering can facilitate the identification of aberrant patterns indicative of potential breakdowns, hence enabling preventive maintenance actions (Amadhe et al. 2024; Mahmoodzadeh et al. 2020).

Before using supplementary machine learning models, data undergoes preprocessing through dimensionality reduction methods such as Principal Component Analysis (PCA). In situations involving high-dimensional data, as is common in pipeline monitoring systems, PCA aids in simplifying datasets while maintaining crucial information (Ossai 2019; Amadhe et al. 2024).

Prior to applying additional machine learning models, data is pre-processed using dimensionality reduction techniques like Principal Component Analysis (PCA). In situations involving high-dimensional data, as is common in pipeline monitoring systems, PCA aids in simplifying datasets while maintaining crucial information (Ossai 2019; Amadhe et al. 2024). PCA can improve subsequent prediction models' performance by streamlining the data structure. Because it may detect outliers or departures from typical operating circumstances without being specifically trained on failure data, unsupervised learning is very helpful for anomaly detection.

- **Clustering Algorithms**: Clustering algorithms group similar data points together, allowing operators to identify patterns and detect outliers (Zimek et al. 2012). In pipeline integrity management, clustering can be used to identify abnormal behavior in sensor data, such as pressure spikes or unusual temperature variations.
- **Dimensionality Reduction**: Dimensionality reduction techniques, such as principal component analysis (PCA), are used to reduce the complexity of large datasets while preserving important features (Hasan and Abdulazeez 2021). This is useful in pipeline integrity management, where operators may be dealing with high-dimensional data from multiple sensors.

10.2.3 Ensemble Learning Models

Multiple individual models are combined in ensemble learning models to increase overall resilience and performance. In pipeline integrity management, strategies including stacking, boosting, and bagging are frequently employed. An ensemble of decision trees called Random Forest has been proven to be very useful in managing the unpredictability of pipeline data. It has been used to forecast corrosion rates and evaluate failure risk (Amadhe et al. 2024; Alobaidi et al. 2022). By concentrating on incorrectly identified occurrences in the training data, boosting techniques like AdaBoost and Gradient Boosting have also been used to improve prediction accuracy (Vanitha et al. 2023; Meda et al. 2021). Because ensemble approaches can take advantage of the capabilities of many models to improve forecast performance, they enable a more thorough examination of pipeline integrity. This approach is especially beneficial in complex environments where multiple factors influence pipeline integrity.

10.2.4 Reinforcement Learning Models

A novel domain in machine learning, known as reinforcement learning (RL), seeks to instruct agents in decision-making through the process of trial and error. Reinforcement learning helps optimize resource allocation and maintenance schedules within pipeline integrity management. Through the simulation of many operational scenarios, reinforcement learning algorithms can identify optimal techniques for risk mitigation and the enhancement of pipeline performance (Amadhe et al. 2024; Mahmoodzadeh et al. 2020). This adaptive learning process allows for continuous improvement in decision-making, making it a valuable tool for dynamic environments where conditions frequently change.

10.2.5 Automated Machine Learning (AutoML)

The advent of Automated Machine Learning (AutoML) has simplified model selection and hyperparameter tuning, hence making machine learning accessible to pipeline integrity management professionals. By utilizing tools such as TPOT and Auto-SKLEARN, users may focus on analyzing results instead of the intricacies of model development, as these tools automate the construction and optimization of machine learning pipelines (Olson and Moore 2016; Olson et al. 2016). The implementation of data-driven methodologies in pipeline integrity monitoring can be expedited using autoML, significantly reducing the time and expertise required to establish effective machine learning systems.

10.2.6 Hybrid Models

Hybrid models that integrate various machine learning strategies are becoming favored for their capacity to utilize the advantages of many methodologies. For instance, integrating ANNs with traditional statistical methods can enhance predictive accuracy while providing interpretable results. Such hybrid approaches have been utilized to model complex relationships in pipeline data, improving the reliability of predictions related to corrosion and structural integrity (Amadhe et al. 2024; Sun and Zhou 2023). The flexibility of hybrid models allows for tailored solutions that can address specific challenges in pipeline integrity management.

10.3 Applications of Machine Learning in Pipeline Integrity Management

Machine learning is being applied across various aspects of pipeline integrity management, from detecting leaks and corrosion to predicting failures and optimizing maintenance schedules. Below are some of the key applications of machine learning in this field.

10.3.1 Leak Detection

Leaks are one of the most significant risks in pipeline operations, leading to environmental damage, financial losses, and safety hazards. Algorithms that use machine learning can analyze sensor data in real time to locate leaks by detecting anomalies in pressure, flow rate, or sound signals (Ullah et al. 2023; Ahmad et al. 2022).

- **Anomaly Detection for Leak Detection**: Anomaly detection algorithms, such as support vector machines (SVM) or autoencoders, can be trained to identify patterns in sensor data that deviate from normal operating conditions (Spandonidis et al. 2022). These deviations may indicate a leak or other structural issues in the pipeline.
- **Acoustic Leak Detection**: Machine learning models can analyze acoustic data from sensors placed along the pipeline to detect the sound of escaping gas or liquid. By comparing the acoustic signature of a leak to normal pipeline sounds, these models can detect leaks early and accurately (Xiao and Li 2023; Quy and Kim 2021).

10.3.2 Corrosion Monitoring and Prediction

Predicting corrosion rates is essential for preserving pipeline integrity because corrosion is a major contributor to pipeline failure. To forecast when and where corrosion is likely to occur, machine learning algorithms can examine past data on corrosion rates, material characteristics, and environmental conditions.

- **Predictive Corrosion Models**: Machine learning algorithms can be trained on historical corrosion data to predict the rate of corrosion in different pipeline segments. These models can consider factors such as operating pressure, temperature, and material type to estimate when a pipeline segment will need to be replaced or repaired (Hussain et al. 2024; Soomro et al. 2022).
- **Example**: A machine learning model could analyze data from corrosion sensors and historical maintenance records to predict that a specific section of pipeline is likely to experience severe corrosion within the next year. Operators can then

plan preventive maintenance to replace or reinforce the affected section before a failure occurs.

10.3.3 Pipeline Failure Prediction

Predicting when a pipeline is likely to fail is one of the most valuable applications of machine learning in pipeline integrity management (Hussain et al. 2023). Machine learning models are capable of anticipating failures by evaluating data from various sources, including as ambient conditions, sensor readings, and inspection reports.

- **Time Series Analysis for Failure Prediction**: Historical sensor data can be analyzed by time series models like ARIMA or recurrent neural networks (RNNs) to find patterns that might point to an imminent failure (Kashpruk et al. 2023). Based on historical performance, these models can forecast when a failure is likely to happen, enabling operators to take preventative action.
- **Remaining Useful Life (RUL) Estimation**: Machine learning models can estimate the remaining useful life (RUL) of pipeline components by analyzing data on wear and degradation (Kahraman et al. 2024). This information helps operators optimize maintenance schedules and avoid costly downtime.

10.3.4 Optimizing Maintenance Schedules

Optimizing maintenance schedules is one of machine learning's main advantages in pipeline integrity management. By predicting when and where maintenance is needed, machine learning algorithms can help operators prioritize repairs and allocate resources more efficiently.

- **Maintenance Prioritization**: Machine learning models can analyze the risk factors for each pipeline segment and prioritize maintenance activities based on the likelihood of failure (Mazumder et al. 2021). This ensures that the most vulnerable sections of the pipeline are addressed first, reducing the overall risk of failure.
- **Dynamic Maintenance Scheduling**: Machine learning models can dynamically adjust maintenance schedules based on real-time data from sensors (Ayvaz and Alpay 2021). For example, if a model detects an increase in pressure fluctuations in a particular section, it can recommend scheduling maintenance earlier than originally planned.

10.4 Benefits of Machine Learning in Pipeline Integrity Management

Numerous advantages that improve pipeline operations' safety, effectiveness, and dependability have resulted from the use of machine learning (ML) into pipeline integrity management. The main benefits of using machine learning approaches to control pipeline integrity are covered in this section, along with pertinent literature to back them up.

10.4.1 Enhanced Predictive Maintenance

Predictive maintenance is one of the most important advantages of machine learning for pipeline integrity management. Scheduled inspections are a common component of traditional maintenance procedures, which can result in needless downtime and higher operating expenses. Machine learning models can anticipate possible failures before they happen, enabling prompt interventions. This is especially true for models that are based on historical data. Studies have demonstrated that, depending on a variety of operating characteristics, machine learning algorithms like Random Forests and Support Vector Machines can accurately forecast corrosion rates and the probability of pipeline failures (Amadhe et al. 2024; Woldesellasse and Tesfamariam 2025). This proactive strategy improves maintenance schedules and lowers the chance of catastrophic failures, which results in significant cost savings.

10.4.2 Improved Risk Assessment

Machine learning enhances risk assessment capabilities in pipeline integrity management by enabling more accurate predictions of potential failure modes. By analyzing vast amounts of data from various sources, including environmental conditions, material properties, and historical failure records, machine learning models can identify critical risk factors associated with pipeline integrity (Vanitha et al. 2023; Woldesellasse and Tesfamariam 2025). For example, the application of relative risk scoring techniques has demonstrated high accuracy in classifying the risk levels of pipelines, allowing operators to prioritize maintenance efforts based on the severity of potential risks (Vanitha et al. 2023). This data-driven approach to risk assessment empowers pipeline operators to make informed decisions regarding resource allocation and risk mitigation strategies.

10.4.3 Anomaly Detection and Leak Identification

Another significant benefit of machine learning is its real-time anomaly detection and leak detection capabilities. Extensive data is generated by advanced sensor technologies, which can be analyzed by machine learning algorithms to identify trends indicative of leaks or structural defects. For pipeline system leak detection, methods like Convolutional Neural Networks (CNNs) have shown effective (Şahin and Yüce 2023). By continuously monitoring pipeline conditions and analyzing sensor data, machine learning models can provide early warnings of potential leaks, enabling swift responses that minimize environmental damage and economic losses.

10.4.4 Data-Driven Decision Making

By fostering a data-driven culture in pipeline integrity management, machine learning empowers operators to make decisions grounded in empirical evidence rather than intuition or past experience. Evaluations of pipeline integrity are enhanced in accuracy and reliability by the integration of machine learning into decision-making processes (Amadhe et al. 2024; Woldesellasse and Tesfamariam 2025). For instance, intricate datasets can be examined by machine learning models to uncover patterns and correlations that traditional analytical methods may overlook. Ultimately, this capability enhances the overall safety and efficiency of pipeline operations by enabling operators to make educated decisions regarding maintenance, repairs, and upgrades.

10.4.5 Automation and Efficiency

The automation of data analysis and decision-making processes through machine learning significantly enhances operational efficiency. Automated Machine Learning (AutoML) tools streamline the development and deployment of machine learning models, reducing the time and expertise required to implement effective solutions (Gijsbers et al. 1801; Celik et al. 2023). By automating routine tasks, such as data preprocessing, model selection, and hyperparameter tuning, pipeline operators can focus on higher-level strategic decisions rather than getting bogged down in technical details (Gijsbers et al. 1801; Olson and Moore 2016). This increased efficiency allows for quicker responses to emerging issues and enhances the overall agility of pipeline management operations.

10.4.6 Cost Reduction

The implementation of machine learning in pipeline integrity management can lead to substantial cost reductions. Operators can drastically reduce repair and operating expenses by reducing the likelihood of unplanned failures and streamlining maintenance plans (Amadhe et al. 2024; Daniel and Olaoye 2025). Predicting and preventing failures also lessens the possibility of expensive environmental catastrophes, which can lead to large financial obligations and legal repercussions. In addition to facilitating more effective resource allocation, machine learning's data-driven insights guarantee that maintenance efforts are concentrated on the most important locations (Amadhe et al. 2024; Woldesellasse and Tesfamariam 2025).

10.4.7 Enhanced Safety and Environmental Protection

Protecting the environment and ensuring operational safety are the main objectives of pipeline integrity management. By enhancing the identification of possible risks and facilitating prompt actions, machine learning helps achieve these goals. For instance, using machine learning algorithms to identify leaks reduces the possibility of environmental contamination while also safeguarding the pipeline infrastructure. Machine learning supports regulatory compliance and corporate social responsibility objectives by improving pipeline operations' safety, protecting the environment and public health.

10.4.8 Continuous Improvement and Adaptation

Continuous increases in operational efficacy and forecast accuracy are made possible by machine learning models' capacity to learn from and adjust to new data. Machine learning algorithms can improve their comprehension of pipeline behavior under different conditions and improve their predictions as more data becomes available. This flexibility helps pipeline operators keep ahead of possible problems and maintain peak performance, which is especially useful in the face of changing climatic circumstances and operational difficulties.

10.4.9 Integration with Advanced Technologies

The benefits of machine learning in pipeline integrity monitoring are amplified by its integration with other technologies, including Big Data Analytics and the Internet of Things (IoT). Real-time data from pipelines can be gathered by IoT devices, and machine learning algorithms can be used to analyze the data and extract useful information (Amadhe et al. 2024; Woldesellasse and Tesfamariam 2025). By combining machine learning and IoT technology, pipeline systems may be monitored and managed more thoroughly, improving overall operational safety and efficiency.

10.4.10 Scalability and Flexibility

Machine learning solutions are inherently scalable and flexible, allowing them to be applied across various pipeline systems and configurations. This scalability is crucial for managing extensive pipeline networks, as machine learning models can process large volumes of data from multiple sources simultaneously (Amadhe et al. 2024; Woldesellasse and Tesfamariam 2025). The adaptability of machine learning algorithms allows them to be tailored to various pipelines, materials, and operational conditions, rendering them appropriate for diverse applications in pipeline integrity monitoring.

10.4.11 Real-Time Monitoring

One important benefit of using machine learning in pipeline integrity control is real-time monitoring. Operators can keep a current awareness of the pipeline's condition by continuously gathering and analyzing data from a variety of sensors positioned along the pipeline system. This capacity is essential for identifying alterations that might point to possible malfunctions or safety risks. In order to detect anomalous patterns that can indicate leaks or structural flaws, machine learning algorithms can be used to assess real-time data on pressure, temperature, and flow rates that is provided by IoT devices (Im et al. 2024). Operators can react quickly to new problems by combining machine learning and real-time data analytics, which reduces the possibility of catastrophic failures and environmental harm (Amadhe et al. 2024). Additionally, machine learning-enabled real-time monitoring improves pipeline systems' operational efficiency. Instead of wasting too much time on manual data processing, operators may concentrate on analyzing outcomes and making wise decisions by automating the data analysis process. In big pipeline networks, where the amount of data created might be overwhelming, this efficiency is especially advantageous. Real-time processing of this data by machine learning algorithms can yield actionable insights that assist operators in allocating resources and optimizing maintenance plans (Amadhe et al. 2024; Nguyen et al. 2020).

10.4.12 Enhanced Anomaly Detection

Improved anomaly detection is a notable advantage of machine learning in pipeline integrity control. Conventional anomaly detection techniques frequently depend on preset thresholds and rules, which might not take into consideration the intricacies and fluctuations of actual pipeline operations. Without the use of explicit rules, machine learning models—especially those that are based on unsupervised learning techniques—can detect anomalies by examining patterns in the data (Putatunda et al. 2005; Li and Zou 2109). In order to identify outliers that can signal to possible problems, clustering algorithms, for instance, might group comparable operational data points (Kubba et al. 2024).

The precision and reliability of anomaly detection in pipeline systems are augmented by advanced machine learning methodologies, including ensemble techniques and deep learning. These models can identify intricate patterns and associations by learning from past data, which allows them to identify little changes that might occur before major failures (Dong and Rekatsinas 2018; Laadan et al. 2020). In contrast to conventional techniques, research has demonstrated that the use of Convolutional Neural Networks (CNNs) for sensor data analysis can result in more precise leak detection (Laux et al. 2020). In addition to increasing pipeline operations' safety, this improved capability lowers the expenses of unforeseen repair and environmental cleanup. Moreover, the amalgamation of machine learning with real-time monitoring systems facilitates ongoing anomaly identification, permitting operators to obtain prompt notifications upon the emergence of prospective concerns. This proactive strategy for anomaly identification is essential in averting catastrophic failures, as it facilitates timely interventions before to the escalation of issues (Doerr et al. 2021). The integration of real-time data analysis and sophisticated machine learning methodologies establishes a strong framework for preserving pipeline integrity and guaranteeing operational safety.

References

Ahmad S et al (2022) A method for pipeline leak detection based on acoustic imaging and deep learning. Sensors 22(4):1562

Alobaidi MH, Meguid MA, Zayed T (2022) Semi-supervised learning framework for oil and gas pipeline failure detection. Sci Rep 12(1):13758

Amadhe FO, Anjorin RO, Uwoghiren FO (2024) Advancements in machine learning for pipeline integrity management: a comprehensive review of predictive and optimization techniques. Cognizance J Multi Stud 4(11):129–138

Ayvaz S, Alpay K (2021) Predictive maintenance system for production lines in manufacturing: a machine learning approach using IoT data in real-time. Expert Syst Appl 173:114598

Celik B, Singh P, Vanschoren J (2023) Online automl: an adaptive automl framework for online learning. Mach Learn 112(6):1897–1921

Cen H et al (2023) Application research on risk assessment of municipal pipeline network based on random forest machine learning algorithm. Water 15(10):1964

Daniel S, Olaoye G (2025) Improving equipment utilization and maintenance using IoT sensors in construction projects https://www.preprints.org/manuscript/202501.1998/v1 [cited 5 March 2025]

Doerr S et al (2021) TorchMD: a deep learning framework for molecular simulations. J Chem Theory Comput 17(4):2355–2363

Dong XL, Rekatsinas T (2018) Data integration and machine learning: a natural synergy. In: Proceedings of the 2018 International Conference on Management of Data

Gijsbers P, Vanschoren J, Olson RS (2018) Layered TPOT: speeding up tree-based pipeline optimization. arXiv preprint arXiv:1801.06007

Hasan BMS, Abdulazeez AM (2021) A review of principal component analysis algorithm for dimensionality reduction. J Soft Comput Data Min 2(1):20–30

Hussain M, Zhang T, Seema M (2023) Adoption of big data analytics for energy pipeline condition assessment. Int J Pressure Vessels Pip 105061

Hussain M et al (2024) Review of prediction of stress corrosion cracking in gas pipelines using machine learning. Machines 12(1):42

Im J et al (2024) Data pipeline for real-time energy consumption data management and prediction. Front Big Data 7:1308236

Kahraman CL et al (2024) Machine learning pipeline for power electronics state of health assessment and remaining useful life prediction. IEEE Access 12:136727–136746

Kashpruk N, Piskor-Ignatowicz C, Baranowski J (2023) Time series prediction in Industry 4.0: a comprehensive review and prospects for future advancements. Appl Sci 13(22):12374

Kubba A, Trabelsi H, Derbel F (2024) Enhanced long-range network performance of an oil pipeline monitoring system using a hybrid deep extreme learning machine model. Future Internet 16(11):425

Laadan D et al (2020) MetaTPOT: enhancing a tree-based pipeline optimization tool using meta-learning. In: Proceedings of the 29th ACM International Conference on Information and Knowledge Management

Laux L et al (2020) Interactive machine learning for fast and robust cell profiling. PLoS ONE 15(9):e0237972

Lei Y et al (2018) Machinery health prognostics: a systematic review from data acquisition to RUL prediction. Mech Syst Signal Process 104:799–834

Li X, Zou B (2021) An automated data engineering pipeline for anomaly detection of IoT sensor data. arXiv preprint arXiv:2109.13828

Lo M, Karuppanan S, Ovinis M (2021) Failure pressure prediction of a corroded pipeline with longitudinally interacting corrosion defects subjected to combined loadings using FEM and ANN. J Mar Sci Eng 9(3):281

Mahmoodzadeh Z et al (2020) Condition-based maintenance with reinforcement learning for dry gas pipeline subject to internal corrosion. Sensors 20(19):5708

Mazumder RK, Salman AM, Li Y (2021) Failure risk analysis of pipelines using data-driven machine learning algorithms. Struct Saf 89:102047

Meda US, Harshitha N, Padubidri AR (2021) Studies on leak detection in process pipelines using artificial neural networks/machine learning. SGS-Eng Sci 1(01)

Nasteski V (2017) An overview of the supervised machine learning methods. Horizons 4(51–62):56

Nguyen T, Gosine RG, Warrian P (2020) A systematic review of big data analytics for oil and gas industry 4.0. IEEE Access 8:61183–61201

Olson RS, Moore JH (2016) TPOT: A tree-based pipeline optimization tool for automating machine learning. In: Workshop on automatic machine learning. PMLR

Olson RS et al (2016) Automating biomedical data science through tree-based pipeline optimization. In: Applications of evolutionary computation: 19th European conference, EvoApplications 2016, Porto, Portugal, March 30–April 1, 2016, proceedings, part I 19. 2016. Springer

Osisanwo F et al (2017) Supervised machine learning algorithms: classification and comparison. Int J Comput Trends Technol (IJCTT) 48(3):128–138

Ossai C (2019) A data-driven machine learning approach for corrosion risk assessment—a comparative study. Big Data Cogn Comput 3(2):28
Putatunda S et al (2020) DriveML: an R package for driverless machine learning. arXiv preprint arXiv:2005.00478
Quy TB, Kim J-M (2021) Real-time leak detection for a gas pipeline using Ak-NN classifier and hybrid AE features. Sensors 21(2):367
Rayhana R et al (2024) Towards next-generation NDE: predictive analysis of pipeline integrity using machine learning. In: NDE 4.0, predictive maintenance, communication, and energy systems: the digital transformation of NDE II. SPIE
Şahin E, Yüce H (2023) Prediction of water leakage in pipeline networks using graph convolutional network method. Appl Sci 13(13):7427
Shaik NB et al (2022) Recurrent neural network-based model for estimating the life condition of a dry gas pipeline. Process Saf Environ Prot 164:639–650
Soomro AA et al (2022) Integrity assessment of corroded oil and gas pipelines using machine learning: a systematic review. Eng Fail Anal 131:105810
Spandonidis C et al (2022) Evaluation of deep learning approaches for oil & gas pipeline leak detection using wireless sensor networks. Eng Appl Artif Intell 113:104890
Sun H, Zhou W (2023) Classification of failure modes of pipelines containing longitudinal surface cracks using mechanics-based and machine learning models. J Infrastruct Preserv Resil 4(1):5
Ullah N, Ahmed Z, Kim J-M (2023) Pipeline leakage detection using acoustic emission and machine learning algorithms. Sensors 23(6):3226
Vanitha C et al (2023) Efficient qualitative risk assessment of pipelines using relative risk score based on machine learning. Sci Rep 13(1):14918
Wang N et al (2023) Multi-parameter maximum corrosion depth prediction model for buried pipelines based on GSCV-XGBoost. IEEE Access 11:116452–116471
Woldesellasse H, Tesfamariam S (2025) Risk assessment of gas pipeline using an integrated Bayesian belief network and GIS: using Bayesian neural networks for external pitting corrosion modelling. Can J Chem Eng 103(1):98–109
Xiao R, Li J (2023) Evaluation of acoustic techniques for leak detection in a complex low-pressure gas pipeline network. Eng Fail Anal 143:106897
Zhang P et al (2023) Machine learning data analytics based on distributed fiber sensors for pipeline feature detection. In: Proceedings Volume 12532, Optical waveguide and laser sensors II. SPIE
Zimek A, Schubert E, Kriegel HP (2012) A survey on unsupervised outlier detection in high-dimensional numerical data. Stat Anal Data Min 5(5):363–387

Chapter 11
Risk Assessment and Big Data Analytics

11.1 Introduction

The utilization of Big Data Analytics has revolutionized risk assessment, particularly within the oil and gas sector. By offering deeper insights into potential dangers that may not be readily obvious through traditional approaches, Big Data utilization enables firms to improve their risk management strategies. For instance, Jung et al. (2022) highlight that the incorporation of big financial data enables risk managers to account for latent risk factors, thereby improving the accuracy of risk assessments and portfolio management through advanced methodologies such as the POET method within a dynamic factor model. This is significant as it handles the prevalent issue of the plague of dimensionality, which can hinder parameter estimation in high-dimensional datasets.

Portfolio management is not the only use of Big Data Analytics in financial risk management. According to El Khatib et al. (2023), Big Data Analytical techniques enable businesses to efficiently identify, evaluate, reduce, track, and communicate a range of risks. This all-encompassing strategy is crucial for preserving business continuity and improving service delivery in a variety of industries, including healthcare and banking. Processing enormous volumes of both structured and unstructured data makes it easier to make evidence-based decisions, which is essential for reducing risks before they become serious problems (Du 2024).

Predicting and preventing financial risk is another important function of Big Data technology breakthroughs. Recent research, for instance, has demonstrated that by examining big datasets to find hidden patterns, deep learning techniques—like Long Short-Term Memory (LSTM) networks—can greatly increase the accuracy of financial risk forecasts (Du 2024). This capability is further supported by the findings of Sun and Huo, who categorize Big Data Analytics into various domains, underscoring its versatility and applicability across different financial contexts (Sun and Huo 2021).

M. Hussain and T. Zhang, *Big Data Analytics in Energy Pipeline Integrity Management*, Lecture Notes in Energy 104, https://doi.org/10.1007/978-981-96-8019-1_11

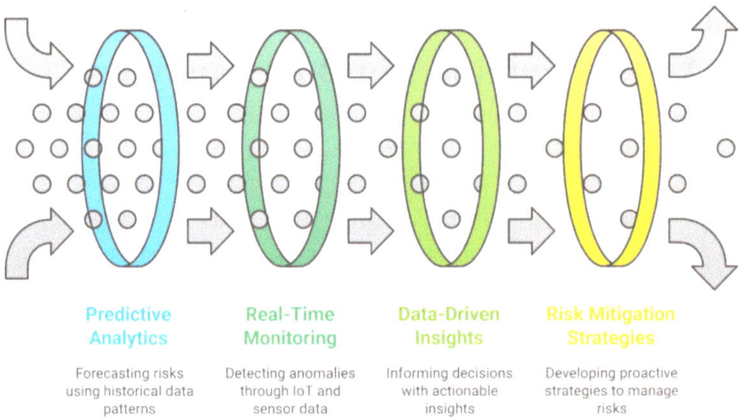

Fig. 11.1 Transforming risk assessment with Big Data

The overall effectiveness of decision-making processes is improved by incorporating Big Data Analytics into risk management frameworks. Figure 11.1 illustrates how Big Data is changing risk assessment. Dicuonzo et al. (2019) contend that efficient risk assessment and management, especially in financial institutions where risk management is vital for survival, depend on the multidirectional flow of information made possible by Big Data Analytics. In addition to protecting company assets, this interconnection lowers the possibility of unforeseen circumstances that could result in losses.

11.2 The Importance of Risk Assessment in Pipeline Integrity Management

Risk assessment plays a key role in pipeline integrity management, acting as a fundamental component for assuring the safety, reliability, and sustainability of pipeline operations. Potential risks related to pipeline systems can be identified, assessed, and mitigated through the integration of risk assessment approaches. This is especially crucial because pipelines carry dangerous compounds that, in the event of a breakdown, could result in serious financial losses, harm to the environment, and dangers to public safety (Jiang and Dong 2021; Fan and Feng 2023).

One of the primary objectives of risk assessment in pipeline integrity management is to estimate the probability of pipeline failures and the potential consequences of such failures. Jiang and Dong emphasize the necessity of establishing a risk-based integrity management model that incorporates both failure probability and consequence assessment, thereby enabling operators to mitigate risks effectively while minimizing costs (Jiang and Dong 2021). Similarly, Dakhel et al. (2023) highlight the importance of experimental data in integrity assessment, which aids in establishing robust integrity management systems and guidelines for safe pipeline operations. This data-driven approach is essential for accurately predicting risks and enhancing the reliability of pipeline systems (Qin et al. 2022).

The application of advanced methodologies, such as Bayesian belief networks and Geographic Information Systems (GIS), has been shown to enhance risk assessment processes. Woldesellasse's and Tesfamariam (2025) research illustrates how these methodologies can serve as early warning systems, facilitating proactive management of pipeline integrity against threats like internal corrosion. The integration of various risk assessment models, such as the one proposed by Qin et al. (2022), further underscores the importance of adaptability and accuracy in predicting risks associated with pipeline operations. These methodologies not only assist in identifying risk factors but also guide maintenance and operational decisions, ensuring that pipeline systems remain safe and efficient (Zhang et al. 2019). In addition to identifying risks, comprehensive risk assessments also consider the socio-economic implications of pipeline failures. Asha et al. (2023) discusses how risk assessment planning is vital for managing pipeline data and preventing incidents that could have severe consequences for communities and the environment. The assessment of potential negative outcomes, including asset loss, environmental degradation, and impacts on human health, is critical for developing effective risk mitigation strategies (Hanafiah et al. 2015). Furthermore, the use of models that incorporate fuzzy logic and analytical hierarchy processes can enhance the understanding of complex risk scenarios, allowing for more nuanced decision-making in pipeline integrity management (Bertuccio and Moraleda 2012; Frefer et al. 2018).

11.2.1 Risk Categories in Pipeline Operations

Pipeline risk assessment typically involves evaluating several categories of risks, including:

- **Operational Risks**: These risks are associated with the day-to-day operation of pipelines, such as pressure fluctuations, material fatigue, and equipment malfunctions.
- **Environmental Risks**: External environmental factors, such as temperature changes, seismic activity, floods, and ground movement, can increase the risk of pipeline failures.
- **Corrosion Risks**: Corrosion is one of the leading causes of pipeline failure. Risk assessment models evaluate the likelihood and severity of internal and external corrosion based on material properties, environmental conditions, and inspection data (Shah et al. 2022; Hussain et al. 2024).
- **Third-Party Risks**: Pipelines are often buried underground, where they can be damaged by third-party activities such as construction, excavation, or drilling. Risk assessment helps identify areas where pipelines are vulnerable to external interference (Hussain et al. 2023).

By categorizing risks, operators can focus on the most pressing threats to pipeline integrity and develop targeted mitigation strategies.

11.2.2 Quantifying Risk in Pipeline Operations

A primary problem in pipeline integrity management is assessing the risk levels associated with various segments of the pipeline. Risk is defined as the multiplication of the probability of a failure and the ramifications of that failure.

$$\text{Risk} = \text{Likelihood} \times \text{Consequence}$$

By assessing both the probability and the ramifications of failure, operators may identify and prioritise the areas that provide the most significant risks to safety, the environment, and operational continuity.

- **Likelihood**: This applies to the likelihood of a failure occurring in a designated segment of the pipeline. Big Data Analytics enables operators to compute this probability using previous failure data, real-time sensor measurements, and external environmental variables.
- **Consequence**: The consequence of failure depends on factors such as the volume of oil or gas transported, the proximity of the pipeline to populated areas or sensitive ecosystems, and the cost of repairs. High-consequence areas, such as those near urban centers or water sources, are given higher priority in risk assessments.

11.3 The Role of Big Data Analytics in Risk Assessment

In a number of industries, such as public safety, healthcare, and finance, the importance of BDA in risk assessment has grown. By utilizing enormous volumes of both organized and unstructured data, BDA makes it easier to identify, assess, and reduce risks. This synthesis looks at how BDA improves decision-making and risk assessment techniques.

In the financial sector, BDA has transformed credit risk assessment and fraud detection. More accurate creditworthiness predictions and the detection of fraudulent activity are made possible by the combination of Big Data and machine learning algorithms. For example, research has demonstrated that sophisticated models, including those that use PSO and neural networks, greatly increase the precision of financial risk management. Furthermore, the application of Big Data Analytics enables financial institutions to analyze real-time data from diverse sources, thereby enhancing their risk management frameworks (Shen 2024).

In healthcare, BDA plays a crucial role in public health risk assessment. By analyzing large datasets, healthcare organizations can predict health risks and monitor disease outbreaks more effectively. For example, Big Data technologies have been employed for early detection of infectious diseases, allowing for timely interventions. This predictive capability not only aids in managing current health risks but also supports the formulation of evidence-based public health policies (Zhang and Pan 2022). The ability to analyze complex datasets contributes to a more nuanced understanding of health risks at both individual and community levels.

BDA is pivotal in enhancing information security risk management. The systematic identification of potential risk factors within network environments has been made possible through Big Data Analytics. This approach allows organizations to evaluate their risk posture in real-time and respond proactively to security incidents. The integration of Big Data Analytics into cybersecurity frameworks has led to improved detection of anomalies and threats, thereby bolstering organizational defenses against cyber risks (Yang 2022).

In construction and project management, BDA helps with risk visualization and modeling. Project managers can improve planning and resource allocation by using Big Data to forecast possible risks related to contracts and procurement procedures. Decision-making is improved and more educated risk management techniques are possible when data trends and patterns can be visualized (Zihao et al. 2023).

11.3.1 Data-Driven Risk Models

Big Data Analytics facilitates the creation of advanced risk models that utilise real-time and historical data to forecast pipeline problems (Mohammadpoor and Torabi 2020; Fedushko et al. 2020). These models utilise machine learning algorithms, statistical techniques, and probabilistic risk assessments that analyze data patterns to predict future dangers.

11.3.2 Integrating Multiple Data Sources

A primary advantage of Big Data Analytics is its capacity to amalgamate data from many sources. In pipeline integrity management, data is gathered from various sources, including:

- **Sensor Data**: Pressure, temperature, and flow rate sensors provide real-time information about the operating conditions of the pipeline.
- **Inspection Data**: In-line inspection (ILI) tools, drones, and field inspections generate data on wall thickness, corrosion, and other physical attributes of the pipeline.
- **Environmental Data**: External factors such as weather conditions, seismic activity, and soil composition are crucial for understanding the external risks to pipeline integrity.

By integrating these diverse datasets, Big Data Analytics provides operators with a more comprehensive view of pipeline health and the associated risks.

- **Example**: A pipeline operator might integrate sensor data showing slight fluctuations with environmental data indicating recent ground movement in the area. By combining these data sources, the operator can assess the likelihood of a rupture and take preventive action before the problem escalates.

11.3.3 Real-Time Risk Monitoring

Conventional risk assessments relied on intermittent inspections and historical data, rendering them incapable of accommodating real-time fluctuations in operational conditions or environmental variables. Big Data Analytics enables operators to conduct real-time risk assessment through the continuous analysis of sensor data and additional inputs. This real-time surveillance allows operators to identify possible problems as they occur and implement prompt corrective measures.

11.4 Techniques for Risk Prioritization and Mitigation

Effective risk management in a variety of fields, including supply chain management and cybersecurity, depends on risk prioritization and mitigation. An organization's capacity to handle possible risks and uncertainties can be greatly improved by having a thorough understanding of risk prioritization and mitigation strategies.

The security risk mitigation framework for cyber-physical systems put out by Zahid et al. (2020) is one of the fundamental frameworks for risk mitigation. Authentication, confidentiality, data integrity, and other critical risks must be addressed at the application layer. This framework also incorporates Safety Integrity Level (SIL) assignments from the IEC 61508 standard, which increases its applicability across different industries that follow comparable risk management procedures. For firms looking to put strong risk management methods into place, this generalizability is essential.

In the context of supply chain management, Ghadge et al. (2012) highlight the necessity of identifying and managing various organizational risks, including inventory and operational risks, to improve overall supply chain resilience. Chang et al. (2015) further elaborate on this by proposing a contingency-based framework that aligns risk mitigation strategies with the specific contexts of supply chain operations, thereby enhancing the effectiveness of risk management efforts. This approach underscores the importance of contextual understanding in the selection of appropriate risk mitigation strategies.

The integration of advanced technologies such as machine learning (ML) into risk evaluation processes is gaining traction. Rizvi's (2024) framework for IoT devices illustrates how ML can streamline risk identification and prioritization, leading to more accurate and timely risk assessments. This technological advancement not only boosts the precision of risk evaluations but also enables organizations to implement more effective control measures.

Structured frameworks are crucial for small and medium-sized businesses (SMEs), which frequently lack the resources necessary for thorough risk management. Comparably, Naude and Chiweshe (2017) offer a framework for operational risk management designed specifically for SMEs, emphasizing the need to include risk management into day-to-day activities in order to increase organizational resilience.

In the realm of environmental and disaster risk management, frameworks such as those discussed by Egbelakin et al. (2013) focus on regulatory mechanisms to prioritize risk mitigation efforts effectively. These frameworks aim to enhance public safety and optimize resource allocation in the face of natural disasters. The incorporation of spatial analysis in environmental risk assessment, as noted by Ferretti and Montibeller (2019), further supports policymakers in making informed decisions regarding risk mitigation strategies.

11.4.1 Risk Prioritization Models

Risk prioritization models use data-driven insights to rank pipeline segments based on the level of risk they pose (Yasseri 2014; Dinh et al. 2021). These models consider factors such as the likelihood of failure, the potential consequences, and the operational importance of each section. By prioritizing high-risk areas, operators can focus on their resources where they are needed most.

- **Example**: A risk prioritization model might rank pipeline segments based on factors such as proximity to sensitive environmental areas, historical failure data, and recent inspection results. Segments that are identified as high risk would be prioritized for immediate inspection or repair, while low-risk segments might be scheduled for routine maintenance at a later date.

11.4.2 Mitigation Strategies

After prioritising risks, operators can formulate mitigation plans to reduce the likelihood of failure and mitigate its potential consequences (Hubbard 2020; Smith 2021). Big Data Analytics can help identify the most effective mitigation strategies based on historical data and predictive models.

- **Preventive Maintenance**: Predictive models can be used to schedule preventive maintenance on high-risk sections of the pipeline before a failure occurs.
- **Pressure Management**: Operators can use real-time data to adjust pressure levels in high-risk areas, reducing the likelihood of a rupture.

- **Material Upgrades**: For pipelines at risk of corrosion, operators may choose to upgrade materials or apply protective coatings to extend the pipeline's lifespan.
- **Example**: An operator using a Big Data risk assessment model might identify that a section of the pipeline is at high risk of failure due to internal corrosion. To mitigate this risk, the operator could schedule preventive maintenance, apply corrosion-resistant coating, and adjust pressure levels to reduce stress on the affected section.

11.5 The Future of Risk Assessment with Big Data Analytics

A strong foundation for recognizing, evaluating, and reducing risks in the oil and gas industry is offered by Big Data Analytics. Kuleshov et al. (2024) highlight the importance of digital technologies and machine learning in risk management, arguing that these tools may greatly enhance risk assessment procedures by facilitating predictive modeling and real-time data analysis. Organizations can make better decisions by using the ability to analyze large amounts of data to find patterns and insights that were previously unavailable.

Big Data Analytics can achieve operational excellence through process optimisation and enhanced safety measures. Tan et al. (2022) assert that the proper utilization of Big Data can yield novel insights that enhance corporate value and profitability, particularly in pipeline and drilling operations. The conclusions of Nguyen et al. (2020), asserting that Big Data Analytics is crucial for improving operational efficiency and mitigating risks in the oil and gas industry, align with this perspective. Utilising contemporary analytics, companies may anticipate potential issues and proactively implement preventive strategies, thus reducing the risk of incidents and operational interruptions. In addition to improving operations, Big Data Analytics is crucial for fostering a culture of safety and accountability within organisations. Sendy and Basaria (2023) promote the incorporation of Total Quality Management (TQM) concepts into risk management techniques, which can be augmented by Big Data Analytics. Through the promotion of continuous improvement and proactive risk management, organisations can foster a safety-centric culture that emphasises risk mitigation at every level.

11.5.1 AI-Driven Risk Assessment

Artificial intelligence (including machine learning) will increasingly play a crucial role in risk assessment due to their capacity to facilitate more dynamic and precise risk models. AI systems conduct real-time analysis of extensive data sets, enabling them to identify trends and frequently revise risk models to reflect changing conditions.

11.5.2 Predictive Risk Analytics

Predictive analytics will enable operators to move from reactive to proactive risk management. By predicting where and when failures are likely to occur, operators can take preventive action before risks escalate. Predictive risk models will become increasingly accurate as they incorporate more data and advanced algorithms.

- **Example**: A predictive risk model might analyze historical data on pipeline failures, environmental conditions, and material properties to predict that a specific section of pipeline is at risk of failure within the next six months. Based on this prediction, operators can schedule repairs or upgrades to prevent failure.

References

Asha LN et al (2023) A quantitative approach of measuring sustainability risk in pipeline infrastructure systems. Sustainability 15(19):14229

Bertuccio I, Moraleda MB (2012) Risk assessment of corrosion in oil and gas pipelines using fuzzy logic. Corros Eng Sci Technol 47(7):553–558

Chang W, Ellinger AE, Blackhurst J (2015) A contextual approach to supply chain risk mitigation. Int J Logistics Manage 26(3):642–656

Dakhel A et al (2023) Fatigue and burst tests of full-scale girth welded pipeline sections for safe operations. Weld World 67(5):1193–1208

Dicuonzo G et al (2019) Risk management 4.0: the role of big data analytics in the bank sector. Int J Econ Financ Issues 9(6):40–47

Dinh N et al (2021) Development and application of a data-driven methodology for validation of risk-informed safety margin characterization models. North Carolina State University, Raleigh, NC (United States); Purdue University, West (United States)

Du Q (2024) Financial risk prediction model in the context of big data-corporate financial risk control based on LSTM deep neural networks. Appl Math Nonlinear Sci 9(1):1–18

Egbelakin T et al (2013) Improving regulatory frameworks for earthquake risk mitigation. Build Res Inf 41(6):677–689

El Khatib M, Al Shehhi H, Al Nuaimi M (2023) How big data and big data analytics mediate organizational risk management. J Financ Risk Manage 12(1):1–14

Fan Y, Feng X-X (2023) Research on integrity management of long distance pipeline. E3S Web Conf

Fedushko S, Ustyianovych T, Gregus M (2020) Real-time high-load infrastructure transaction status output prediction using operational intelligence and big data technologies. Electronics 9(4):668

Ferretti V, Montibeller G (2019) An integrated framework for environmental multi-impact spatial risk analysis. Risk Anal 39(1):257–273

Frefer AA, Matoug MM, Haddada FL (2018) An investigation of corrosion risks in the oil and gas pipelines using analytical hierarchy process and fuzzy analytical hierarchy process. In: Proceedings of First Conference for Engineering Sciences and Technology (CEST-2018), September 25-27, 2018, vol. 2. AIJR Publisher

Ghadge A, Dani S, Kalawsky R (2012) Supply chain risk management: present and future scope. Int J Logistics Manage 23(3):313–339

Hanafiah NM et al (2015) Comparison of human health and safety loss due to corroded gas pipeline failure in rural and urban areas: a case study in Malaysia. Solid State Phenom 227:221–224

Hubbard DW (2020) The failure of risk management: why it's broken and how to fix it. Wiley

Hussain M, Zhang T, Seema M (2023) Adoption of big data analytics for energy pipeline condition assessment. Int J Pressure Vessels Pip 105061
Hussain M et al (2024) Review of prediction of stress corrosion cracking in gas pipelines using machine learning. Machines 12(1):42
Jiang F, Dong S (2021) Risk-based integrity model for offshore pipelines subjected to impact loads from falling objects. Proc Inst Mech Eng Part M: J Eng Marit Environ 235(2):623–641
Jung K, Kim D, Yu S (2022) Next generation models for portfolio risk management: an approach using financial big data. J Risk Insur 89(3):765–787
Kuleshov V et al (2024) Assessing and managing risks in the oil and gas industry with machine learning and digital technologies. E3S Web Conf
Mohammadpoor M, Torabi F (2020) Big data analytics in oil and gas industry: an emerging trend. Petroleum 6(4):321–328
Naude MJ, Chiweshe N (2017) A proposed operational risk management framework for small and medium enterprises. S Afr J Econ Manage Sci 20(1):1–10
Nguyen T, Gosine RG, Warrian P (2020) A systematic review of big data analytics for oil and gas industry 4.0. IEEE Access 8:61183–61201
Qin G et al (2022) An information entropy-based risk assessment method for multiple-media gathering pipelines. J Infrastruct Preserv Resil 3(1):19
Rizvi S, Govere E (2024) A risk management framework for IoT devices and networks. Preprints. https://doi.org/10.20944/preprints202410.1083.v1
Sendy S, Basaria FT (2023) A comparative analysis of hazard analysis methods for sustainable energy development. E3S Web Conf
Shah V et al (2022) Big data analytics in oil and gas industry. Emerg Technol Sustain Smart Energy 37–55
Shen Z (2024) Financial risk management based on big data technology. Adv Econ Manage Polit Sci 124:227–233
Smith DJ (2021) Reliability, maintainability and risk: practical methods for engineers. Butterworth-Heinemann
Sun Z, Huo Y (2021) The spectrum of big data analytics. J Comput Inf Syst 61(2):154–162
Tan V, Aziz KA, Razavi SH (2022) Data analytics for effective project management in the oil and gas industry. In: Proceedings of the international conference on technology and innovation management (ICTIM 2022)
Woldesellasse H, Tesfamariam S (2025) Risk assessment of gas pipeline using an integrated Bayesian belief network and GIS: using Bayesian neural networks for external pitting corrosion modelling. Can J Chem Eng 103(1):98–109
Yang M (2022) Information security risk management model for big data. Adv Multimedia 2022(1):3383251
Yasseri SF (2014) Transmission pipeline risk filtering and ranking. J Pip Eng 13(2)
Zahid M et al (2020) A security risk mitigation framework for cyber physical systems. J Softw 32(2):e2219
Zhang H-Y, Pan T (2022) Public health risk assessment and prevention based on big data. J Environ Public Health 2022(1):7965917
Zhang P, Qin G, Wang Y (2019) Risk assessment system for oil and gas pipelines laid in one ditch based on quantitative risk analysis. Energies 12(6):981
Zihao D et al (2023) The nexus between construction contract risk assessment and big data in China. Development 33(1):441–448

Chapter 12
Data Visualization and Reporting for Pipeline Integrity

Reporting and data visualization are essential elements in pipeline integrity management, especially in the oil and gas industry. Stakeholders can make better decisions and improve the general safety and dependability of pipeline operations by using effective visualization techniques to understand complicated datasets. In the framework of pipeline integrity, this chapter summarizes the many approaches and tools that support data reporting and visualization.

The integration of many data sources, such as geometric, temporal, and semantic information, is one of the main obstacles in pipeline integrity management. Shen et al. (2024) highlight the significance of a geographical data fusion model that successfully integrates multi-feature information to improve the accuracy and completeness of pipeline data. This approach not only improves the efficiency of data sharing but also supports rapid modeling and maintenance of underground pipelines in urban environments. Such models are essential for visualizing the spatial relationships and conditions of pipelines, which can be crucial for identifying potential integrity issues.

Corrosion remains a significant threat to pipeline integrity, necessitating continuous monitoring and analysis. Usuoyibo and Obogai (2020) highlight the need for a systematic approach to Pipeline Integrity Management (PIM), which includes the collection and analysis of data to predict potential failure points. This predictive capability is further supported by advanced data visualization techniques that can illustrate corrosion trends and anomalies over time, allowing operators to prioritize maintenance efforts effectively.

In-line inspection (ILI) tools play a pivotal role in assessing pipeline conditions, particularly in identifying defects such as dents and corrosion. Lei et al. (2024) discusses the importance of strain analysis and the assessment of dents detected by ILI tools, which are critical for ensuring safe pipeline operations. The visualization of strain data and defect locations can provide operators with a clearer understanding of the structural integrity of pipelines, enabling them to make informed decisions regarding repairs or replacements. Lei et al. (2017) also emphasize the significance of

M. Hussain and T. Zhang, *Big Data Analytics in Energy Pipeline Integrity Management*, Lecture Notes in Energy 104, https://doi.org/10.1007/978-981-96-8019-1_12

strain analysis in the context of dented pipelines, highlighting the need for thorough assessments to ensure safe operations.

The use of Geographic Information Systems (GIS) has been proposed as a robust framework for managing pipeline integrity data. The integration of GIS with decision support platforms allows for comprehensive visualization of pipeline networks, facilitating risk assessment and management. This approach not only enhances the visualization of spatial data but also aids in the identification of high-consequence areas that require closer monitoring.

Notable is also that artificial intelligence (AI) plays in managing pipeline integrity. According to Macdonald et al. (2022), real-time monitoring systems can be used to view pipeline ruptures, which can be detected using AI classifiers. Understanding pipeline integrity and the efficacy of maintenance plans can be greatly improved by being able to display AI-generated predictions alongside historical data.

12.1 The Importance of Data Visualization in Pipeline Integrity Management

Data visualisation is important in pipeline integrity management (PIM) as it improves the comprehension and communication of intricate data sets concerning pipeline operations. Effective data visualisation enables stakeholders to identify possible risks, evaluate the integrity of pipeline systems, and make educated decisions about maintenance and operational plans.

The capacity of data visualization in PIM to integrate and present data from multiple sources in a logical way is one of its main advantages. For example, integrating the geometric, temporal, and semantic properties of subterranean pipes can greatly improve the completeness and accuracy of data, which will increase the overall effectiveness of pipeline management systems. In metropolitan settings, where pipes are frequently exposed to a variety of external pressures and hazards, this integration makes it easier to develop and maintain pipeline systems quickly. Additionally, pipeline data can be spatially represented through the use of Geographic Information Systems (GIS) and visualization tools, which helps operators identify vulnerabilities and evaluate risks (Feng and Xu 2011).

Analyzing historical data and identifying failure patterns are made easier with the help of data visualization. Operators can identify patterns and probable reasons for pipeline failures by visualizing failure data, which is essential for risk identification and mitigation tactics (Dai et al. 2017). A more intuitive comprehension of the elements causing pipeline integrity problems, such as corrosion and external damage, is made possible by the capacity to visualize complicated datasets (Fan and Feng 2023). The application of sophisticated analytical models, like Bayesian survival analysis, which can forecast probable failures based on operational conditions and historical data, significantly improves this understanding (Feng et al. 2019).

Data visualization aids condition-based maintenance plans in addition to risk assessment. Operators can gain instant insights into the operational state of pipelines by seeing the massive amounts of real-time data generated by the use of Internet of Things (IoT) technologies in pipeline monitoring (Mahmoodzadeh et al. 2020). By facilitating preventative maintenance choices, this real-time visualization lowers the risk of catastrophic failures and the resulting environmental effects (Feng et al. 2019). Furthermore, predictive maintenance capabilities can be improved by combining machine learning algorithms with visualization tools, which enables more precise pipeline integrity issue forecasting.

Lastly, improving stakeholder communication is another function of data visualization. By facilitating talks between engineers, management, and regulatory agencies, effective visual representations of data can guarantee that everyone involved is aware of pipeline integrity issues and the steps being taken to resolve them (El-Akruti et al. 2016). Creating thorough integrity management plans that adapt to the changing needs of pipeline operations requires this cooperative approach.

12.1.1 Enhanced Decision-Making

The capacity of data visualization to improve decision-making is among its most significant advantages. Operators can rapidly understand important trends, anomalies, and hazards when data is presented visually, such as through graphs, charts, heatmaps, or geographic information systems (GIS). This enables them to decide on pipeline upkeep, repairs, and operating modifications more quickly and intelligently.

12.1.2 Improved Communication

Effective data visualization improves communication between different stakeholders, including pipeline operators, engineers, and executives. By presenting complex data in an accessible format, visualizations help ensure that all stakeholders can understand the current state of the pipeline, the risks involved, and the actions that need to be taken. This is particularly important when presenting data to non-technical audiences (Ahmad et al. 2022).

12.1.3 Faster Detection of Anomalies

Real-time data visualization allows operators to detect anomalies more quickly. Visualization dashboards that display real-time data streams from sensors enable operators to monitor pipeline conditions continuously and identify potential issues, such as pressure drops or temperature spikes, as soon as they occur (Al Mubarak 2022; Hussain et al. 2023).

12.2 Tools for Data Visualization in Pipeline Integrity

Several tools and platforms are available to help pipeline operators visualize and interpret their data. These tools range from basic charting software to more advanced platforms that support real-time monitoring and geographic data visualization. Various tools and methodologies have been developed to enhance the visualization of data related to pipeline integrity, each contributing uniquely to the field.

One significant advancement in this area is the integration of Geographic Information Systems (GIS) into pipeline integrity management. A decision support platform that combines GIS with data acquisition processes during construction and operation phases has been proposed. This platform facilitates visual management of construction quality, risk assessment, and integrity evaluation, thereby enhancing the overall decision-making process in pipeline management. The use of GIS allows for spatial analysis and visualization of pipeline data, which is essential for identifying high-consequence areas and managing risks associated with pipeline integrity (Xie and Tian 2018).

The development of specialized software tools has further improved the visualization capabilities in pipeline integrity management. For instance, the integration of multi-feature information through spatial data fusion models has shown to enhance data accuracy and completeness, which is vital for effective pipeline monitoring and maintenance (Shen et al. 2024). Such tools enable the rapid modeling of underground pipelines, allowing for efficient data sharing and improved work efficiency in urban environments where pipelines are often buried. In addition to GIS-based solutions, there are also software packages designed for specific data types, such as RNA-Seq and proteomics, which can be adapted for pipeline integrity data visualization. For example, the TRAPR R package provides functions for data management, filtering, normalization, and visualization, which can be crucial for ensuring the integrity of data processed through various pipelines (Lim et al. 2017). Similarly, tools like PANDA-view and iMAP have been developed to facilitate statistical analysis and visualization of complex biological data, which can also be applied to pipeline integrity data (Chang et al. 2018; Buza et al. 2019). These tools allow researchers to visualize data interactively, making it easier to identify trends and anomalies that could indicate potential integrity issues.

Furthermore, the use of cloud computing and automated pipelines has emerged as a significant trend in data visualization for pipeline integrity. The MS-PyCloud pipeline, for example, leverages cloud resources to enhance the speed and efficiency of data analysis, which is critical for real-time monitoring of pipeline conditions (Chang et al. 2018). This approach not only improves the processing capabilities but also allows for scalable and flexible data visualization solutions that can adapt to the evolving needs of pipeline integrity management (Wang et al. 2024). Below are some of the most commonly used tools for pipeline data visualization:

12.2.1 Geographic Information Systems (GIS)

GIS platforms are essential for visualizing the geographic layout of pipeline networks and understanding how external environmental factors affect pipeline integrity (Shekargoftar et al. 2022; Li 2020). GIS tools allow operators to overlay pipeline data with geographic maps, making it easier to identify risks related to weather, terrain, population density, and other external factors (Li and Yang 2021).

12.2.2 Dashboards for Real-Time Monitoring

Dashboards offer a real-time, visual platform for observing pipeline status. These platforms consolidate data from sensors, inspections, and many sources into a unified interface, enabling operators to monitor critical parameters such as pressure, temperature, flow rate, and corrosion levels. Dashboards may feature customisable charts, graphs, and alarms that inform operators upon the detection of a problem.

12.2.3 Heatmaps and Risk Maps

Heatmaps are a powerful visualization tool for displaying the severity of risks, such as corrosion, leaks, or external damage, across different sections of the pipeline. Risk maps use color gradients to indicate which areas are at the highest risk, helping operators prioritize maintenance and repairs.

12.2.4 Time Series Visualizations

Sensors that continuously detect variables like pressure, temperature, and flow rate provide pipelines with vast amounts of time series data. Operators can monitor changes over time, spot patterns, and spot departures from standard operating conditions by visualizing time series data.

12.3 Best Practices for Data Visualization in Pipeline Integrity

Operators should adhere to best practices while creating reports and visualizations to guarantee that data visualizations are efficient and actionable. By following these guidelines, visualizations are made to be understandable, clear, and insightful. Standardizing metrics, integrating various data sources, and utilizing interactive visualization technologies are some of the main categories into which best practices in data visualization for pipeline integrity can be divided.

First, for data visualization to be effective, quality control measures must be standardized. This is demonstrated by the SCTK-QC pipeline, which simplifies the creation and display of quality control metrics for single-cell RNA sequencing data by combining multiple techniques into a unified framework that makes the visible data easily accessible (Hong et al. 2022). This approach can be mirrored in pipeline integrity assessments, where standardized metrics can help in the consistent evaluation of pipeline conditions across different inspections. Moreover, the integration of deterministic and probabilistic models for evaluating pipeline integrity can enhance the accuracy of visualizations, as demonstrated by the analysis of error distributions in in-line inspection data. Such models enable the visualization of potential defect growth and the overall integrity of pipelines over time.

Second, it's crucial to incorporate many data sources into a single visualization framework. For example, thorough analysis and visualization of complicated datasets are made possible by the integration of genomic and phenotypic data in a multimodal framework (Das et al. 2018). A comprehensive picture of pipeline conditions can be obtained by integrating data from multiple inspection techniques (such as visual inspections, in-line inspections, and environmental evaluations) in the context of pipeline integrity. Effective visualization of pipeline integrity data can be achieved by adapting tools such as the Bio-Docklets framework, which combines interactive data visualizations with pipelines (Kim et al. 2017). Better maintenance and repair decision-making procedures are made possible by this integration, which also improves the data's interpretability.

Finally, the use of interactive visualization tools is paramount in making complex data accessible to stakeholders. The development of user-friendly visualization engines, such as genomeview (Spies et al. 2018), allows researchers to extend and integrate visualizations into existing analysis pipelines. In the pipeline integrity context, such tools can enable engineers and decision-makers to interact with the data, exploring various scenarios and outcomes based on different inspection results. Additionally, the usability of data visualizations in pipeline integrity monitoring can be greatly enhanced by utilizing frameworks that follow the FAIR principles (Findable, Accessible, Interoperable, and Reusable) (Lee et al. 2024).

12.3.1 Keep Visualizations Simple and Focused

Effective visualizations should be simple, clear, and focused on the most important information. Overloading a chart or dashboard with too many data points or complex visual elements can make it difficult for operators to interpret the information and take action.

- **Tip**: Use minimal colors, avoid unnecessary chart elements, and focus on the key metrics that are most relevant to decision-making. For example, a heatmap showing corrosion levels should use a simple color scale (e.g., green to red) to indicate the severity of corrosion, without adding unnecessary details that could clutter the visualization.

12.3.2 Use Real-Time Data for Dynamic Decision-Making

For pipeline integrity management to make dynamic decisions, real-time data visualization is crucial. Operators can spot possible issues as soon as they appear and take prompt action thanks to dashboards that provide real-time data from sensors and other monitoring equipment.

- **Tip**: Implement real-time alert systems that trigger notifications when data exceeds predefined thresholds. For example, if pressure in a certain section of the pipeline exceeds safe limits, the dashboard should issue an alert, allowing operators to address the issue quickly.

12.3.3 Make Visualizations Interactive

Interactive visualisations enable users to delve into data more thoroughly and reveal insights that may not be readily discernible from static charts or graphs. Interactive dashboards allow operators to magnify individual pipeline parts, filter data by date or location, and modify visualisation parameters to emphasise particular indicators.

- **Tip**: Design dashboards that allow users to interact with the data. For example, an operator might use a slider to adjust the time range displayed on a time series chart or click on a section of the pipeline map to view detailed inspection data for that area.

12.3.4 Ensure Data Accuracy and Consistency

Reliable and current data is essential for efficient decision-making and visualization. Data that is out-of-date or inconsistent can result in poor decision-making and inaccurate conclusions, which raises the possibility of pipeline failures. Before data is visualized, it is crucial to make sure that it is regularly updated, verified, and trustworthy.

- **Tip**: Implement regular data validation processes to ensure that sensor data, inspection records, and external data sources are accurate and up to date. This will improve the quality of the visualizations and the decisions that are made based on them.

12.4 Reporting for Pipeline Integrity Management

Effective PIM requires the integration of many management systems and compliance frameworks since it enables businesses to manage the intricacies of regulatory obligations while preserving operational effectiveness. The requirement for strong compliance systems that can adjust to changing laws and technical developments is one of PIM's main obstacles. It has been demonstrated that the incorporation of standardized management systems (MSs) improves the risk management of pipeline operations. For example, Rebelo et al. point out that incorporating standardized MSs can improve risk management process coordination, which is essential for preserving pipeline integrity and guaranteeing adherence to safety rules (Rebelo et al. 2017). Furthermore, the application of automation and systematic approaches to integrating security standards into DevOps pipelines can significantly improve compliance and operational efficiency, particularly in environments where industrial control systems (ICS) are employed, as they must adhere to stringent security regulations (Moyón et al. 2020).

The role of knowledge management in PIM cannot be overstated. Effective knowledge management practices facilitate the sharing of critical information regarding compliance and operational standards, which can enhance decision-making processes within organizations (Masuin et al. 2019). By guaranteeing that all stakeholders are aware of compliance standards and best practices, the integration of PIM with knowledge management systems can enhance organizational performance (Olan et al. 2016). This is corroborated by Ikram et al.'s (2020) findings, which highlight how integrating management systems through sophisticated techniques can improve overall organizational effectiveness and expedite compliance procedures.

Technological improvements significantly contribute to the enhancement of PIM. Big Data Analytics and artificial intelligence have been recognised as tools to enhance compliance monitoring and risk management in pipeline operations (Muslim 2024). These tools empower organisations to scrutinize extensive data sets, discern possible dangers, and execute proactive strategies to minimize them. AI-driven compliance training has demonstrably decreased compliance infractions, consequently improving the overall integrity of pipeline management systems.

12.4.1 Components of an Effective Pipeline Integrity Report

An effective pipeline integrity report should include the following components:

- **Executive Summary**: A high-level overview of the pipeline's current condition, key risks, and recommended actions. This section is particularly important for communicating with non-technical stakeholders.
- **Risk Assessment**: A detailed analysis of the risks associated with different sections of the pipeline, including the likelihood of failure and the potential consequences.
- **Data Visualizations**: Charts, graphs, and maps that illustrate key trends, risks, and anomalies. These visualizations should be designed to be easily interpreted by the intended audience.
- **Recommendations**: Specific recommendations for addressing the risks identified in the report, such as scheduling maintenance, adjusting operating conditions, or conducting further inspections.
- **Example**: A pipeline integrity report might include a heatmap showing areas of high corrosion, followed by recommendations to schedule maintenance on the most severely affected sections. The report would also include an executive summary outlining the overall health of the pipeline and the actions needed to ensure its continued integrity.

12.4.2 Automated Reporting Systems

Many pipeline operators are adopting automated reporting systems that generate regular reports based on real-time data from sensors and inspections. These systems can automatically compile data, create visualizations, and produce reports that provide a snapshot of pipeline integrity at any given time. Automated reports help ensure that key stakeholders have up-to-date information and can make decisions more quickly.

- **Example**: An automated reporting system might generate a weekly report on pipeline pressure levels, including a time series chart that shows pressure trends over the past seven days. The report could also include alerts for any pressure anomalies that occurred during that time.

References

Ahmad A, Leifler O, Sandahl K (2022) Data visualisation in continuous integration and delivery: information needs, challenges, and recommendations. IET Software 16(3):331–349

Al Mubarak OMM (2022) Design and implementation of an industrial internet of things based Scada system: case study the petroleum pipeline systems. İstanbul Gelişim Üniversitesi Lisansüstü Eğitim Enstitüsü

Buza TM et al (2019) IMAP: an integrated bioinformatics and visualization pipeline for microbiome data analysis. BMC Bioinform 20:1–18

Chang C et al (2018) PANDA-view: an easy-to-use tool for statistical analysis and visualization of quantitative proteomics data. Bioinformatics 34(20):3594–3596

Dai L et al (2017) Analysis and comparison of long-distance pipeline failures. J Pet Eng 2017(1):3174636

Das S et al (2018) Integration of "omics" Data and phenotypic data within a unified extensible multimodal framework. Front Neuroinform 12:91

El-Akruti K, Zhang T, Dwight R (2016) Maintaining pipeline integrity through holistic asset management. Eur J Ind Eng 10(5):618–638

Fan Y, Feng X-X (2023) Research on integrity Management of long distance pipeline. E3S Web Conf

Feng XL, Xu XJ (2011) Hydraulic calculation and visualization of long distance pipeline based on GIS. Appl Mech Mater 63:502–506

Feng Q, Sha S, Dai L (2019) Bayesian survival analysis model for girth weld failure prediction. Appl Sci 9(6):1150

Hong R et al (2022) Comprehensive generation, visualization, and reporting of quality control metrics for single-cell RNA sequencing data. Nat Commun 13(1):1688

Hussain M, Zhang T, Seema M (2023) Adoption of big data analytics for energy pipeline condition assessment. Int J Pressure Vessels Pip 105061

Ikram M, Zhang Q, Sroufe R (2020) Developing integrated management systems using an AHP-Fuzzy VIKOR approach. Bus Strateg Environ 29(6):2265–2283

Kim B et al (2017) Bio-Docklets: virtualization containers for single-step execution of NGS pipelines. GigaScience 6(8):gix048

Lee KH et al (2024) Spyglass: a framework for reproducible and shareable neuroscience research. bioRxiv

Lei ZQ et al (2017) Strain analysis of oil and gas pipeline with dents based on ILI data. Appl Mech Mater 853:372–376

Lei Z et al (2024) Modeling and strain analysis of pipeline dent based on geometric in-line-inspection data. In: Ninth international conference on energy materials and electrical engineering (ICEMEE 2023). SPIE

Li Z (2020) Pipeline spatial data modeling and pipeline WebGIS. Springer

Li Z, Yang L (2021) Pipeline real-time data integration and pipeline network virtual reality system: digital oil & gas pipeline: research and practice. Springer Nature

Lim JH, Lee SY, Kim JH (2017) TRAPR: R package for statistical analysis and visualization of RNA-Seq data. Genomics Inform 15(1):51

MacDonald C et al (2022) Transferable pipeline rupture detection using multiple artificial intelligence classifiers during transient operations. J Pressure Vessel Technol 144(4):041802

Mahmoodzadeh Z et al (2020) Condition-based maintenance with reinforcement learning for dry gas pipeline subject to internal corrosion. Sensors 20(19):5708

Masuin R, Latief Y, Zagloel TY (2019) Development of knowledge management in integration management systems in order to increase the organisational performance of construction companies. MATEC Web Conf

Moyón F et al (2020) Integration of security standards in devops pipelines: an industry case study. In: Product-focused software process improvement: 21st international conference, PROFES 2020, Turin, Italy, November 25–27, 2020, proceedings 21. Springer

Muslim M (2024) Managerial finance tactics in the era of enhanced regulation following financial Scandals. Adv Manage Financ Rep 2(1):1–10

Olan F et al (2016) How knowledge sharing and business process contribute to organizational performance: using the fsQCA approach. J Bus Res 69(11):5222–5227

Rebelo MF, Silva R, Santos G (2017) The integration of standardized management systems: managing business risk. Int J Qual Reliab Manage 34(3):395–405
Shekargoftar A et al (2022) An integrated framework for operation and maintenance of gas utility pipeline using BIM, GIS, and AR. J Perform Constr Facil 36(3):04022023
Shen T et al (2024) Spatial data fusion model design and research for an underground pipeline in urban environment scene modeling. Appl Sci 14(15):6760
Spies N et al (2018) Genomeview-an extensible python-based genomics visualization engine. BioRxiv 355636
Usuoyibo A, Obogai L (2020) Corrosion analysis of API 5L Gr. Three phase crude. NIPES-J Sci Technol Res 2(3)
Wang X et al (2024) ViromeFlowX: a comprehensive nextflow-based automated workflow for mining viral genomes from metagenomic sequencing data. Microb Genomics 10(2):001202
Xie M, Tian Z (2018) A review on pipeline integrity management utilizing in-line inspection data. Eng Fail Anal 92:222–239

Chapter 13
The Role of the Internet of Things (IoT) in Pipeline Integrity Management

In pipeline integrity management, the Internet of Things (IoT) has revolutionized the area by providing innovative methods for monitoring, maintenance, and operating efficiency. The integration of IoT technology into pipeline systems facilitates real-time data collecting and analysis, which is crucial for ensuring the reliability and safety of oil and gas transportation.

The capacity of IoT to monitor continuously is one of its main benefits for pipeline integrity management. Singh et al. stress the value of strong, dependable systems that can function in demanding conditions by highlighting the usage of Zigbee and long-range architectures for remote oil pipeline monitoring (Singh et al. 2021). In a similar vein, Liu et al. talk about how low-power routing methods can improve wireless sensor networks' (WSNs') ability to identify leaks and other issues in gas and oil pipelines (Liu et al. 2022). These systems are made to work well in isolated locations, guaranteeing that possible problems may be found and fixed quickly, reducing risks to the environment and the economy (Rahmani et al. 2022).

IoT applications extend beyond mere monitoring; they encompass decision support systems that utilize data analytics to protect integrity. The authors of a study on lifecycle-based decision support systems assert that the amalgamation of Geographic Information Systems (GIS) with Internet of Things (IoT) data facilitates comprehensive assessments of pipeline conditions. This strategy facilitates the management of various phases of pipeline operation, ensuring that integrity assessments are both proactive and responsive to emerging concerns.

The role of WSNs in pipeline integrity management cannot be overstated. These networks provide high flexibility and low-cost solutions for monitoring pipeline conditions, as highlighted by Liu et al. (2022). The ability to deploy multiple sensors across extensive pipeline networks enables the collection of diverse data points, which can be analyzed to detect anomalies such as corrosion or leaks. Furthermore, the use of advanced algorithms, such as those discussed by Rahmani et al. (2022), enhances the area coverage and energy efficiency of monitoring systems, ensuring that even the most remote sections of pipelines are adequately supervised.

M. Hussain and T. Zhang, *Big Data Analytics in Energy Pipeline Integrity Management*, Lecture Notes in Energy 104, https://doi.org/10.1007/978-981-96-8019-1_13

In addition to monitoring and data collection, IoT technologies also facilitate the automation of maintenance processes. For instance, Kadar et al. (2018) illustrate how IoT can automate water level management and leakage detection in water distribution systems, which parallels the needs of pipeline integrity management. The automation of these procedures enhances operating efficiency and mitigates the danger of human error, which may result in catastrophic failures in pipeline systems.

13.1 What is the Internet of Things (IoT)?

The Internet of Things (IoT) is a transformative technology that enables a diverse array of physical items, sensors, and systems to autonomously collect, disseminate, and analyze data via the Internet. Since Kevin Ashton coined the phrase "Internet of Things" in 1999, it has expanded to include a broad variety of applications in a number of industries, such as industrial automation, smart cities, healthcare, and education (Perera et al. 2015; Cao and Wachowicz 2020). The Internet of Things' fundamental function is to enable communication between devices, also known as "things," which can function separately or in tandem to deliver improved services and efficiencies (Perera et al. 2015; Fisher et al. 2015).

IoT is fundamentally defined by its wide-ranging connectivity, which enables objects to communicate "anytime, anywhere, with anything and anyone" via a variety of communication networks (Perera et al. 2015; Said and Zolkipli 2022). This connectivity is essential for developing smart surroundings, where gadgets can serve as sensors integrated into commonplace items, helping to build smart homes and cities (Cao and Wachowicz 2020). In industries like agriculture, logistics, and healthcare, where real-time data collecting and analysis may optimize operations and improve decision-making processes, the incorporation of IoT technology has resulted in notable breakthroughs (Chen et al. 2014; Zhao et al. 2011). A number of enabling technologies, such as cloud computing, wireless sensor networks, and Radio Frequency Identification (RFID), support the Internet of Things ecosystem and make it easier for devices to communicate with one another (Chen and Guo 2013; Razzaque et al. 2015). As the Internet of Things grows, it also brings up important security and privacy issues, which calls for strong frameworks to preserve the integrity of the linked systems (Said and Zolkipli 2022; Vermesan et al. 2020). The importance of IoT in the current digital landscape is highlighted by its wide range of applications, which include improving industrial processes and developing intelligent learning environments (Uspabayeva et al. 2022).

IoT devices can interact with central systems and with each other, allowing for remote control and ongoing data collecting. IoT technology is utilized in pipeline operations to track variables including temperature, pressure, flow rate, vibration, and corrosion, giving operators real-time information about the pipeline's condition.

13.1.1 How IoT Works in Pipeline Management

The primary application of IoT in pipeline management is the utilization of smart sensors to monitor water pipeline flow rates and detect leaks. Systems utilizing water flow sensors can quantify both the volume of water traversing pipelines and its flow rate. Subsequently, the data is transmitted to the cloud for further processing and analysis, facilitating immediate responses if anomalies are detected (Kumar and Jagadeep 2022). By promptly addressing any waste, these monitoring systems facilitate leak detection and promote water conservation.

The Internet of Things is crucial for preserving the integrity and safety of pipelines within the oil and gas sector. Wireless sensor networks are employed to incessantly monitor several parameters, including temperature, pressure, and flow rates. These sensors provide adaptability and expandability in operational monitoring, especially under challenging conditions (Liu et al. 2022). The collected information is crucial for anomaly detection, enhancing the overall safety of pipeline operations and preventing catastrophic failures (Priyanka et al. 2021). Predictive maintenance is facilitated by the integration of IoT and advanced data analytics, allowing for the identification of potential issues prior to their escalation. The establishment of dependable data pipelines facilitates the management of the substantial volumes of data generated by IoT devices. These pipelines provide data integrity and reliability by facilitating the interpretation of various data streams from several sensors. Formulating informed decisions in pipeline management necessitates the proficient management of extensive data collections, facilitated by the use of Big Data technology (de Haro-Olmo et al. 2023).

13.2 Types of IoT Devices Used in Pipeline Integrity

In the oil and gas sector, the Internet of Things (IoT) has emerged as a transformative instrument for monitoring and maintaining pipeline integrity. A range of IoT devices are utilized to enhance the reliability, efficiency, and safety of pipeline operations. These technologies enable predictive maintenance, data collection, and real-time monitoring, which are essential for preventing leaks and other faults.

A principal kind of IoT devices utilized in pipeline integrity is wireless sensor networks (WSNs). These sensors are systematically positioned along pipes to measure factors including pressure, temperature, and flow rates. Karam et al. (2024) emphasize that IoT devices can be deployed in difficult environments, facilitating ongoing monitoring and accurate data collecting for leak detection. Singh et al. (2021) assert that real-time monitoring via IoT-enabled devices facilitates ongoing oversight of oil pipelines, substantially reducing environmental and human consequences during incidents. The implementation of these sensors is essential for the prompt identification of anomalies, facilitating quick interventions and minimizing risks.

Energy-efficient routing protocols are vital for the effective operation of these IoT devices, especially in remote areas where power supply may be limited. Liu et al. (2022) discuss the importance of low-energy consumption routing protocols that enhance the performance of IoT systems in oil and gas pipelines, addressing challenges related to energy limitations of sensor nodes. This is further supported by Ayeni and Ayogu (2020), who note that IoT can help monitor pipeline facilities efficiently, particularly in regions prone to sabotage and theft. The integration of energy-efficient technologies ensures that these monitoring systems remain operational over extended periods, which is critical for maintaining pipeline integrity. In addition to WSNs, advanced monitoring systems utilizing Narrowband IoT (NB-IoT) technology are also being developed. Xi et al. (2024) describe a low-power monitoring system designed for gas wells that incorporates temperature and humidity sensors, allowing for real-time data transmission to IoT platforms. This capability is essential for monitoring the conditions around pipelines and ensuring that any potential hazards, such as gas leaks, are promptly addressed.

Pipeline integrity management greatly benefits from predictive analytics driven by IoT technology. Zhang and Xie (2024) propose an IoT-based failure prediction mechanism model that can improve safety management through the creation of extensive lifetime databases and algorithmic pipeline monitoring models. This predictive skill is essential for foreseeing malfunctions before they happen and averting disastrous events. Finally, pipeline monitoring systems are more effective overall when machine learning and data analytics are integrated with IoT devices. In order to optimize sensor deployment in pipeline monitoring, Rahmani et al. present an energy-aware area coverage method that uses Q-learning (Rahmani et al. 2022). This method guarantees that the data gathered is applicable and actionable for decision-making processes, in addition to increasing the effectiveness of monitoring activities.

13.2.1 Pressure Sensors

Pressure sensors are one of the most commonly used IoT devices in pipeline integrity management. They continuously monitor the internal pressure within the pipeline, providing critical data that can help detect leaks, ruptures, or blockages (Zaman et al. 2020; Korlapati et al. 2022). Pressure sensors are carefully positioned at critical locations throughout the pipeline to provide operators with a comprehensive understanding of pressure dynamics.

13.2.2 Temperature Sensors

Temperature sensors quantify the internal temperature of the fluid or gas conveyed through the pipeline. Changes in temperature can signal problems such as inefficiencies in flow, material fatigue, or external environmental conditions affecting the pipeline (Korlapati et al. 2022).

13.2.3 Vibration Sensors

Vibration sensors are used to detect mechanical vibrations along the pipeline, which can indicate potential issues such as material fatigue, external disturbances, or nearby construction activities (Sharma et al. 2021; Razi and Taheri 2014). Excessive vibrations can lead to cracks or structural damage over time, making vibration monitoring essential for pipeline integrity.

13.2.4 Flow Meters

Flow meters measure the flow rate of liquids or gases through the pipeline, ensuring that the flow remains within safe and efficient operating ranges (LaNasa and Upp 2014; Thorn et al. 2023). Any deviation from normal flow rates can indicate blockages, leaks, or operational inefficiencies.

13.2.5 Corrosion Sensors

Corrosion is a leading cause of pipeline failures, making corrosion sensors critical for monitoring the condition of pipeline materials (Reddy et al. 2021). These sensors detect signs of corrosion, rust, or material degradation, allowing operators to take preventive action before the integrity of the pipeline is compromised (Wright et al. 2019).

13.2.6 Acoustic Sensors

Acoustic sensors are used for leak detection by capturing sound waves generated by escaping gases or liquids (Meribout et al. 2020). These sensors can detect the unique acoustic signature of a leak and alert operators to its location.

13.3 Integration of IoT with Big Data and AI

The integration of the Internet of Things (IoT), Big Data, and Artificial Intelligence (AI) signifies a transformative shift across numerous industries, enhancing operational efficiency and facilitating data-driven decision-making. This synthesis analyzes the interconnections among these technologies, their applications, and the challenges they present.

The integration of IoT, Big Data, and AI is essential for deriving actionable insights from the extensive data produced by IoT devices. Wong (2021) observes that the interaction among these technologies enhances sophisticated analytics, which can greatly benefit IoT applications, although it also presents challenges related to power consumption and security. The dependence on AI is essential, as human abilities are inadequate to handle the vast data quantities generated by IoT devices. This necessity is emphasized by the reality that, in the absence of AI, the created data remains predominantly unexploited and unused.

Emerging trends in AI-driven data analysis highlight the potential for enhanced privacy protection and intelligent data processing. Marengo (2024) emphasizes the importance of AI in developing sophisticated data analysis techniques that can detect anomalies in sensor data, thereby improving the reliability of IoT systems. The integration of AI with IoT not only optimizes data handling but also enhances communication protocols, leading to more robust IoT frameworks. This synergy is particularly evident in sectors such as agriculture, where AI and IoT technologies are utilized to monitor crop conditions and optimize resource use, thereby promoting sustainable practices (Senoo et al. 2024; Issa et al. 2024).

According to research, artificial intelligence (AI) can significantly improve IoT security by using machine learning techniques to detect and eliminate risks. Advanced security measures are necessary due to the complexity of IoT systems, especially as the number of devices increases due to 5G technology improvements (Abed and Anupam 2023). Significant thought must be given to the ethical ramifications of AI in IoT applications, especially with regard to data protection and decision-making transparency.

13.3.1 Real-Time Data Analysis

IoT devices generate continuous streams of data, which are processed in real time by Big Data platforms. These platforms analyze the data to detect anomalies, track performance metrics, and provide actionable insights to operators (Kumar et al. 2022).

13.3.2 Predictive Maintenance

The Internet of Things (IoT) data is essential for predictive maintenance, utilizing both historical and real-time information to forecast the timing and location of necessary maintenance activities. Through the analysis of data from IoT devices, predictive maintenance models can anticipate probable failures and suggest maintenance measures prior to the emergence of issues (Ayvaz and Alpay 2021; Zonta et al. 2020).

13.3.3 Autonomous Monitoring and Control

IoT devices, when integrated with AI, can enable autonomous monitoring and control of pipeline operations. AI algorithms may evaluate IoT data instantaneously and autonomously modify operating parameters, like as flow rates or pressure levels, to enhance performance and avert failures.

13.4 Challenges of IoT in Pipeline Integrity Management

Numerous challenges must be addressed when integrating the Internet of Things (IoT) with pipeline integrity management to provide effective monitoring and maintenance of pipeline systems. The energy efficiency of IoT devices placed throughout vast pipeline networks is one of the main obstacles. In order to manage and restore malfunctioning devices dispersed throughout the pipeline, energy-efficient and fault-tolerant communication protocols must be developed, as Karam et al. (2024) pointed out that the energy consumption of a single IoT node can have a substantial impact on the monitoring operation as a whole. This is further corroborated by Liu et al. (2022), who highlight how sensor node energy limits impede IoT system performance in oil and gas pipelines, highlighting the urgent need for routing protocols with low energy consumption.

The robustness and dependability of IoT-based monitoring systems present another major obstacle. When implementing industrial IoT systems for pipeline monitoring, Ahmed et al. (2020) stress the significance of solving concerns related to scalability, maintainability, and reliability. The integrity management process is made more difficult by the intricacy of overseeing numerous IoT devices, each of which has the potential to malfunction. Since IoT technology is still developing and many current systems rely more on data visualization than on reliable algorithmic models for predictive maintenance, this complexity is exacerbated by the requirement for real-time data analysis and decision-making (Zhang and Xie 2024).

A significant obstacle in the area of IoT applications for pipeline monitoring is cybersecurity. According to Diaz et al. (2019), IoT devices may be vulnerable to cyberattacks due to systemic flaws, thus strong cybersecurity measures must be put in place to guard against possible assaults. IoT systems' dispersed architecture, which frequently combines cloud, edge, and fog computing, makes cybersecurity management even more challenging and makes it challenging to guarantee uniform protection across all endpoints.

Along with these technical difficulties, there are operational barriers to IoT integration with pipeline integrity control. Zhang and Xie (2024) emphasize that accurate algorithmic models and a thorough lifespan database are essential for failure prediction, and that present research frequently lacks the support these models require. The implementation of IoT solutions must consider the environmental and geographic aspects that impact pipeline operations, as well as the requirement for real-time monitoring capabilities to quickly identify problems like corrosion or leaks (Rahmani et al. 2022).

13.4.1 Data Overload

IoT devices produce substantial quantities of data, which can inundate operators if not effectively managed. Without effective data processing and analytics tools, operators may struggle to interpret the data and make timely decisions.

- **Solution**: Implementing advanced Big Data platforms that can filter, organize, and analyze IoT data is essential for ensuring that operators can focus on the most relevant information and respond to issues quickly.

13.4.2 Connectivity Issues

Pipelines often operate in remote or challenging environments where reliable internet connectivity may be limited. IoT devices require stable connections to transmit data to central systems, and connectivity issues can lead to delays in data transmission or even data loss.

- **Solution**: To overcome connectivity challenges, operators can use satellite communication systems, low-power wide-area networks (LPWAN), or other long-range communication technologies to ensure continuous data transmission from remote pipeline locations.

13.4.3 Cybersecurity Risks

The increasing reliance on IoT devices also introduces cybersecurity risks, as connected devices can be vulnerable to hacking and unauthorized access. A cyberattack on pipeline IoT systems could disrupt operations or cause data breaches, leading to safety and financial risks.

- **Solution**: Implementing strong cybersecurity protocols, such as encryption, authentication, and intrusion detection systems, is critical for protecting IoT devices and networks from cyber threats.

13.5 Future of IoT in Pipeline Integrity Management

The integration of smart technologies and data analytics significantly enhances pipeline integrity control through the Internet of Things (IoT). IoT technology enables real-time pipeline monitoring and predictive maintenance, which are crucial for preventing leaks and ensuring operational safety. IoT systems may continuously collect and analyze data from pipeline infrastructures using advanced sensors and communication protocols. This enhances the decision-making process for overseeing pipeline integrity.

Leak detection constitutes a primary use of IoT in the maintenance of pipeline integrity. Zhang and Xie (2024) assert that IoT technology offers innovative methods for pipeline leak detection, enabling the development of comprehensive lifespan datasets and precise algorithmic models for failure prediction. Proactive maintenance techniques, which can greatly lower the danger of catastrophic breakdowns and environmental harm, depend on these predictive capabilities. Karam et al. (2024) further emphasize the significance of energy-efficient routing algorithms in guaranteeing dependable low-latency communication for oil and gas pipeline monitoring, which is especially difficult because of low infrastructure and hazardous settings.

Pipeline monitoring is seeing an increase in the use of low-power wide-area networks (LPWANs), such LoRa. Long-distance data transmission is made possible by LoRa technology, which is essential for overseeing vast pipeline networks, according to Narayana et al. (2022). This feature makes it possible to install many sensor nodes that can keep an eye on different parameters, such as flow and pressure, and guarantee the integrity of the pipeline system.

IoT technologies help with pipeline operational optimization in addition to monitoring. Mekki et al. (2023) propose a printed circuit board design that improves the overall management of oil pipeline traffic by enabling real-time monitoring and control of pipeline cathodic protection systems via cloud and Internet of Things platforms. In addition to increasing operational effectiveness, this integration helps ensure that pipeline management techniques are sustainable.

The implementation of IoT technologies for pipeline integrity management still faces difficulties despite these developments. To guarantee the resilience of IoT systems in industrial applications, concerns about energy consumption, scalability, and dependability must be resolved. According to Ahmed et al. (2020), IoT-based monitoring systems' resilience is essential for efficiently identifying and addressing pipeline faults. Therefore, to improve these technologies and go past current constraints, continuous research and development is required.

13.5.1 5G Connectivity

The deployment of 5G networks would markedly enhance the speed and reliability of data transfer from IoT devices, facilitating swifter and more responsive pipeline monitoring systems (Esenogho et al. 2022). 5G will facilitate real-time data processing with negligible latency, enabling operators to respond promptly to abnormalities.

13.5.2 Edge Computing

Rather than transmitting data to a centralized cloud platform, edge computing processes data at the source of generation, such as at the IoT device itself. In areas with inadequate connectivity, operators can reduce latency and expedite decision-making by utilizing edge computing for data processing.

13.5.3 AI-Driven Autonomous Systems

As AI technologies continue to advance, IoT systems will become increasingly autonomous, with AI-driven algorithms capable of making real-time decisions based on sensor data. These systems will be able to detect risks, adjust operating conditions, and even initiate repairs without human intervention.

References

Abed AK, Anupam A (2023) Review of security issues in Internet of Things and artificial intelligence-driven solutions. Secur Privacy 6(3):e285

Ahmed S, Le Mouël F, Stouls N (2020) Resilient IoT-based monitoring system for crude oil pipelines. In: 2020 7th International conference on Internet of Things: systems, management and security (IOTSMS). IEEE

Ayeni TP, Ayogu BA (2020) Intelligent pipeline monitoring system based on internet of things. Scient Res J 8(8):44–49

Ayvaz S, Alpay K (2021) Predictive maintenance system for production lines in manufacturing: a machine learning approach using IoT data in real-time. Expert Syst Appl 173:114598

Cao H, Wachowicz M (2020) A holistic overview of anticipatory learning for the internet of moving things: research challenges and opportunities. ISPRS Int J Geo Inf 9(4):272

Chen C, Guo H (2013) Design method of database system based on DPT and P2P point cloud for Internet of Things. Adv Mater Res 760:2077–2080

Chen J et al (2014) Applications of Internet of Things in facility agriculture. Appl Mech Mater 685:517–523

de Haro-Olmo FJ et al (2023) ELI: an IoT-aware big data pipeline with data curation and data quality. PeerJ Comp Sci 9:e1605

Diaz J et al (2019) Self-service cybersecurity monitoring as enabler for DevSecOps. IEEE Access 7:100283–100295

Esenogho E, Djouani K, Kurien AM (2022) Integrating artificial intelligence Internet of Things and 5G for next-generation smartgrid: a survey of trends challenges and prospect. IEEE Access 10:4794–4831

Fisher R et al (2015) Open hardware: a role to play in wireless sensor networks? Sensors 15(3):6818–6844

Issa AA et al (2024) Farming in the digital age: smart agriculture with AI and IoT. In: E3S Web of conferences. EDP Sciences

Kadar HH, Sameon SS, Rusli ME (2018) SMART2L: smart water level and leakage detection. Int J Eng Technol (UAE) 7(4.36):448–452

Karam SN et al (2024) Energy-efficient routing protocol for reliable low-latency Internet of Things in oil and gas pipeline monitoring. PeerJ Comp Sci 10:e1908

Korlapati NVS et al (2022) Review and analysis of pipeline leak detection methods. J Pipeline Sci Eng 2(4):100074

Kumar DM, Jagadeep T (2022) Water pipeline leakage detection and monitoring system using smart sensor with IOT. J Phys: Conf Ser

Kumar H, Soh PJ, Ismail MA (2022) Big data streaming platforms: a review. Iraqi J Comp Sci Math 3(2):95–100

LaNasa PJ, Upp EL (2014) Fluid flow measurement: a practical guide to accurate flow measurement. Butterworth-Heinemann

Liu M et al (2022) Low energy consumption routing protocol for oil and gas pipeline Internet of Things. Preprints. Research Square

Marengo A (2024) Future of AI-driven IoT: identifying emerging trends in intelligent data analysis and privacy protection. Prepints.org

Mekki I et al (2023) Design of a printed circuit board for real-time monitoring and control of pipeline's cathodic protection system via IoT and a cloud platform. Int J Eng 36(9):1667–1676

Meribout M et al (2020) Leak detection systems in oil and gas fields: present trends and future prospects. Flow Meas Instrum 75:101772

Narayana CL, Singh R, Gehlot A (2022) Performance evaluation of LoRa based sensor node and gateway architecture for oil pipeline management. Int J Elec Comp Eng (IJECE) 12(1):974–982

Perera C, Liu CH, Jayawardena S (2015) The emerging internet of things marketplace from an industrial perspective: a survey. IEEE Trans Emerg Top Comput 3(4):585–598

Priyanka EB, Maheswari C, Thangavel S (2021) A smart-integrated IoT module for intelligent transportation in oil industry. Int J Numer Model Electron Netw Devices Fields 34(3):e2731

Rahmani A et al (2022) An energy-aware and Q-learning-based area coverage for oil pipeline monitoring systems using sensors and Internet of Things. Sci Rep 12(1):9638

Razi P, Taheri F (2014) A vibration-based strategy for health monitoring of offshore pipelines' girth-welds. Sensors 14(9):17174–17191

Razzaque MA et al (2015) Middleware for internet of things: a survey. IEEE Internet Things J 3(1):70–95

Reddy MSB et al (2021) Sensors in advancing the capabilities of corrosion detection: a review. Sens Actuators, A 332:113086

Said ZM, Zolkipli MF (2022) Internet of Things (IoT): a study of security issues and challenges. Int J Recent Contrib Eng Sci IT 10(2):16–31

Senoo EEK et al (2024) IoT solutions with artificial intelligence technologies for precision agriculture: definitions, applications, challenges, and opportunities. Electronics 13(10):1894

Sharma VB et al (2021) Review of structural health monitoring techniques in pipeline and wind turbine industries. Appl Syst Innov 4(3):59

Singh R et al (2021) Zigbee and long-range architecture based monitoring system for oil pipeline monitoring with the internet of things. Sustainability 13(18):10226

Thorn R et al (2023) Flow measurement. In: Mechanical variables measurement-solid, fluid, and thermal. CRC Press, pp 9-1–9-132

Uspabayeva A et al (2022) Evaluation of high school students' new trends in education: Internet of Things. Int J Emerg Technol Learn (iJET) 17(19):159–175
Vermesan O et al (2020) Internet of robotic things intelligent connectivity and platforms. Front Robot AI 7:104
Wong YK (2021) Understanding the features of Internet of Things (IoT) and Big Data analysis. In: CS & IT conference proceedings
Wright RF et al (2019) Corrosion sensors for structural health monitoring of oil and natural gas infrastructure: a review. Sensors 19(18):3964
Xi J et al (2024) Design of a low-power monitoring system for gas wells with temperature and humidity compensation. In: 4th International conference on Internet of Things and smart city (IoTSC 2024). SPIE
Zaman D et al (2020) A review of leakage detection strategies for pressurised pipeline in steady-state. Eng Fail Anal 109:104264
Zhang X, Xie X (2024) Model for the failure prediction mechanism of in-service pipelines based on IoT technology. Processes 12(8):1642
Zhao C, Li XS, Chen JS (2011) Study on the application of Internet of Things in the logistics in forest industry. Appl Mech Mater 97:664–668
Zonta T et al (2020) Predictive maintenance in the Industry 4.0: a systematic literature review. Comp Industr Eng 150:106889

Chapter 14
Leveraging Blockchain for Pipeline Integrity Management

It is nearly hard to change historical data covertly thanks to blockchain's immutable ledger structure, which guarantees that all transactions and pipeline integrity data are safely recorded. This quality is essential for pipeline management since leaks and failures caused by integrity problems can have serious negative effects on the environment and the economy (Fan and Feng 2023).

The improvement of accountability and transparency is a primary benefit of utilizing blockchain technology for pipeline integrity monitoring. According to Rahayu et al. (Rahayu et al. 2021), blockchain enables stakeholders to track down the origin of supply chain issues, which is especially important in pipeline operations as several parties are involved throughout the pipeline's lifecycle. Frameworks like SIGNORA, which integrates blockchain technology with a chain of signatures to offer dataflow integrity and guarantee that every data sent through the pipeline is verifiable and impenetrable, further improve this capability. Such mechanisms not only bolster the integrity of the data but also facilitate reliable identity management and accountability among participants, which is essential in an industry where trust is paramount (Oktian et al. 2022).

Two of the main reasons for pipeline failures, corrosion and third-party damage, can be considerably reduced by using blockchain technology in pipeline integrity management. Operators can keep a thorough record of the pipeline's state and the steps required to maintain its integrity by putting in place a decentralized system that logs all maintenance operations, inspections, and incidents. Because it offers a clear picture of the pipeline's operational history, this historical data can be extremely helpful for regulatory compliance and auditing (Fan and Feng 2023).

Blockchain integration has the potential to improve cooperation amongst pipeline ecosystem players. According to Fan and Feng (2023), a blockchain-enabled collaborative approach to integrity management can result in more efficient monitoring and reaction plans for possible integrity risks. Blockchain's decentralized structure makes it possible for all stakeholders to share data in real time, which speeds up and improves pipeline integrity decision-making.

M. Hussain and T. Zhang, *Big Data Analytics in Energy Pipeline Integrity Management*, Lecture Notes in Energy 104, https://doi.org/10.1007/978-981-96-8019-1_14

14.1 What is Blockchain?

Blockchain technology is a decentralized and distributed digital ledger system that enables secure and transparent transaction recording among multiple parties without necessitating a central authority. Essentially, blockchain consists of a series of blocks, each containing a collection of transactions that are cryptographically linked. This structure ensures data integrity and immutability; once a block is added to the chain, it cannot be altered without changing all subsequent blocks, requiring consensus from network participants (Ruoti et al. 2019; Yaga et al. 2019).

The fundamental characteristics of blockchain include decentralization, transparency, and security. Decentralization enables a peer-to-peer network in which each participant (or node) has access to the entire ledger, hence eliminating the need for intermediaries like as banks or governments (Yaga et al. 2019). Transparency is achieved as all transactions are visible to all participants, hence fostering confidence among them (Rizal Batubara et al. 2019). Security is enhanced using cryptographic techniques that protect data and consensus protocols that verify transactions, hence complicating attempts by malicious actors to undermine the system (Wang et al. 2022; Conti et al. 2018).

Despite its potential, the application of blockchain technology faces significant challenges. This includes technical issues such as scalability and interoperability, together with regulatory and governance limitations (Reshi and Sholla 2024). The distinction between public and private blockchains raises questions about their suitability for different applications, particularly in contexts where trust among participants already exists (Alles and Gray 2023). As technology progresses, ongoing research is essential to address these challenges and explore innovative applications that leverage blockchain's unique characteristics (Honey et al. 2022).

14.1.1 How Blockchain Works

Blockchain operates by a consensus mechanism, necessitating many nodes within the network to agree on the validity of data before its incorporation into the ledger. Upon verification, data is irrevocably inscribed on the blockchain and is available to authorized parties (Praveena Anjelin and Ganesh Kumar 2021). This decentralized architecture reduces the need for middlemen and improves the security and transparency of data exchanges.

Blockchain technology operates as a decentralized digital ledger that records transactions across multiple computers, ensuring the security, transparency, and permanence of the data. Each transaction is aggregated into a block, which is then linked to the prior block, forming a sequential chain. This structure is crucial for blockchain functionality, as it guarantees that once data is recorded, it cannot be altered without the consensus of network participants (Tijan et al. 2019).

The process begins when a user starts a transaction, which is then transmitted to a network of nodes (computers) participating in the blockchain. Each node independently verifies the transaction using proven consensus mechanisms, such as Proof of Work (PoW) or Proof of Stake (PoS). These processes require nodes to either solve complex mathematical problems or verify transactions based on their interest in the network, respectively. Upon transaction confirmation, it is integrated into a new block with additional transactions validated within the same time frame (Kh-Madhloom 2022).

After the accumulation of transactions within a block is finalized, it undergoes a hashing process, during which a cryptographic hash function generates a unique identifier for the block. This not only signifies the contents of the block but also integrates the hash of the preceding block, so forming a linkage between them. This cryptographic linkage ensures that any attempt to alter a block requires the recalculation of hashes for all subsequent blocks, making it computationally unfeasible in a large network (Mohite and Kale 2022). When a new block is created, it is distributed around the network, compelling all nodes to update their copies of the blockchain to include the new block. This decentralized approach eliminates the need for a central authority, as all participants have access to the same version of the ledger, hence enhancing trust and accountability among users (Zhang et al. 2024). Blockchain technology employs smart contracts—self-enforcing agreements with terms embedded directly in software. These contracts independently enforce and execute transactions when specified conditions are met, thus streamlining processes and reducing the need for intermediaries (Shrestha et al. 2021).

14.2 Applications of Blockchain in Pipeline Integrity Management

The application of blockchain technology in pipeline integrity monitoring is rising owing to its essential attributes of decentralization, immutability, and transparency. These qualities are particularly beneficial in managing the complex data flows and interactions characteristic of pipeline operations, ensuring that all transactions and alterations are securely recorded and validated.

A primary benefit of blockchain in pipeline integrity management is its ability to enhance data security and integrity. Blockchain's distributed ledger technology (DLT) ensures that all data entries are immutable, meaning that once a transaction is recorded, it cannot be altered or deleted without the agreement of network participants (Yaga et al. 2019). This functionality is essential in pipeline management, where accurate recording of inspections, maintenance activities, and operational data is crucial for compliance and safety. The immutable properties of blockchain can prevent data tampering, hence enhancing trust among stakeholders (Bhutta et al. 2021).

Blockchain enhances traceability and transparency in pipeline operations. Every transaction or occurrence can be timestamped and connected to preceding records, establishing a thorough audit trail accessible to authorized individuals (Qureshi et al. 2025). This skill is especially beneficial in situations where accountability is crucial, such as in environmental monitoring and regulatory compliance. Blockchain facilitates a clear overview of the pipeline's operational history, enabling stakeholders to swiftly identify and address concerns effectively (Zhu et al. 2023).

The incorporation of smart contracts significantly improves the efficacy of blockchain in pipeline integrity control. Smart contracts can automate numerous operations, such the initiation of maintenance alerts based on sensor data or the execution of payments upon the fulfillment of designated tasks (Popchev et al. 2023). This automation minimizes human mistake and optimizes operations, facilitating more efficient resource and time management. In pipeline inspections, smart contracts can autonomously confirm the completion of inspections and the submission of all requisite documentation prior to permitting operations to proceed (Cheng et al. 2024).

The implementation of blockchain technology in pipeline integrity management not only provides operational advantages but also results in considerable cost reductions. Organizations can decrease administrative expenses and enhance operational efficiency by minimizing the reliance on intermediaries and manual record-keeping (Iskamto and Juariyah 2023). Moreover, the augmented security and transparency offered by blockchain might mitigate the risks linked to data breaches and fraud, which can be especially expensive in the energy sector (Bhutta et al. 2021).

14.2.1 Data Security and Integrity

A primary benefit of blockchain is its ability to protect data and preserve its integrity (Idrees et al. 2021). In pipeline operations, where real-time data is collected from diverse IoT sensors and monitoring systems, blockchain can be utilized to verify this data and protect it from alteration.

14.2.2 Asset Tracking and Maintenance Verification

Blockchain can be used to track the lifecycle of pipeline assets, from installation to decommissioning (Chen et al. 2022). By recording all maintenance activities, repairs, and inspections on a blockchain, operators can create an immutable record of asset history. This helps ensure that maintenance is performed on time and according to regulatory requirements (Demir and Yomralioglu 2024).

14.2.3 Regulatory Compliance and Reporting

Blockchain's ability to create an immutable record of transactions makes it ideal for ensuring compliance with regulatory standards. By recording inspection results, maintenance activities, and operational data on a blockchain, pipeline operators can provide regulators with verifiable proof of compliance.

14.2.4 Smart Contracts for Automated Compliance

Blockchain can be combined with smart contracts, which are self-executing agreements encoded in software (Kaur et al. 2022). Smart contracts can automate specific compliance duties, including scheduling maintenance or activating notifications upon the fulfillment of specified requirements (Mohanta et al. 2018).

14.2.5 Supply Chain Management

Blockchain technology can be utilized to monitor the supply chain for materials used in pipeline construction and maintenance. By establishing a visible and immutable record of materials from suppliers, operators may verify that components adhere to safety and quality criteria prior to installation.

14.3 Benefits of Blockchain in Pipeline Integrity Management

Blockchain provides numerous advantages for pipeline operators, especially regarding data security, transparency, and efficiency. These advantages can result in enhanced safety, decreased expenses, and increased assurance among stakeholders.

A key advantage of blockchain in pipeline integrity management is its capacity to furnish an immutable record of all transactions and data entries. This feature guarantees that once data is recorded, it cannot be modified or erased without the agreement of network members, hence augmenting data integrity and confidence among stakeholders (Noviani and Muda 2021). In pipeline management, this entails the secure storage of documents pertaining to inspections, maintenance activities, and operational data, hence mitigating the danger of data manipulation or fraud (Olatoye et al. 2024). The unchangeable characteristics of blockchain enhance adherence to regulatory mandates by offering a verifiable audit trail accessible to pertinent authorities.

A significant benefit is the increased transparency provided by blockchain technology. Every transaction on a blockchain is timestamped and interconnected with preceding entries, establishing a thorough and readily verifiable record of all operations associated with the pipeline (Olatoye et al. 2024). This transparency is crucial for stakeholders, especially regulatory authorities, as it facilitates real-time oversight of pipeline operations and prompt detection of possible problems. Moreover, the decentralized characteristic of blockchain obviates the necessity for middlemen, hence optimizing procedures and diminishing operational expenses (Cao et al. 2022).

The implementation of smart contracts enhances the advantages of blockchain in pipeline integrity control. Smart contracts are autonomous agreements that automatically enforce and implement stipulations based on predetermined situations. In pipeline management, smart contracts can automate numerous operations, like the initiation of maintenance alerts based on sensor data or the execution of payments upon the fulfillment of designated tasks. This automation reduces the likelihood of human error and improves operational efficiency by decreasing the time needed for manual interventions.

Blockchain technology can dramatically enhance data security. The encryption and decentralized storage of data on a blockchain render it more resilient against cyber-attacks and unlawful access. Access and decryption of the data are restricted to authorized individuals possessing the requisite private keys, thereby safeguarding sensitive information. This degree of security is especially vital in pipeline management, where the integrity of operating data is essential for safety and compliance.

14.3.1 Enhanced Data Security

The decentralized and tamper-proof nature of blockchain ensures that pipeline data, such as sensor readings and inspection reports, cannot be altered or deleted once it is added to the blockchain (Chanson et al. 2019). This provides an additional layer of security for sensitive operational data.

14.3.2 Improved Transparency and Accountability

Blockchain establishes a transparent, communal record of data accessible to all parties, including regulators, contractors, and maintenance personnel. This enhances accountability, as all stakeholders have access to identical data and can confirm that safety and compliance standards are being fulfilled.

14.3.3 Streamlined Compliance Processes

By providing a permanent, verifiable record of inspections, maintenance, and repairs, blockchain can streamline compliance processes and reduce the administrative burden of regulatory reporting. This ensures that operators can quickly demonstrate compliance with safety standards and avoid penalties for non-compliance.

14.3.4 Reduced Operational Costs

Blockchain can help reduce operational costs by automating compliance tasks, streamlining supply chain management, and reducing the need for manual data verification (Narayanaswami et al. 2019). Smart contracts can automate routine tasks such as scheduling inspections or triggering alerts, freeing up resources for more critical tasks.

14.4 Challenges of Implementing Blockchain in Pipeline Operations

The integration of blockchain technology in pipeline operations poses numerous problems that must be resolved to fully realize its potential. A primary difficulty is maintaining data integrity and security, especially in contexts where data is vulnerable to manipulation or illegal access. The immutable nature of blockchain provides a solution by offering a decentralized ledger that securely records transactions, hence boosting data integrity (Oktian et al. 2022). The shift to blockchain necessitates substantial modifications to current operational frameworks, which may encounter opposition from stakeholders familiar with conventional systems (Ioannou and Demirel 2022; Lu et al. 2019).

A notable problem is the integration of blockchain with current technologies and processes. In the oil and gas sector, the amalgamation of blockchain with Internet of Things (IoT) systems is essential for guaranteeing data integrity and real-time transparency throughout the supply chain (Ioannou and Demirel 2022; Kariri 2023). This integration is frequently obstructed by insufficient comprehension and proficiency in blockchain technology among industry participants (Lu et al. 2019). The necessity for regulatory compliance introduces additional complication, as current legal frameworks may not sufficiently include the intricacies of blockchain operations (Ioannou and Demirel 2022; Kariri 2023). This requires a cooperative strategy among stakeholders to amend regulatory frameworks and create standards that promote blockchain implementation (Ioannou and Demirel 2022).

Operational risks provide a considerable hurdle in the deployment of blockchain inside pipeline operations. The dependence on smart contracts, which automate activities according to predetermined circumstances, presents hazards associated with coding errors and the possibility of malicious attacks (Zheng et al. 2019). The scalability of blockchain solutions is a significant challenge, as numerous existing systems struggle to manage the transaction volume characteristic of pipeline operations without sacrificing performance. Resolving these operational problems necessitates continuous research and development to improve the resilience and scalability of blockchain systems (Zheng et al. 2019).

14.4.1 Scalability Issues

Blockchain networks may have scalability issues, especially when managing substantial data quantities (Khan et al. 2021). The generation of substantial data volumes from IoT devices through pipelines may render the storage and verification of all this data on a blockchain resource-intensive, thus impeding system performance.

- **Solution**: To address scalability challenges, operators can use hybrid blockchain models that store critical data on-chain while offloading larger datasets to external storage systems. This approach ensures that only the most important data is stored on the blockchain, while reducing the strain on the network.

14.4.2 Regulatory Uncertainty

While blockchain holds promises to improve regulatory compliance, its use in pipeline operations may face uncertainty due to a lack of established legal frameworks. Regulators may not yet have clear guidelines on how blockchain-based records should be treated in compliance audits or legal disputes.

- **Solution**: Pipeline operators should work closely with regulators to develop best practices and standards for using blockchain in compliance processes. Engaging in industry initiatives and pilot projects can help create regulatory clarity and promote the adoption of blockchain in the energy sector.

14.4.3 Collaboration Across the Industry

For blockchain to be effective in pipeline integrity management, collaboration between different stakeholders—such as regulators, suppliers, and maintenance contractors—is essential. Achieving industry-wide collaboration can be challenging, as it requires standardization and the alignment of goals.

- **Solution**: Industry consortia and working groups can play a key role in promoting collaboration and developing common standards for blockchain use in the energy sector. By working together, stakeholders can ensure that blockchain solutions are interoperable and meet the needs of the entire industry.

14.5 The Future of Blockchain in Pipeline Integrity Management

The future use of blockchain in pipeline integrity management holds the potential to revolutionize the industry by enhancing transparency, security, and efficiency in the oversight and upkeep of pipeline systems. The integration of blockchain technology offers a decentralized and immutable record that significantly improves the traceability of pipeline operations, thereby addressing integrity issues that have historically plagued the industry.

A key benefit of blockchain in pipeline integrity management is its capacity to furnish an immutable record of all transactions and events associated with the pipeline. This attribute is essential for identifying the origin of issues, such as leaks or failures, which may result from multiple reasons, including third-party damage or corrosion (Fan and Feng 2023). Blockchain records every transaction within a distributed network, guaranteeing that all stakeholders access identical data, hence augmenting responsibility and confidence among participants in pipeline operations (Oktian et al. 2022). This transparency is especially crucial in contexts involving numerous untrusted entities, as it reduces the risks linked to data modification and fraud (Lin et al. 2020).

Blockchain enables real-time surveillance and documentation of pipeline conditions, crucial for preventative maintenance and swift reaction to possible integrity violations. Smart contracts in blockchain frameworks can automate procedures like inspections and maintenance schedules, ensuring timely actions based on predetermined criteria (Olatoye et al. 2024). This automation decreases operational expenses and minimizes human mistake, hence strengthening the integrity of the pipeline management system.

Besides enhancing operational efficiency, blockchain technology can facilitate the integration of new technologies like the Internet of Things (IoT) and artificial intelligence (AI). For example, IoT devices can perpetually monitor pipeline conditions and transmit data to the blockchain, where it can be securely stored and evaluated (Xue et al. 2023). This integration facilitates advanced data analytics capable of forecasting possible failures and enhancing maintenance plans, therefore prolonging the lifespan of pipeline assets and mitigating environmental concerns linked to leaks (Liu et al. 2023).

The utilization of blockchain in pipeline integrity monitoring corresponds with the increasing focus on sustainability and regulatory adherence. Blockchain enables a transparent and immutable record of all pipeline activities, enabling firms to exhibit compliance with environmental rules and industry norms, thus bolstering their reputation and stakeholder trust (Götz et al. 2020). This is especially pertinent given the growing examination of environmental effects and the necessity for companies to implement more accountable procedures.

14.5.1 Integration with AI and IoT

The integration of blockchain with advanced technologies, like as artificial intelligence (AI) and the Internet of Things (IoT), will provide strong new systems for monitoring and preserving pipeline integrity. By combining AI's predictive capabilities with blockchain's data security and transparency, operators may create fully automated systems for detecting and mitigating pipeline risks.

14.5.2 Smart Grids and Decentralized Energy Systems

Blockchain possesses the capacity to significantly contribute to the advancement of smart grids and decentralized energy systems (Hasankhani et al. 2021; Pop et al. 2018). In the future, pipelines may become part of a broader energy ecosystem that leverages blockchain to securely manage the flow of energy and data between different stakeholders.

14.5.3 Standardization and Industry Adoption

As blockchain technology matures, industry-wide standards for its use in pipeline operations will likely emerge (Zehir et al. 2022). Standardization will make it easier for operators to adopt blockchain solutions and ensure that these solutions are compatible across different regions and regulatory environments.

References

Alles M, Gray GL (2023) Hope or hype? Blockchain and accounting. Int J Digit Acc Res 23:19–45

Bhutta MNM et al (2021) A survey on blockchain technology: evolution, architecture and security. IEEE Access 9:61048–61073

Cao B et al (2022) Blockchain systems, technologies, and applications: a methodology perspective. IEEE Commun Surv Tutorials 25(1):353–385

Chanson M et al (2019) Blockchain for the IoT: privacy-preserving protection of sensor data. J Assoc Inf Syst 20(9):1274–1309

Chen B-Q, Videiro PM, Guedes Soares C (2022) Opportunities and challenges to develop digital twins for subsea pipelines. J Mar Sci Eng 10(6):739

Cheng M, Chong H-Y, Xu Y (2024) Blockchain-smart contracts for sustainable project performance: bibliometric and content analyses. Environ Dev Sustain 26(4):8159–8182
Conti M et al (2018) A survey on security and privacy issues of bitcoin. IEEE Commun Surv Tutorials 20(4):3416–3452
Demir S, Yomralioglu T (2024) Bridging geo-data and natural gas pipeline design standards: a systematic review of BIM-GIS integration for natural gas pipeline asset management. Energies 17(10):2306
Fan Y, Feng X-X (2023) Research on integrity management of long distance pipeline. In: E3S Web of conferences. EDP Sciences
Götz CS, Karlsson P, Yitmen I (2020) Exploring applicability, interoperability and integrability of Blockchain-based digital twins for asset life cycle management. Smart Sustain Built Environ 11(3):532–558
Hasankhani A et al (2021) Blockchain technology in the future smart grids: a comprehensive review and frameworks. Int J Electr Power Energy Syst 129:106811
Honey K, Robb A, Rohde F (2022) Are we contributing? The who, when, where, & what of the Blockchain research landscape. In: Proceedings of the 56th Hawaii International Conference on System Sciences
Idrees SM et al (2021) Security aspects of blockchain technology intended for industrial applications. Electronics 10(8):951
Ioannou I, Demirel G (2022) Blockchain and supply chain finance: a critical literature review at the intersection of operations, finance and law. J Bank Fin Technol 6(1):83–107
Iskamto D, Juariyah L (2023) Blockchain technology challenge in the future: data security and efficiency. Int J Law, Policy, Govern 2(2):65–76
Kariri MH (2023) Leveraging blockchain technology to create a more resilient supply chain for energy industry. Int J Supply Chain Manage 8(1):1–76
Kaur R, Ali A, Faisal M (2022) Smart contracts: the self-executing contracts, in Blockchain. Chapman and Hall/CRC, p 35–49
Khan D, Jung LT, Hashmani MA (2021) Systematic literature review of challenges in blockchain scalability. Appl Sci 11(20):9372
Kh-Madhloom J (2022) Dynamic cryptography integrated secured decentralized applications with blockchain programming. Wasit J Comp Math Sci 1(2):21–33
Lin W et al (2020) Blockchain technology in current agricultural systems: from techniques to applications. IEEE Access 8:143920–143937
Liu X et al (2023) FP-BFT: a fast pipeline Byzantine consensus algorithm. IET Blockchain 3(3):123–135
Lu H et al (2019) Blockchain technology in the oil and gas industry: a review of applications, opportunities, challenges, and risks. IEEE Access 7:41426–41444
Mohanta BK, Panda SS, Jena D (2018) An overview of smart contract and use cases in blockchain technology. In: 2018 9th International conference on computing, communication and networking technologies (ICCCNT). IEEE
Mohite JS, Kale V (2023) Blockchain technology based supply chain management system. In: International conference on applications of machine intelligence and data analytics (ICAMIDA 2022). Atlantis Press
Narayanaswami C et al (2019) Blockchain anchored supply chain automation. IBM J Res Dev 63(2/3):7:1–7:11
Noviani N, Muda I (2021) The evolution of accounting software: review of blockchain technology in audit. In: Proceedings of the 1st international conference on social, science, and technology, ICSST
Oktian YE, Heo S, Kim H (2022) SIGNORA: a blockchain-based framework for dataflow integrity provisioning in an untrusted data pipeline. IEEE Access 10:89714–89731
Olatoye FO et al (2024) Blockchain in asset management: an extensive review of opportunities and challenges. Int J Sci Res Archive 11(1):2111–2119

Pop C et al (2018) Blockchain based decentralized management of demand response programs in smart energy grids. Sensors 18(1):162

Popchev I, Radeva I, Doukovska L (2023) Oracles integration in blockchain-based platform for smart crop production data exchange. Electronics 12(10):2244

Praveena Anjelin D, Ganesh Kumar S (2021) Blockchain technology for data sharing in decentralized storage system. In: Intelligent computing and applications: proceedings of ICICA 2019. Springer

Qureshi JN et al (2025) Exploring the integration of blockchain and distributed DevOps for secure, transparent, and traceable software development. Access 13:15489–15502

Rahayu SB et al (2021) Military supply chain management and blockchain development. Comp Sci Inf Technol 11(1):93–105

Reshi IA, Sholla S (2024) The blockchain conundrum: an in-depth examination of challenges, contributing technologies, and alternatives. Concurr Comput: Pract Exp 36(8):e7987

Rizal Batubara F, Ubacht J, Janssen M (2019) Unraveling transparency and accountability in blockchain. In: Proceedings of the 20th annual international conference on digital government research

Ruoti S et al (2019) SoK: Blockchain technology and its potential use cases. arXiv preprint arXiv:1909.12454

Shrestha AK et al (2021) Augmenting the technology acceptance model with trust model for the initial adoption of a blockchain-based system. PeerJ Comp Sci 7:e502

Tijan E et al (2019) Blockchain technology implementation in logistics. Sustainability 11(4):1185

Wang X et al (2022) Bft in blockchains: from protocols to use cases. ACM Comput Surv (CSUR) 54(10s):1–37

Xue H et al (2023) Integration of blockchain and edge computing in internet of things: a survey. Futur Gener Comput Syst 144:307–326

Yaga D et al (2019) Blockchain technology overview. arXiv preprint arXiv:1906.11078

Zehir C et al (2022) Tailored blockchain applications for the natural gas industry: the case study of SOCAR. Energies 15(16):6010

Zhang X et al (2024) Blockchain implementation and shareholder value: a complex adaptive systems perspective. Int J Oper Prod Manag 44(3):666–698

Zheng Z, Dai H-N, Wu J (2019) Blockchain intelligence: when blockchain meets artificial intelligence. arXiv preprint arXiv:1912.06485

Zhu J et al (2024) Fusion diagnostics based on blockchain technology. In: International conference on computer graphics, artificial intelligence, and data processing (ICCAID 2023). SPIE

Chapter 15
Future Trends and Innovations in Pipeline Integrity Management

Traditional approaches to pipeline condition assessment often rely on periodic inspections and conventional analysis, which is time-consuming, costly, and prone to human error. However, with the advancements in data analytics and the increasing availability of large volumes of data, adopting Big Data Analytics for energy pipeline condition assessment has emerged as a promising solution. By leveraging the massive volumes of data generated by sensors, meters, and other monitoring devices installed on these pipelines. Operators can gain valuable insights into the pipeline's health and performance. Predictive maintenance is one of the primary applications of Big Data in pipeline condition assessment. Traditional maintenance approaches are often based on predetermined schedules or reactive responses to failures. Big Data Analytics is all about managing and analyzing large amounts of data to improve asset performance, predict anomalies, understand the current condition of the asset, reduce maintenance costs, make better decisions, and reduce workplace risks.

Big Data, defined in the context of the oil and gas industry, encompasses vast and complex datasets generated from various sources such as geological surveys, production data, inspections (ILI, direct, CP, DCVG etc.) and real-time monitoring systems. These data sources provide a foundation for analytics, enabling pipeline companies to enhance operational efficiency, safety, and decision-making processes. Technological advancements in the energy pipeline sector in recent years have had an enormous impact and generated a significantly large number of datasets. It requires the development of advanced technologies to process, analyze, and apply the data to dig out the embedded values.

Big Data Analytics is important and useful in the life-cycle management of energy pipelines. This covers the condition assessment, the prediction of the remaining useful life, replacement, retrofit planning, optimal exploitation, etc. Big Data Analytics is designed to handle a large amount of data corresponding to the pipeline, making it a powerful asset management tool. The core application of Big Data Analytics is a user-friendly environment that acts as an on-the-fly management of right-of-the-use data

M. Hussain and T. Zhang, *Big Data Analytics in Energy Pipeline Integrity Management*, Lecture Notes in Energy 104, https://doi.org/10.1007/978-981-96-8019-1_15

and efficiently retrieves design and quality documents. Meaningful insights extraction, such as patterns and themes, as well as professional preferences, market trends, and unknown correlations, make it more important to use in energy pipeline asset management. Integrating Big Data Analytics with IoT and other digital technologies has enabled more precise and timely condition monitoring, improved pipeline integrity and reduced failure risk.

Overall, this chapter summarizes the achievements presented in the above chapters. By acknowledging the significant strides made in adopting Big Data Analytics in the oil and gas industry while also pointing out existing challenges such as data silos, security concerns, and the need for skilled personnel, it calls for continued investment in technology and training to fully leverage the potential of Big Data Analytics, paving the way for further innovations and advancements in the sector.

The focus in the literature has been on the benefits of Big Data Analytics and its application to pipeline integrity management, while the concept of Big Data Analytics has not yet been fully utilized or understood in the oil and gas sector. As such, the research proposition addresses the problem that organizations are not sufficiently in control of the required integrity management activities, and relationships between integrity management activities, to properly utilize Big Data Analytics to make appropriate decisions to enhance pipeline integrity by predicting anomalies and aligning them to assist with competitive strategy development and implementation.

Pipeline Integrity Management (PIM) refers to the systematic process of ensuring pipeline systems' safety, reliability, and functionality for transporting energy materials. The primary goal of PIM is to prevent pipeline failures, leaks, or ruptures that could lead to environmental damage, safety hazards, or product loss. Beyond theoretical examination, this study proposes a conceptual framework for applying Big Data Analytics to pipeline integrity management. This conceptual framework for integrating Big Data Analytics into energy pipeline asset management provides a strategic approach for optimizing both the business value chain and the data value chain. The proposed framework, represented in Fig. 6.10, aligns the Big Data value chain (horizontal axis) with the business value chain (vertical axis) to create a synergistic interaction between data collection, analysis, processing, and visualization, ultimately leading to actionable business insights for energy pipeline integrity management.

The novelty of this framework lies in its layered approach, which supports a holistic policy development process through efficient management of resources, processes, competencies, and technologies. This structured approach ensures that businesses can target critical assets, determining when and how to manage, maintain, or decommission them to achieve maximum value. By integrating Big Data Analytics into the lifecycle of assets, from acquisition to utilization, maintenance, and renewal, the framework paves the way for more effective asset management, aligning business goals with the entire lifecycle of infrastructure assets.

A key feature of the framework is its focus on proactive decision-making, which is enabled by real-time data collection from various sensors and field inspections. Applying artificial intelligence (AI) techniques enhances predictive maintenance and anomaly detection, allowing pipeline operators to preempt potential failures. For example, AI can detect structural issues like weld cracks, ensuring timely intervention and avoiding costly breakdowns. As depicted in Fig. 6.11, integrating real-time data and advanced analytics not only enhances pipeline safety and efficiency but also reduces environmental risks and operational downtime.

The framework proposes a stepwise adoption strategy for Big Data Analytics in pipeline integrity management. This approach bridges the gap between theoretical research and practical implementation, offering a detailed roadmap for businesses to enhance safety, optimize maintenance, and improve decision-making through data-driven insights. A deeper examination of data collection, storage, and application reveals how Big Data Analytics can transform the pipeline industry. Data collection is achieved through diverse sensors, measuring pressure, flow, temperature, vibration, and sound, creating a holistic view of pipeline conditions. Integrating internal sensor data with external influences, historical records, and public data enables a comprehensive understanding of pipeline health. The framework emphasizes the importance of standardization and data quality controls to ensure that the resulting analyses are accurate and actionable.

Data applications are central to the framework's effectiveness. Real-time anomaly detection, powered by continuous sensor analysis, enables immediate responses to leak or blockages. Predictive maintenance models help optimize resource allocation and prevent costly downtime. Furthermore, data-driven inspection planning prioritizes inspections based on risk, enhancing both efficiency and effectiveness. Long-term trend analysis and scenario planning inform strategic decision-making, while network and geospatial analysis provide a comprehensive view of pipeline interactions with external environmental factors, thus improving risk management. Data visualization is critical to transforming raw data into actionable insights. The framework advocates for the use of interactive dashboards and customizable visualizations that cater to different user needs, supporting better collaboration and decision-making across all levels of the organization. Ensuring that these visualizations are accessible and effectively convey complex information is essential for inclusive and informed decision-making.

The challenges in implementing Big Data Analytics for pipeline integrity management have been addressed in this work. Data quality and integration remain significant obstacles, necessitating the development of robust data governance frameworks and standardization efforts. Infrastructure and expertise pose further challenges, as handling the volume and complexity of Big Data requires both advanced technology and skilled personnel. Cybersecurity is another key concern, with secure data storage and access protocols being vital to protect sensitive operational information. Despite these challenges, the framework offers substantial benefits for the oil and gas industry. It enhances pipeline safety through proactive threat identification and mitigation, optimizes maintenance by enabling data-driven insights, and supports informed decision-making with clear, actionable visualizations. In doing so, it positions the industry to achieve greater operational efficiency, cost-effectiveness, and environmental sustainability.

Chapter 7 focuses on data quality issues, such as inconsistencies, missing values, defect mismatching, long-distance mismatching, girth weld mismatching, and outliers, that pose significant challenges in the accurate assessment and interpretation of inspection results in the oil and gas pipeline industry. These issues can lead to severe consequences, including missed anomaly detection, misinterpretation of defects, safety hazards, and operational disruptions. Therefore, addressing these data quality concerns is paramount for ensuring pipeline integrity and operational efficiency.

In this book, the critical problem of outlier detection in in-line inspection (ILI) data has been addressed, which holds substantial value for pipeline integrity management. By identifying and analyzing outliers in raw ILI data, it is possible to improve anomaly detection and prevent the misinterpretation of potential defects, ultimately contributing to the safety and reliability of pipeline operations. This work compares machine learning models and traditional statistical analysis techniques for detecting outliers in ILI data. The findings indicate a significant overlap in the results produced by both approaches, with machine learning models demonstrating a comparable level of accuracy to statistical methods. While machine learning models are often considered complex and require substantial computational resources, they are proved effective in detecting outliers, matching the performance of established statistical techniques. This convergence highlights the potential of machine learning models to serve as either robust alternatives or complements to statistical methods, especially when dealing with large and complex datasets commonly encountered in pipeline inspections.

A key aspect of outlier detection is the selection of appropriate thresholds, as the effectiveness of a method depends on the nature and characteristics of the dataset being analyzed. For instance, this research work identified that the conventional interquartile range (IQR) method was not effective in detecting outliers in the ILI datasets under study. To address this limitation, we proposed a modification of the IQR method, calculating the upper fence by using the 95th percentile plus three times of the IQR. This modified method was tested and demonstrated similar performance to other outlier detection techniques, offering a more tailored approach for this specific application.

The purpose of testing multiple approaches in this study is to propose a comprehensive framework for outlier detection that balances robustness and efficiency. On the one hand, it is crucial to take a conservative approach and thoroughly investigate all detected outliers to confirm whether they are genuine or false measurements. However, excessive verification can result in unnecessary labor, particularly when many of the identified points are not true outliers. Therefore, the selection of the outlier detection method should be guided by the dataset's characteristics, the specific requirements of the analysis, and the available resources.

Looking toward future research, the exploration of hybrid approaches that combine the strengths of both machine learning and statistical methods offers exciting potential. These hybrid models could create more robust and reliable outlier detection frameworks, further enhancing data quality and improving the safety and efficiency of pipeline infrastructure management. Additionally, incorporating more advanced machine learning techniques, such as deep learning or anomaly detection algorithms, may yield even more accurate results, particularly in detecting subtle or complex patterns in ILI data. Future studies could also investigate the application of these outlier detection methods across various industries, expanding their impact beyond the oil and gas sector and providing broader contributions to infrastructure safety and operational excellence.

Chapter 8 addresses a critical challenge in pipeline integrity management by focusing on the prediction of corrosion defects using In-line Inspection (ILI) data. Through a systematic evaluation of multiple mathematical models, the study explores the applicability and accuracy of Exponential, Logarithmic, Polynomial (Order 2), Linear, and Power Law models in predicting corrosion defects in pipelines. ILI data from three distinct oil and gas pipelines were used to rigorously test and validate these models, leading to significant findings that advance the current understanding of corrosion prediction in the oil and gas sector.

The analysis demonstrated that Logarithmic and Power Law models consistently outperformed other models in general. Using a comprehensive dataset of 47 identified defects, the research highlights the practicality of using the simple models within the pipeline integrity management system (PIMS) for forecasting defect growth over time.

A notable contribution of this research lies in its emphasis on the adaptability of models based on pipeline-specific characteristics. The study underscores that no single model fits all pipeline scenarios, urging operators to tailor their choice of models according to the unique conditions of each pipeline. This adaptive approach is crucial for enhancing predictive accuracy and effectively managing corrosion growth over time. The results indicate that, while the Logarithmic and Power Law models are widely applicable, operators should still consider other models under certain conditions to capture specific defect growth patterns with sufficient accuracy.

In terms of practical applications, this research provides an invaluable framework for improving pipeline monitoring and maintenance strategies. By recommending the use of these simple yet highly effective models, the study equips pipeline operators with a reliable tool to predict corrosion defects, enabling more proactive maintenance and reducing the likelihood of pipeline failures. The insights gained from this research have the potential to reduce operational costs, extend the lifespan of pipelines, and enhance overall safety in the oil and gas industry.

The novelty of this research lies not only in the comparative analysis of the models but also in its methodological rigor, which provides a benchmark for future studies aiming to refine or develop new predictive models. The findings suggest that future research could explore the integration of advanced data analytics, such as machine learning algorithms, to further enhance the predictive power of simple models such as Logarithmic and Power Law models. Additionally, investigating the application of these models to other forms of pipeline anomalies, beyond corrosion, could offer broader insights into pipeline integrity management.

In conclusion, this research offers a significant advancement in predicting pipeline corrosion defects, with Logarithmic, Power Law, Linear and Exponential models being highly recommended for widespread use in the oil and gas industry. Their applicability, combined with the model's demonstrated accuracy, positions them as a key tool for future pipeline integrity management. Future research should build upon these findings by exploring model refinement, incorporating real-time data analytics, and expanding the scope of prediction models to other defect types, thereby contributing to more robust and reliable pipeline maintenance strategies.

The role of AI (including machine learning) in PIM is becoming increasingly prominent. These technologies enable the development of predictive models that can assess the likelihood of pipeline failures based on historical data and real-time monitoring (Amadhe et al. 2024). For instance, machine learning algorithms can analyze patterns in pipeline performance and identify potential failure points, allowing for timely interventions (Zhang and Xie 2024). The application of AI in conjunction with IoT technology will further enhance the monitoring capabilities of pipeline systems, providing operators with real-time insights into pipeline conditions and facilitating rapid response to emerging issues (Liu et al. 2024). This integration of advanced technologies is expected to lead to more efficient and effective pipeline management practices.

Finally, the obstacles related to standardization and data sharing in PIM must be resolved to fully harness the promise of these technologies. The creation of standardized inspection protocols and cooperative data-sharing frameworks among industry participants is crucial for improving the safety and dependability of pipeline systems. As the sector advances towards a more cohesive strategy for PIM, collaboration among academia, industry, and regulatory entities will be essential for fostering ongoing enhancements in pipeline safety and integrity management.

15.1 Artificial Intelligence

Artificial Intelligence (AI) including Machine Learning are transformative technologies that are reshaping various sectors, including education, healthcare, and industry. The progression of AI technology is recognized as a vital factor in fostering sustainable growth and enhancing human capabilities. The establishment of guidelines for AI systems is crucial to ensure their safe and beneficial integration into society. This involves the creation of equilibrium-safe intelligent systems that emphasize human well-being and ethical principles in their design and execution (Bryndin 2020).

The utilization of AI in the educational sector is notably substantial. Artificial intelligence technology facilitates personalized learning experiences, optimizes assessment processes, and enhances educational management systems. These enhancements increase educational outcomes while simultaneously optimizing teacher training and resource allocation (Akgun and Greenhow 2022). AI-driven platforms can tailor educational content to address specific student requirements, hence enhancing engagement and understanding. The integration of AI in second language education demonstrates how these technologies can create new options for learners, improving accessibility and effectiveness in a globalized context.

The swift progress in AI is accompanied by considerable ethical and social ramifications. Concerns including data privacy, algorithmic bias, and the risk of job displacement from automation require meticulous attention and regulation. As

AI systems gain autonomy and the ability to execute intricate tasks, the necessity for comprehensive legal frameworks to tackle these challenges grows progressively imperative. The designation of AI as a legal entity prompts inquiries regarding accountability and intellectual property rights, especially in creative domains where AI-generated content is increasingly common.

The advancement of AI, especially in deep learning and neural networks, has been driven by the accessibility of extensive data and considerable computational resources (Shawahna et al. 2018). This has resulted in advancements across multiple sectors, including healthcare diagnostics and autonomous transportation systems. The ramifications of these developments surpass simple efficiency improvements; they contest conventional concepts of creativity and intelligence, necessitating a reassessment of human–machine interactions.

15.1.1 AI-Driven Predictive Maintenance

AI-driven predictive maintenance will utilize sophisticated machine learning algorithms to forecast impending breakdowns with unparalleled precision (Hari et al. 2021; Ahmad et al. 2021). AI systems will evaluate past data and learn from real-time sensor inputs, constantly modifying models to reflect changing situations. This will enable pipeline operators to foresee problems prior to their escalation and arrange maintenance with little disturbance.

15.1.2 AI for Autonomous Monitoring and Control

As AI technology advances, the role of human operators in monitoring and controlling pipelines will diminish. AI systems will take over the day-to-day management of pipelines, autonomously adjusting pressure levels, flow rates, and other operational parameters to optimize performance and prevent failures (Hari et al. 2021).

15.2 Advanced Robotics for Inspection and Maintenance

The integration of advanced robotics in inspection and maintenance across various sectors has become increasingly significant, driven by the need for enhanced safety, efficiency, and cost-effectiveness. In the railway sector, for instance, the employment of robotic systems can streamline maintenance operations by automating tasks such as track inspection and rolling stock servicing. Vithanage et al. emphasize the potential of collaborative robots to assist maintenance technicians, thereby reducing physical strain and enhancing operational efficiency in railway depots (Vithanage et al. 2019). The use of autonomous systems in railway maintenance can significantly improve the accuracy and reliability of inspections, as highlighted by Rahman et al. (2022), who note that robotic systems can effectively combine autonomous sensing and decision-making capabilities for track maintenance.

In maritime applications, robotics has revolutionized the inspection and maintenance of offshore structures. Sahoo et al. (2024) discuss how autonomous underwater vehicles (AUVs) have transformed oceanographic research and the inspection of maritime assets by collecting critical data while minimizing risks to human divers. Similarly, AI-powered robots are being utilized for routine inspections and repairs in hazardous offshore environments, as noted by Ukato et al. (2024), who point out that these technologies enhance operational safety and efficiency. The deployment of robotic systems in maritime contexts not only improves safety but also reduces operational costs associated with human intervention in dangerous settings.

The offshore wind energy sector has also seen significant advancements through the use of robotic technology. Jiang et al. (2023) describe a multi-robot system designed for the autonomous deployment and recovery of maintenance robots for wind turbine blades, which not only lowers maintenance costs but also mitigates health and safety risks for technicians. The development of specialized robots for blade inspection and maintenance, such as BladeBug, exemplifies the trend towards automation in the operation and maintenance of renewable energy assets. This trend is echoed in the broader context of subsea maintenance, where AI-enhanced robots are increasingly capable of navigating complex underwater environments to perform inspections and repairs, thereby improving safety and operational efficiency (Nauert and Kampmann 2023).

The use of mobile manipulators in railway track maintenance is gaining traction, as they can effectively replace human labor in hazardous tasks, enhancing both safety and efficiency (Rahman et al. 2023). The potential for robotic systems to handle routine maintenance tasks across various sectors is further supported by the development of specialized robots for ship hull maintenance, which automate processes that are traditionally labor-intensive and risky for human workers (Prabakaran et al. 2020; Abdulkader et al. 2020). These advancements indicate a clear trajectory towards greater reliance on robotics for inspection and maintenance tasks across multiple industries, driven by the need for improved safety, efficiency, and cost-effectiveness.

15.2.1 In-Line Inspection Robots

In-line inspection (ILI) robots, commonly known as "smart pigs," are extensively utilized for internal pipeline inspection (Parlak and Yavasoglu 2023). These robots possess sensors that identify corrosion, fractures, and other imperfections (Wu 2024). In the future, ILI robots will evolve to undertake real-time data processing and make repair decisions while within the pipeline.

15.2.2 Drones for External Inspections

Drones equipped with advanced sensors and imaging technologies will continue to play a critical role in inspecting above-ground and hard-to-access sections of pipelines (Ndehedehe 2022). Future drones will be able to perform more complex tasks, such as detecting leaks through thermal imaging, assessing corrosion, and even delivering repair materials to remote locations (Tarr et al. 2021).

15.2.3 Autonomous Repair Robots

While robots are already used for inspections, future innovations will focus on robots capable of performing repairs autonomously (Wu 2024). These machines could fix minor issues, such as welding cracks or applying corrosion-resistant coatings, without the need for human intervention.

15.3 Quantum Computing for Complex Data Analysis

Quantum computing possesses the capacity to revolutionize data processing within the energy sector, especially in the realm of pipeline integrity management (AlSuwaidan 2021). Unlike classical computers, quantum computers can simultaneously process vast amounts of data, enabling them to tackle complex problems that surpass the capabilities of traditional systems. The implications of these characteristics are substantial, particularly in the domains of machine learning, optimization, and data analysis.

Quantum computing exhibits considerable potential in enhancing complex data analysis tasks. Quantum algorithms, such as those employing quantum walks and variational methods, may provide exponential enhancements in efficiency for specific scenarios. Quantum walks are linked to route integrals, enabling innovative techniques for simulating quantum physical systems and enhancing large-scale data analysis capabilities (Feng et al. 2021). The progression of quantum kernel methods has shown that quantum computers can proficiently handle data in high-dimensional spaces, a challenge often insurmountable for classical approaches (Park et al. 2020). These advancements suggest that quantum computing could significantly enhance the effectiveness of machine learning algorithms, particularly in scenarios involving large datasets and complex feature interactions.

Quantum machine learning (QML) is a nascent discipline that integrates quantum computing with machine learning methodologies to address intricate data analysis problems. QML possesses the capacity to transform data processing and analysis by using the intrinsic parallelism of quantum systems (Schuld et al. 2015). Quantum principal component analysis (QPCA) has been suggested as a potentially more computationally efficient alternative to classical PCA; however, current evidence indicates that while QPCA may present certain advantages, it does not ensure exponential speedup relative to classical methods (Daskin et al. 2023). Additionally, hybrid quantum–classical algorithms are being formulated to use the advantages of both paradigms, facilitating more advanced data processing frameworks (McClean et al. 2016).

Besides machine learning applications, quantum computing is being investigated for its potential in topological data analysis and complicated network analysis. Quantum algorithms developed for calculating persistent Betti numbers can elucidate the fundamental structure of data, proving especially beneficial in domains such as biology and social network analysis (Hayakawa 2022). The capacity to examine complex networks using quantum navigation and rating methodologies exemplifies the adaptability of quantum computing in managing intricate data interrelations (Sánchez-Burillo et al. 2012).

Notwithstanding the encouraging progress, obstacles persist in the practical application of quantum computing for intricate data analysis. Challenges such as error rates in quantum calculations and the necessity for fault-tolerant quantum systems must be resolved to fully actualize the potential of quantum algorithms (Bordoni et al. 2023). Nevertheless, the swift advancement of quantum hardware and algorithms suggests that we are approaching a future in which quantum computing can be regularly utilized for intricate data processing jobs across multiple fields, such as finance, genomics, and artificial intelligence.

15.3.1 Quantum Algorithms for Risk Assessment

Quantum computing could significantly enhance risk assessment models by processing large datasets more efficiently. By applying quantum algorithms, operators could evaluate multiple risk factors simultaneously, considering variables such as environmental conditions, pipeline material properties, and operational data to provide more accurate risk assessments (Azeez et al. 2024).

15.3.2 Optimizing Maintenance Scheduling

Quantum computing could also optimize maintenance schedules by evaluating multiple potential maintenance scenarios simultaneously (Bayerstadler et al. 2021). This would allow operators to determine the best times to perform maintenance based on predicted risks, operational demands, and resource availability, minimizing downtime and maximizing efficiency.

- **Example**: A quantum algorithm could simulate thousands of potential maintenance schedules, considering factors such as predicted weather conditions, equipment availability, and pipeline demand, to recommend the optimal maintenance window for each section of the pipeline.

15.4 Enhanced IoT Networks and Edge Computing

Advanced Internet of Things (IoT) networks, combined with edge computing, signify a substantial advancement in the administration and processing of data produced by numerous gadgets. The incorporation of edge computing into IoT frameworks mitigates significant issues such latency, bandwidth constraints, and security risks, therefore improving overall system efficacy and user satisfaction.

Edge computing enables data processing near its source, which is especially advantageous in IoT settings marked by substantial data generation and the necessity for real-time analytics. Transferring processing from centralized cloud servers to the network's edge enables apps to attain reduced latency and enhanced responsiveness. This is crucial for applications necessitating instantaneous data processing, such as remote monitoring and telematics, where delays may result in considerable operational inefficiencies (Jin et al. 2020; Muñoz et al. 2018). Furthermore, edge computing enhances the scalability of IoT networks by facilitating localized data processing, which diminishes the volume of data transported to the cloud, thereby mitigating bandwidth limitations (Wang et al. 2019; Zhao et al. 2018).

Enhanced IoT network architecture frequently integrates diverse edge computing paradigms, such as fog computing and microservices. Fog computing enhances cloud functionalities at the edge, facilitating decentralized processing and storage, which is especially beneficial for resource-limited IoT devices (Kokate and Shrawankar 2023; Ahanger et al. 2018). The implementation of microservices at the edge significantly improves flexibility and efficiency by enabling the dynamic allocation of resources according to real-time demands (Xu et al. 2021; Souza et al. 2018). This architecture optimizes the performance of IoT applications while also improving data privacy and security by reducing the visibility of sensitive information during transmission (Su et al. 2018; Jin et al. 2023).

Security is a major concern in the deployment of IoT networks. The decentralized framework of edge computing helps mitigate risks associated with centralized data processing, such as single points of failure. Blockchain methodologies can be integrated into edge computing architectures to provide secure and dependable data transmissions, hence ensuring data integrity (Muthanna et al. 2019; Zhao et al. 2019). Moreover, edge computing facilitates the implementation of privacy-preserving techniques, crucial for maintaining user trust in IoT applications.

15.4.1 5G-Enabled IoT Networks

The implementation of 5G networks will markedly improve the speed and dependability of IoT devices utilized in pipeline monitoring. The low latency and large bandwidth of 5G will provide real-time data transmission from the most remote pipeline locations, offering operators immediate insights into pipeline status (Mukherjee et al. 2022; Alsamhi et al. 2021).

15.4.2 Edge Computing for Real-Time Decision-Making

Edge computing is processing data in proximity to its source, rather than transmitting it to a centralized cloud for analysis. This minimizes latency and facilitates real-time decision-making, essential for swiftly identifying and addressing pipeline irregularities.

15.5 The Impact of Renewable Energy on Pipeline Infrastructure

The incorporation of renewable energy sources into current pipeline infrastructure offers both advantages and obstacles that profoundly affect the energy sector. As countries transition from fossil fuels to greener energy sources, the implications for pipeline systems, which were historically built for transmission of oil and natural gas, are profound.

A significant effect of renewable energy on pipeline infrastructure is the possibility of repurposing current natural gas pipes for hydrogen transportation. Research suggests that incorporating hydrogen into natural gas pipes can ease the transition to a hydrogen economy by utilizing existing infrastructure and reducing the necessity for new construction (Bdioui et al. 2023). The viability of this blending has been examined in multiple scenarios, emphasizing both technical obstacles and economic advantages. This transformation is essential, as hydrogen is anticipated to be pivotal in future energy systems, especially in areas that are challenging to electrify.

The emergence of renewable energy sources, like wind and solar, requires the establishment of new pipeline infrastructure for the transportation of hydrogen and other renewable fuels. The development of these new infrastructures must account for the stability and resilience of the systems, particularly with the implications of climate change, which can compromise both the physical integrity of pipelines and the reliability of energy delivery. The interconnection of energy infrastructures, particularly the integration of electrical grids with gas pipelines, complicates the situation, as variations in renewable energy production can cause pressure fluctuations in gas pipelines.

The economic ramifications of incorporating renewable energy into pipeline systems are substantial, alongside technical and operational factors. The implementation of renewable energy can stimulate economic growth, as demonstrated by comparative studies of rich and developing countries. Investments in renewable energy infrastructure bolster energy security, provide employment, and invigorate local economies (Asha et al. 2023). The transformation necessitates meticulous planning and investment to guarantee that current pipeline systems can integrate new energy types without jeopardizing safety and efficiency.

The environmental consequences of pipeline infrastructure require attention. The building and operation of pipelines may result in habitat disruption and biological alterations, especially in marine ecosystems. As renewable energy initiatives proliferate, it is imperative to adopt solutions that alleviate these effects, guaranteeing that the advantages of cleaner energy do not compromise environmental integrity.

15.5.1 Hydrogen Pipelines

With the growing emphasis on hydrogen as a clean energy source, pipelines will need to be retrofitted or designed to transport hydrogen safely. Hydrogen molecules are smaller and more prone to leaking than natural gas, so advanced materials and monitoring technologies will be required to ensure the integrity of hydrogen pipelines (Ahad et al. 2023; Erdener et al. 2023; Shehata and El-Shamy 2023).

15.5.2 Carbon Capture and Storage (CCS) Pipelines

As carbon capture and storage (CCS) technologies become more prevalent, pipelines will play a critical role in transporting captured carbon dioxide to storage facilities. New monitoring and integrity management systems will be needed to handle the unique challenges associated with transporting CO_2.

15.6 Limitation of Research Findings

The limitations of using Big Data Analytics on pipeline integrity management have been found. One of the primary limitations of this research is the dependency on the quality and availability of data. The effectiveness of Big Data Analytics in pipeline integrity management heavily relies on the accuracy, completeness, and timeliness of the data collected from different inspection, sensors, and monitoring systems. The data may contain noise, errors, or missing values, which could affect the precision and reliability of the analysis.

The study assumes the availability of advanced computational resources and technologies for processing large datasets, which may not be universally accessible to all organizations involved in pipeline management. Some of the companies with a limited IT infrastructure might face challenges in adopting the recommended Big Data Analytics techniques due to high initial costs, lack of technical expertise, or insufficient computational power, thereby limiting the practical application of the research outcomes. The models and algorithms developed and applied in this research are based on certain assumptions and simplifications to manage the complexity of real-world pipeline integrity scenarios. For instance, prediction models may not fully account for all potential external factors influencing pipeline integrity, such as unpredictable environmental conditions, third-party interference, or human error. These assumptions may lead to deviations between the model predictions and actual outcomes, potentially affecting the robustness and applicability of the research findings.

This research primarily focuses on specific use cases of Big Data Analytics in pipeline integrity management, such as anomaly detection, predictive maintenance, and risk assessment. While these use cases are significant, the study does not cover all possible applications of Big Data Analytics in the pipeline industry. Big Data Analytics and pipeline integrity management technologies are rapidly evolving fields. The findings of this research are based on the current state of technology, which may become outdated as new methods, tools, and techniques emerge. Future advancements in artificial intelligence, machine learning, and data processing technologies could offer more sophisticated solutions for pipeline integrity management, potentially surpassing the approaches discussed in this study.

Finally, the generalizability of the research findings is limited by the specific context in which the study was conducted. The data and scenarios used in this research are drawn from geographical locations and pipeline types, which may not be representative of all pipeline systems globally. The unique characteristics of different pipelines, such as varying materials, environmental conditions, and operational practices, mean that the findings may not be universally applicable without further validation in different contexts.

15.7 Future Work and Recommendations

The future research potential and development using Big Data in oil and gas pipelines offer several exciting possibilities. Current literature indicates a lack of research on real-time monitoring and alert systems specifically designed for pipelines. These systems are crucial for quickly detecting anomalies, leaks, or security breaches. There is a need for research to develop robust algorithms and techniques that enable real-time data processing, anomaly detection, and automated alert generation. Such advancements would allow for proactive responses to potential incidents, predict anomalies, significantly reducing the risks associated with pipeline operations.

While some predictive models have been developed to detect corrosion and other anomalies in pipelines, there are few practical applications where these models have been implemented by pipeline owners for condition assessment. Existing models often consider only one or two parameters, typically focusing on corrosion. To enhance the reliability of predictive analytics, researchers should aim to develop more comprehensive models that can forecast pipeline failures, identify maintenance needs, and optimize operational performance by incorporating multiple parameters. This would involve exploring advanced machine learning algorithms, integrating real-time data, and leveraging historical patterns to improve prediction accuracy and reliability. The current literature lacks detailed information on data integration specifically for the oil and gas pipeline industry. There is a significant opportunity to develop comprehensive and integrated data analytics platforms tailored to this sector. These platforms could streamline the use of Big Data Analytics by combining various data sources, offering advanced data visualization capabilities, and providing a suite of analytics tools and algorithms designed for pipeline operations. Such platforms would enhance decision-making processes and operational efficiency.

Integrating Big Data Analytics with the Internet of Things (IoT) and sensor networks presents a promising area for research and development. Effective collection, processing, and integration of sensor data with other data sources could enhance real-time monitoring, enable condition-based maintenance, and improve overall pipeline performance. Research also needs to focus on developing new methodologies for optimizing data flow from sensors and enhancing the synergy between IoT devices and data analytics systems. Research opportunities exist in developing advanced data fusion and integration techniques that combine data from multiple sources, including legacy systems, Supervisory Control and Data Acquisition (SCADA) systems, satellite imagery, and external data providers. By focusing on these techniques, researchers can aim to provide more accurate and comprehensive insights for decision-making in pipeline management. To utilize machine learning models, data quality is another area that needs more work to explore best solutions for reading and analyzing the data for analysis and decision making. Further research could focus on the integration of hybrid models that combine machine learning techniques with traditional statistical methods. This approach could enhance the robustness and interpretability of outlier detection systems. For instance, machine learning models could be employed to preprocess data, reducing noise, and isolating potential outliers, which can then be further validated using statistical analysis.

Comparative studies between machine learning models and traditional statistical methods should be expanded to assess their performance across various pipeline conditions and data types. These comparisons could help identify the strengths and limitations of each approach, guiding the selection of appropriate techniques for specific pipeline integrity challenges. Interdisciplinary collaboration between data scientists, pipeline engineers, and industry experts will be crucial in advancing the practical implementation of machine learning-based outlier detection systems. Such collaborations could lead to the development of standardized protocols and best practices, facilitating the widespread adoption of these technologies in the pipeline industry.

References

Abdulkader RE et al (2020) Sparrow: a magnetic climbing robot for autonomous thickness measurement in ship hull maintenance. J Mar Sci Eng 8(6):469

Ahad MT et al (2023) An overview of challenges for the future of hydrogen. Materials 16(20):6680

Ahanger TA, Tariq U, Nusir M (2018) Mobility of Internet of Things and fog computing: concerns and future directions. Int J Commun Netw Inf Security 10(3):534

Ahmad T et al (2021) Artificial intelligence in sustainable energy industry: status Quo, challenges and opportunities. J Clean Prod 289:125834

Akgun S, Greenhow C (2022) Artificial intelligence in education: addressing ethical challenges in K-12 settings. AI Ethics 2(3):431–440

Alsamhi SH et al (2021) Green internet of things using UAVs in B5G networks: A review of applications and strategies. Ad Hoc Netw 117:102505

AlSuwaidan L (2021) The role of data management in the Industrial Internet of Things. Concurr Comput: Pract Exp 33(23):e6031

Amadhe F, Anjorin R, Uwoghiren F (2024) Advancements in machine learning for pipeline integrity management: a comprehensive review of predictive and optimization techniques. Cognizance J Multi Stud 4(11):129–138

Asha LN et al (2023) A quantitative approach of measuring sustainability risk in pipeline infrastructure systems. Sustainability 15(19):14229

Azeez M et al (2024) Developing intelligent cyber threat detection systems through quantum computing. International Journal of Science and Research Archive 12(02):1297–1307

Bayerstadler A et al (2021) Industry quantum computing applications. EPJ Quantum Technol 8(1):25

Bdioui H et al (2023) Feasibility study on hydrogen blending into Tunisian natural gas distributing system. Preprints. https://doi.org/10.20944/preprints202312.0662.v1

Bordoni S et al (2023) Long-lived particles anomaly detection with parametrized quantum circuits. Particles 6(1):297–311

Bryndin E (2020) Standardization of artificial intelligence for the development and use of intelligent systems. Adv Wirel Commun Netw 6(1):1–9

Daskin A, Gupta R, Kais S (2023) Dimension reduction and redundancy removal through successive Schmidt decompositions. Appl Sci 13(5):3172

Erdener BC et al (2023) A review of technical and regulatory limits for hydrogen blending in natural gas pipelines. Int J Hydrogen Energy 48(14):5595–5617

Feng Y, Miceli R, McGuigan M (2021) Quantum walks, Feynman propagators and graph topology on an IBM quantum computer. arXiv preprint arXiv:2104.06458

Hari SKK et al (2021) Operation of natural gas pipeline networks with storage under transient flow conditions. IEEE Trans Control Syst Technol 30(2):667–679

Hayakawa R (2022) Quantum algorithm for persistent Betti numbers and topological data analysis. Quantum 6:873

Jiang Z et al (2023) A multirobot system for autonomous deployment and recovery of a blade crawler for operations and maintenance of offshore wind turbine blades. J Field Robot 40(1):73–93

Jin W et al (2020) Secure edge computing management based on independent microservices providers for gateway-centric IoT networks. IEEE Access 8:187975–187990

Jin W et al (2023) Transparent access to heterogeneous IoT based on virtual resources. Comp Mater Continua 74(3):4983–4999

Kokate S, Shrawankar U (2023) An efficient approach for secured data transmission between IoT and Cloud. Research Reports Comp Sci:35–44

Liu H et al (2024) Research on PID control method for predictive modeling of oil pipeline infrastructure under Industrial Internet of Things driving. In: International conference on algorithms, high performance computing, and artificial intelligence (AHPCAI 2024). SPIE

McClean JR et al (2016) The theory of variational hybrid quantum-classical algorithms. New J Phys 18(2):023023

Mukherjee S et al (2022) Leveraging big data analytics in 5G-enabled IoT and industrial IoT for the development of sustainable smart cities. Trans Emerg Telecommun Technol 33(12):e4618
Muñoz R et al (2018) Integration of IoT, transport SDN, and edge/cloud computing for dynamic distribution of IoT analytics and efficient use of network resources. J Lightwave Technol 36(7):1420–1428
Muthanna A et al (2019) Secure and reliable IoT networks using fog computing with software-defined networking and blockchain. J Sens Actuator Netw 8(1):15
Nauert F, Kampmann P (2023) Inspection and maintenance of industrial infrastructure with autonomous underwater robots. Front Robot AI 10:1240276
Ndehedehe C (2022) Remotely piloted aircraft systems. Satellite remote sensing of terrestrial hydrology. Springer, pp 177–207
Park DK, Blank C, Petruccione F (2020) The theory of the quantum kernel-based binary classifier. Phys Lett A 384(21):126422
Parlak BO, Yavasoglu HA (2023) A comprehensive analysis of in-line inspection tools and technologies for steel oil and gas pipelines. Sustainability 15(3):2783
Prabakaran V et al (2020) Hornbill: a self-evaluating hydro-blasting reconfigurable robot for ship hull maintenance. IEEE Access 8:193790–193800
Rahman M et al (2022) Challenges for a railway inspection and repair system from railway infrastructure. In: 2022 10th International conference on control, mechatronics and automation (ICCMA). IEEE
Rahman M et al (2023) A review on the prospects of mobile manipulators for smart maintenance of railway track. Appl Sci 13(11):6484
Sahoo SK, Choudhury BB, Dhal PR (2024) Exploring the role of robotics in maritime technology: innovations, challenges, and future prospects. Spect Mech Eng Oper Res 1(1):159–176
Sánchez-Burillo E et al (2012) Quantum navigation and ranking in complex networks. Sci Rep 2(1):605
Schuld M, Sinayskiy I, Petruccione F (2015) An introduction to quantum machine learning. Contemp Phys 56(2):172–185
Shawahna A, Sait SM, El-Maleh A (2018) FPGA-based accelerators of deep learning networks for learning and classification: a review. IEEE Access 7:7823–7859
Shehata M, El-Shamy A (2023) Hydrogen-based failure in oil and gas pipelines a review. Gas Sci Eng 115:204994
Souza A et al (2018) Osmotic monitoring of microservices between the edge and cloud. In: 2018 IEEE 20th international conference on high performance computing and communications; IEEE 16th international conference on smart city; IEEE 4th international conference on data science and systems (HPCC/SmartCity/DSS). IEEE
Su X et al (2018) Distribution of semantic reasoning on the edge of internet of things. In: 2018 IEEE international conference on pervasive computing and communications (PerCom). IEEE
Tarr AA et al (2021) Drone law and policy: global development, risks, regulation and insurance. Routledge
Ukato A et al (2024) Optimizing maintenance logistics on offshore platforms with AI: current strategies and future innovations. World J Adv Res Rev 22(1):1920–1929
Vithanage RK, Harrison CS, DeSilva AK (2019) Importance and applications of robotic and autonomous systems (RAS) in railway maintenance sector: a review. Computers 8(3):56
Wang Y et al (2019) Satellite edge computing for the internet of things in aerospace. Sensors 19(20):4375
Wu K (2024) A survey on wireless in-pipe inspection robotics. Int J Intell Robot Appl:1–23
Xu R, Jin W, Kim D (2021) Enhanced service framework based on microservice management and client support provider for efficient user experiment in edge computing environment. IEEE Access 9:110683–110694
Zhang X, Xie X (2024) Model for the failure prediction mechanism of in-service pipelines based on IoT technology. Processes 12(8):1642

Zhao Z et al (2018) Deploying edge computing nodes for large-scale IoT: a diversity aware approach. IEEE Internet Things J 5(5):3606–3614

Zhao Y et al (2019) Blockchain based privacy-preserving software updates with proof-of-delivery for internet of things. J Parallel Distrib Comput 132:141–149

MIX
Papier aus verantwortungsvollen Quellen
Paper from responsible sources
FSC® C105338

If you have any concerns about our products,
you can contact us on
ProductSafety@springernature.com

In case Publisher is established outside the EU,
the EU authorized representative is:
Springer Nature Customer Service Center GmbH
Europaplatz 3, 69115 Heidelberg, Germany

Printed by Libri Plureos GmbH
in Hamburg, Germany